SCHAUM'S OUTLINE SERIES
THEORY AND PROBLEMS OF

ESSENTIAL COMPUTER MATHEMATICS

SEYMOUR LIPSCHUTZ

INCLUDING 840 SOLVED PROBLEMS

SCHAUM'S OUTLINE SERIES IN COMPUTERS

McGRAW-HILL BOOK COMPANY

SCHAUM'S OUTLINE OF

THEORY AND PROBLEMS

of

ESSENTIAL COMPUTER MATHEMATICS

•

by

SEYMOUR LIPSCHUTZ, Ph.D.

Professor of Mathematics
Temple University

SCHAUM'S OUTLINE SERIES

McGRAW-HILL BOOK COMPANY

New York   St. Louis   San Francisco   Auckland   Bogotá   Guatemala   Hamburg   Johannesburg
Lisbon   London   Madrid   Mexico   Montreal   New Delhi   Panama   Paris
San Juan   São Paulo   Singapore   Sydney   Tokyo   Toronto

SEYMOUR LIPSCHUTZ, who is presently on the mathematics faculty of Temple University, formerly taught at the Polytechnic Institute of Brooklyn and was a visiting professor in the Computer Science Department of Brooklyn College.   He received his Ph.D. in 1960 at the Courant Institute of Mathematical Sciences of New York University.   His other books in the Schaum's Outline Series include *Fortran* (with Arthur Poe), *Discrete Mathematics*, and *Probability*.

Schaum's Outline of Theory and Problems of
ESSENTIAL COMPUTER MATHEMATICS

3 4 5 6 7 8 9 10 11 12 13 14 15 16 17 18 19 20 SH SH 8 6 5 4 3 2

ISBN 0-07-037990-4

Sponsoring Editor, David Beckwith
Production Manager, Nick Monti
Editing Supervisor, Marthe Grice

Library of Congress Cataloging in Publication Data

Lipschutz, Seymour.
    Schaum's outline of theory and problems of
essential computer mathematics.

    (Schaum's outline series)
    Includes index.
        1.   Mathematics - - Problems, exercises, etc.
2.   Electronic data processing - - Mathematics - -
Problems, exercises, etc.   I.   Title.   II.   Series.
QA43.L669   519.4   81-15574
ISBN 0-07-037990-4 (pbk.)   AACR2

Preface

Computers appear in almost all fields of human endeavor, wherever data are collected and analyzed. Furthermore, with the development of inexpensive microcomputers, more and more individuals are buying and operating their own computer. For these reasons, certain mathematical topics related to the computer and information sciences—in particular, the binary number system, logic circuits, graph theory, linear systems, probability and statistics—are now being more widely studied. This book is designed to present these and associated topics in an elementary yet comprehensive form.

The material has been divided into fourteen chapters, so written that they are mostly independent of one another. Therefore one can change the order of many chapters, or can even omit certain chapters, without difficulty and without loss of continuity. The only prerequisite for most of the book is a minimal amount of high school mathematics.

Each chapter begins with clear statements of pertinent definitions and principles, together with illustrative and other descriptive material. This is followed by graded sets of solved and supplementary problems. The solved problems apply and amplify the theory, as well as provide the repetition of basic principles so vital to effective learning. The supplementary problems serve as a complete review of the material of each chapter.

I wish to thank many of my friends and colleagues for invaluable suggestions and critical review of the manuscript. I also wish to express my gratitude to the staff of the Schaum's Outline Series, particularly to David Beckwith, for their very helpful cooperation.

SEYMOUR LIPSCHUTZ

Contents

CONTENTS

CONTENTS

CONTENTS

Chapter 1

Binary Number System

1.1 INTRODUCTION

Many of the electronic components of a computer are bistable in nature; that is, they can be in either of two states (such as on/off, or clockwise magnetized/counterclockwise magnetized). These two possible states are usually denoted by 0 and 1, which are also the symbols for the digits of the binary number system. Moreover, an individual unit of information is usually represented in the computer by a sequence of these binary digits (called *bits* for short). Such sequences of bits may be viewed as binary numbers, and many computers use the binary number system not only to represent quantities but to perform calculations, using binary arithmetic.

The binary number system and the familiar decimal number system are examples of *positional numeration systems*. Any such system requires only a finite number of symbols, called the *digits* of the system, to represent arbitrarily large numbers. In terms of these digits, the execution of numerical calculations is relatively simple. The number b of digits in the system is called its *base* or *radix*. As we shall see, any number can be represented as a sum of powers of the base b, where each power is weighted by one of the digits.

Although this chapter is mainly concerned with the binary number system and its arithmetic, we begin with a review of the decimal system, since such a review will simplify the analogous topics of the binary system.

1.2 DECIMAL SYSTEM

The decimal system contains ten digits denoted by the symbols

$$0, 1, 2, 3, 4, 5, 6, 7, 8, 9$$

and representing the integers from zero to nine, respectively. Thus, the base of the decimal system is $b = 10$.

Any positive integer N, represented in the decimal system as a string of decimal digits, may also be expressed as a sum of powers of 10, with each power weighted by a digit. For example, $N = 8253$ can be expressed as follows:

$$8253 = 8 \times 10^3 + 2 \times 10^2 + 5 \times 10^1 + 3 \times 10^0 = 8 \times 1000 + 2 \times 100 + 5 \times 10 + 3 \times 1$$

This is called the *expanded notation* for the integer. Observe that

$$8253 = 8000 + 200 + 50 + 3$$

Thus, the digit 3 in the integer represents three 1s, the 5 represents five 10s, the 2 represents two 100s, and the 8 represents eight 1000s. The powers of ten,

$$10^0 = 1 \qquad 10^1 = 10 \qquad 10^2 = 100 \qquad 10^3 = 1000 \qquad \cdots$$

which correspond respectively to the digits in a decimal integer as read from right to left, are called the *place values* of the digits.

Any fractional value M, represented in the decimal system by a string of decimal digits together with an embedded decimal point, may also be expressed in expanded notation by using negative powers of 10. Specifically, the place values of the digits in M to the right of the decimal point are respectively

$$10^{-1} = \frac{1}{10} \qquad 10^{-2} = \frac{1}{100} \qquad 10^{-3} = \frac{1}{1000} \qquad \cdots$$

For example, $M = 837.526$ is expressed in expanded notation as follows:

$$837.526 = 8 \times 10^2 + 3 \times 10 + 7 \times 10^0 + 5 \times 10^{-1} + 2 \times 10^{-2} + 6 \times 10^{-3}$$

$$= 800 + 30 + 7 + \frac{5}{10} + \frac{2}{100} + \frac{6}{1000}$$

This decimal fraction is said to have three *decimal places*, the number of digits to the right of the decimal point.

The arithmetic of decimal fractions is not very complicated provided one keeps track of the decimal points.

Addition

One must vertically align the decimal points when adding decimal fractions. For example, $34.215 + 513.48 + 2.1326$ is obtained as follows:

$$
\begin{array}{r}
34.2150 \\
513.4800 \\
+ \quad 2.1326 \\
\hline
549.8276
\end{array}
$$

Observe that additional 0s are introduced so that the numbers have the same number of decimal places.

Subtraction

As with addition, one must vertically align the decimal points. For example, $45.217 - 23.64$ and $123.45 - 75.168$ are obtained as follows:

$$
\begin{array}{r}
45.217 \\
- 23.640 \\
\hline
21.577
\end{array}
\qquad
\begin{array}{r}
123.450 \\
- \quad 75.168 \\
\hline
48.282
\end{array}
$$

Again additional 0s are introduced to give the numbers the same number of decimal places.

Multiplication

The number of decimal places in the product is the sum of the decimal places in the numbers being multiplied. For example, 2.35×43.162 is obtained as follows:

$$
\begin{array}{r}
43.1\,62 \\
\times \quad 2.35 \\
\hline
2\ 15\ 8\ 10 \\
12\ 94\ 8\ 6 \\
86\ 32\ 4 \quad\ \\
\hline
101.43\ 0\ 70
\end{array}
$$

The product is given 5 decimal places because 43.162 has 3 decimal places and 2.35 has 2 decimal places.

Division

In dividing one decimal fraction by another, one moves the decimal point in the divisor to the right to make the divisor an integer. This is compensated by also moving the decimal point in the dividend to the right the same number of places. For example, $387.167 \div 2.55$ is obtained as follows:

```
          1 51.83
2×55.)387×16.70
      255
      132 1
      127 5
        4 66
        2 55
        2 11 7
        2 04 0
          7 70
          7 65
            5
```

Observe that we moved the decimal point in 2.55 two places to the right, so that the divisor became 255. Next we moved the decimal point in 387.167 two places to the right, so that the dividend became 38 716.7. We then performed the long division. We can add terminal zeros to the dividend to perform the division to as many places as we want. Here, we added one zero, and carried out the division to 2 decimal places.

1.3 BINARY SYSTEM

The binary system is the positional numeration system to the base $b = 2$. Its two digits, denoted 0 and 1, are called *bits* for short. Any *binary number* is therefore a sequence of bits, possibly with an embedded *binary point*. Those binary numbers that have no fractional part, i.e. are without an embedded binary point, are called *binary integers*.

The place values in the binary system are the powers of the base $b = 2$, just as the place values in the decimal system are the powers of ten. Specifically, the place values of the integral part of a binary number are the nonnegative powers of two,

$$2^0 \qquad 2^1 \qquad 2^2 \qquad 2^3 \qquad \ldots$$

and the place values of the fractional part of a binary number are the negative powers of two,

$$2^{-1} \qquad 2^{-2} \qquad 2^{-3} \qquad \ldots$$

Table 1-1 gives the values of some of the powers of two.

Binary-to-Decimal Conversion

Any binary number can be written in expanded notation as the sum of each digit times that digit's place value. For example,

$$110101 = 1 \times 2^5 + 1 \times 2^4 + 0 \times 2^3 + 1 \times 2^2 + 0 \times 2 + 1 \times 1$$

$$101.1101 = 1 \times 2^2 + 0 \times 2 + 1 \times 1 + 1 \times 2^{-1} + 1 \times 2^{-2} + 0 \times 2^{-3} + 1 \times 2^{-4}$$

Since each power of two is weighted by either 0 or 1, the binary number is simply the sum of those place values in which the bit 1 appears. This sum at once gives us the decimal equivalent of the binary number. (For another method, restricted to binary integers, see Problem 1.12.)

Table 1-2 lists the binary representations of the integers from 0 to 25, with the place values of the bits shown at the top of the table. Sometimes a subscript 2 is used to distinguish a binary number, e.g. one may write 101011_2 if it is not clear from the context that 101011 is a binary number rather than a decimal number. Also, for easier reading, one sometimes separates a binary number into 4-bit groups, to the left and right of the binary point; e.g.

10110100.011010   might be written   1011 0100.0110 10

Table 1-1

Power of Two	Decimal Value
2^{10}	1024
2^{9}	512
2^{8}	256
2^{7}	128
2^{6}	64
2^{5}	32
2^{4}	16
2^{3}	8
2^{2}	4
2^{1}	2
2^{0}	1
2^{-1}	$1/2 = 0.5$
2^{-2}	$1/4 = 0.25$
2^{-3}	$1/8 = 0.125$
2^{-4}	$1/16 = 0.062\ 5$
2^{-5}	$1/32 = 0.031\ 25$
2^{-6}	$1/64 = 0.015\ 625$

Table 1-2

Decimal Number	Binary Number				
	16s	8s	4s	2s	1s
0					0
1					1
2				1	0
3				1	1
4			1	0	0
5			1	0	1
6			1	1	0
7			1	1	1
8		1	0	0	0
9		1	0	0	1
10		1	0	1	0
11		1	0	1	1
12		1	1	0	0
13		1	1	0	1
14		1	1	1	0
15		1	1	1	1
16	1	0	0	0	0
17	1	0	0	0	1
18	1	0	0	1	0
19	1	0	0	1	1
20	1	0	1	0	0
21	1	0	1	0	1
22	1	0	1	1	0
23	1	0	1	1	1
24	1	1	0	0	0
25	1	1	0	0	1

EXAMPLE 1.1

(*a*) To convert 110101_2 to its decimal equivalent, write the appropriate place value over each bit and then add up those powers of two which are weighted by 1:

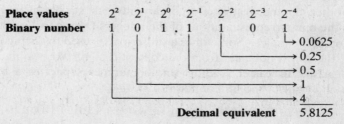

Place values 2^5 2^4 2^3 2^2 2^1 2^0
Binary number   1   1   0   1   0   1

→ 1
→ 4
→ 16
→ 32

Decimal equivalent   53

(*b*) To convert 101.1101_2 to its decimal equivalent, use Table 1-1 for the decimal values of the negative powers of two:

Place values 2^2 2^1 2^0 2^{-1} 2^{-2} 2^{-3} 2^{-4}
Binary number   1   0   1 . 1   1   0   1

→ 0.0625
→ 0.25
→ 0.5
→ 1
→ 4

Decimal equivalent   5.8125

Decimal-to-Binary Conversion

We find the binary representation of a decimal number N by converting its integral part, N_I, and its fractional part, N_F, separately.   We illustrate with the decimal number $N = 109.781\,25$.

EXAMPLE 1.2

(a) To convert $N_I = 109$ to its binary equivalent, divide N_I and each successive quotient by 2, noting the remainders, as follows:

Divisions	Quotients	Remainders
$109 \div 2$	54	1
$54 \div 2$	27	0
$27 \div 2$	13	1
$13 \div 2$	6	1
$6 \div 2$	3	0
$3 \div 2$	1	1
$1 \div 2$	0	1

The zero quotient indicates the end of the calculations.   Observe that the remainders can only be 0 or 1, since the divisions are by 2.   The sequence of remainders from the bottom up, as indicated by the arrow, yields the required binary equivalent.   That is, $N_I = 109 = 1101101_2$.

In actual practice, the above divisions may be condensed as follows:

$$
\begin{array}{ll}
 & \text{Remainders} \\
2\overline{)109} & \\
2\overline{)54} & 1 \\
2\overline{)27} & 0 \\
2\overline{)13} & 1 \\
2\overline{)6} & 1 \\
2\overline{)3} & 0 \\
1 & 1 \\
\end{array}
$$

Here we stop when the quotient, 1, is less than the divisor, 2, since this last quotient will be the next and last remainder.   Again the arrow indicates the sequence of bits that gives the binary equivalent of the number.   (Another method of converting a decimal integer into its binary equivalent is given in Problem 1.14.)

(b) To convert $N_F = 0.781\,25$ to its binary equivalent, multiply N_F and each successive fractional part by 2, noting the integral part of the product, as follows:

Multiplications	Integral parts
$0.781\,25 \times 2 = 1.562\,50$	1
$0.5625 \times 2 \;\; = 1.1250$	1
$0.125 \times 2 \;\;\; = 0.250$	0
$0.25 \times 2 \;\;\;\; = 0.50$	0
$0.50 \times 2 \;\;\;\; = 1.00$	1

The zero fractional part indicates the end of the calculations.   Observe that the integral part of any product can only be 0 or 1, since we are doubling a number less than one.   The sequence of integral-part digits from the top down, as indicated by the arrow, yields the required binary equivalent.   That is, $N_F = 0.781\,25 = 0.11001_2$.

In actual practice, the above multiplications may be condensed as follows:

$$
\begin{array}{r}
0.781\ 25 \\
\times 2 \\
\hline
\underline{1}.562\ 50 \\
\times 2 \\
\hline
\underline{1}.125\ 00 \\
\times 2 \\
\hline
\underline{0}.250\ 00 \\
\times 2 \\
\hline
\underline{0}.500\ 00 \\
\times 2 \\
\hline
\underline{1}.000\ 00
\end{array}
$$

Observe that the integral part of each product is underlined and does not figure in the next multiplication. Again the arrow indicates the sequence of integral-part digits that gives the required binary representation.

We have found the binary equivalents of the integral and fractional parts of the decimal number $N = 109.781\ 25$. The binary equivalent of N is simply the sum of these two equivalents:

$$N = N_I + N_F = 110\ 1101.1100\ 1$$

EXAMPLE 1.3 Let $N = 13.6875$. We convert the integral part, $N_I = 13$, and the fractional part, $N_F = 0.6875$, into binary form as above:

	Remainders		Integral parts
2)13			0.6875
2)6	1		$\times 2$
2)3	0		$\underline{1}.3750$
1	1		$\times 2$
			$\underline{0}.7500$
			$\times 2$
			$\underline{1}.5000$
			$\times 2$
			$\underline{1}.0000$

Thus, $N = 13.6875 = 1101.1011_2$.

Remark: The binary equivalent of a terminating decimal fraction does not always terminate. For example, we convert $N = 0.6$ as above:

Multiplications	Integral parts
$0.6 \times 2 = \underline{1}.2$	1
$0.2 \times 2 = \underline{0}.4$	0
$0.4 \times 2 = \underline{0}.8$	0
$0.8 \times 2 = \underline{1}.6$	1

At this point in the procedure, we again multiply 0.6 by 2. This means the above four steps will repeat and hence we will obtain the above four bits again and again. That is,

$$N = 0.6 = 0.1001\ 1001\ 1001\ \ldots_2$$

(See Problem 1.15. The number of bits which repeat is not always four; nor does the repeating block necessarily begin at the binary point: it all depends on the given N.)

1.4 BINARY ADDITION AND MULTIPLICATION

The execution of numerical calculations is essentially the same in all positional numeration systems. This section investigates addition and multiplication of binary numbers; subtraction and division will be covered in Section 1.5.

Binary Addition

First we illustrate addition in the familiar decimal system. The addition of two decimal numbers is accomplished according to the following three-step algorithm:

STEP 1. Add the first (rightmost) column.

STEP 2. Record the units digit of the column sum. If the sum exceeds nine, carry the tens digit, 1, to the next column.

STEP 3. If there are additional columns or if there is a carry from Step 2, add the next column and repeat Step 2. Otherwise stop.

(As we are adding only two decimal numbers, no column sum, even with a carry, can exceed nineteen; so the tens digit of the sum cannot exceed 1.)

EXAMPLE 1.4 Consider the following decimal sum:

$$
\begin{array}{rl}
34\,573 & \textbf{Addend} \\
+\,52\,861 & \textbf{Augend}
\end{array}
$$

We add the numbers using the above algorithm.

STEP 1. $3+1=4$.

STEP 2.
$$
\begin{array}{rl}
34\,573 & \textbf{Addend} \\
+\,52\,861 & \textbf{Augend} \\
\hline
4 &
\end{array}
$$

STEP 3. $7+6=13$.

STEP 2.
$$
\begin{array}{rl}
1 & \textbf{Carries} \\
34\,573 & \textbf{Addend} \\
+\,52\,861 & \textbf{Augend} \\
\hline
34 &
\end{array}
$$

STEP 3. $1+5+8=14$.

STEP 2.
$$
\begin{array}{rl}
1\ 1 & \textbf{Carries} \\
34\,573 & \textbf{Addend} \\
+\,52\,861 & \textbf{Augend} \\
\hline
434 &
\end{array}
$$

STEP 3. $1+4+2=7$.

STEP 2.
$$
\begin{array}{rl}
1\ 1 & \textbf{Carries} \\
34\,573 & \textbf{Addend} \\
+\,52\,861 & \textbf{Augend} \\
\hline
7\,434 &
\end{array}
$$

STEP 3. $3+5=8$.

STEP 2.
$$
\begin{array}{rl}
1\ 1 & \textbf{Carries} \\
34\,573 & \textbf{Addend} \\
+\,52\,861 & \textbf{Augend} \\
\hline
87\,434 & \textbf{Sum}
\end{array}
$$

STEP 3. Stop.

In actual practice, Steps 1 and 3 are done in one's head. Also, each Step 2 is applied to the same scheme, to yield

$$
\begin{array}{r}
1\ 1 \\
34\,573 \\
+\,52\,861 \\
\hline
87\,434
\end{array}
$$

The basic property of the algorithm is that any two numbers, however large, can be added if one merely knows (i) the addition of (any) two digits and (ii) the addition of two digits and a carry of 1. The required addition facts are usually presented in an addition table of decimal digits, which is memorized early in our education.

Happily, the three-step algorithm for the addition of decimal numbers holds also for the addition of binary numbers if, in Step 2, we replace "nine" by "one" and "tens" by "twos". The addition table for the binary digits 0 and 1 appears as Table 1-3; the only addition facts needed for binary addition appear in Table 1-4.

Table 1-3. Binary Addition Table

+	0	1
0	0	1
1	1	10

Table 1-4. Binary Addition Facts

$0 + 0 = 0$
$0 + 1 = 1$
$1 + 0 = 1$
$1 + 1 = 0$, with a carry of 1
$1 + 1 + 1 = 1$, with a carry of 1

EXAMPLE 1.5 We evaluate the binary sum

$$
\begin{array}{rl}
111 & \textbf{Addend} \\
+\,101 & \textbf{Augend}
\end{array}
$$

by means of the three-step algorithm.

STEP 1. $1 + 1 = 0$, with a carry of 1.

STEP 2.

$$
\begin{array}{rl}
1 & \textbf{Carries} \\
111 & \textbf{Addend} \\
+\ \ 101 & \textbf{Augend} \\
\hline
0 &
\end{array}
$$

STEP 3. $1 + 1 = 0$, with a carry of 1.

STEP 2.

$$
\begin{array}{rl}
1\,1 & \textbf{Carries} \\
111 & \textbf{Addend} \\
+\ \ 101 & \textbf{Augend} \\
\hline
00 &
\end{array}
$$

STEP 3. $1 + 1 + 1 = 1$, with a carry of 1.

STEP 2.

$$
\begin{array}{rl}
1\,1\,1 & \textbf{Carries} \\
111 & \textbf{Addend} \\
+\ \ 101 & \textbf{Augend} \\
\hline
100 &
\end{array}
$$

STEP 3. $1 + 0 = 1$.

STEP 2.

$$
\begin{array}{rl}
1\,1\,1 & \textbf{Carries} \\
111 & \textbf{Addend} \\
+\ \ 101 & \textbf{Augend} \\
\hline
1101 & \textbf{Sum}
\end{array}
$$

STEP 3.   Stop.

Again, in actual practice, we do Steps 1 and 3 in our head, using Table 1-4. Also, when adding the last (leftmost) column, we need not carry the last 1 but simply place it below the line. Thus, the sum would actually appear as

$$
\begin{array}{r}
11 \\
111 \\
+\ \ 101 \\
\hline
1100
\end{array}
$$

In fact, often we don't even write down the 1s that are carried.

EXAMPLE 1.6

(a) Calculate the binary sum $110011101 + 10110111$. We have

$$
\begin{array}{r}
1\ \ 111111 \\
110011101 \\
+\ \ \ \ 10110111 \\
\hline
1001010100
\end{array}
$$

(b) Calculate the sum $1001 + 1101 + 110 + 1011$. We add the first two numbers and then add the others one at a time to a running total, as follows:

$$
\begin{array}{rl}
1001 & \textbf{First number} \\
+\ \ \ \ 1101 & \textbf{Second number} \\
\hline
10110 & \textbf{Sum} \\
+\ \ \ \ \ 110 & \textbf{Third number} \\
\hline
11100 & \textbf{Sum} \\
+\ \ \ \ 1011 & \textbf{Fourth number} \\
\hline
100111 & \textbf{Final sum}
\end{array}
$$

(c) Calculate $11011.01 + 101.1101$. As with decimal fractions, first write the binary numbers so that the binary points are vertically aligned, and then add:

$$
\begin{array}{r}
11011.01 \\
+\ \ \ \ \ 101.1101 \\
\hline
100001.0001
\end{array}
$$

Binary Multiplication

Recall that the multiplication of decimal numbers can be reduced to multiplying numbers by digits and addition.

EXAMPLE 1.7 To calculate 1234×263, first multiply 1234 by the digits 3, 6, and 2, as follows:

$$
\begin{array}{r}
1234 \\
\times\ \ 263 \\
\hline
3702 \\
7404 \\
2468 \ \ \ \ \
\end{array}
$$

Then add the three bottom rows of numbers:

$$
\begin{array}{r}
1234 \\
\times\ \ 263 \\
\hline
3702 \\
7404 \\
2468 \ \ \ \ \ \\
\hline
324542
\end{array}
$$

The sum of the three rows is the required product.

The rule for decimal multiplication also holds for binary multiplication. In fact, binary multiplication is simpler, since multiplying a number by the bit 0 or 1 yields respectively 0 or the number itself.

EXAMPLE 1.8   To calculate the binary product 1101011×10110 multiply 1101011 by the digits 0, 1, 1, 0, and 1, as follows:

$$
\begin{array}{r}
1101011 \\
\times \quad 10110 \\
\hline
0000000 \\
1101011 \\
1101011 \\
0000000 \\
1101011 \\
\hline
\end{array}
$$

Then add the five bottom rows of numbers. In actual practice, we do not write down any zero products.   We bring down initial zeros, if any, and form a running total, adding one nonzero row after another:

110101 1	**Initial zero**
$\times$ 1011⓪	
1101011⓪	**First nonzero product**
1101011	**Second nonzero product**
101000001 0	**Sum**
1101011	**Third nonzero product**
10010011001 0	**Final sum**

The final sum is the required product.   We emphasize that *it is extremely important to line up the numbers in the correct columns.*

EXAMPLE 1.9   Calculate the binary product 11.01×101.1.   We have

$$
\begin{array}{r}
1\,1.0\,1 \\
\times\,1\,0\,1.1 \\
\hline
1\,1\,0\,1 \\
1\,1\,0\,1 \\
100\,1\,1\,1 \\
1101 \\
\hline
10001.1\,1\,1 \\
\end{array}
$$

The number of binary places in the product is three, which is the sum of the binary places in the numbers being multiplied.   This is the same rule that is applied in decimal multiplication.

1.5   BINARY SUBTRACTION AND DIVISION

This section investigates the other two basic arithmetic operations, subtraction and division. Once again, in discussing each operation, we first show the analogous steps followed in the decimal system.

Binary Subtraction

Subtraction in the decimal system can be performed using the following two-step algorithm:

STEP 1.   If the lower (subtrahend) digit is greater than the upper (minuend) digit, borrow from the next column to the left.   (The value borrowed is equal to ten.)

STEP 2.   Subtract the lower value from the upper value.

In Step 1 "borrowing" means appropriating, with no intention of paying back.   Although the value borrowed is ten, the column from which one borrows is only decreased by one.

EXAMPLE 1.10   Consider the following decimal difference:

$$
\begin{array}{r}
548 \qquad \textbf{Minuend} \\
-263 \qquad \textbf{Subtrahend} \\
\hline
\end{array}
$$

We subtract the numbers using the above algorithm.

STEPS 1 and 2.
$$\begin{array}{r} 548 \\ -263 \\ \hline 5 \end{array}$$
Minuend
Subtrahend

STEP 1.
$$\begin{array}{r} 4\,14 \\ \not5\,\not4\,8 \\ -2\,6\,3 \\ \hline 5 \end{array}$$
Borrowings
Minuend
Subtrahend

STEP 2.
$$\begin{array}{r} 4\,14 \\ \not5\,\not4\,8 \\ -2\,6\,3 \\ \hline 8\,5 \end{array}$$
Borrowings
Minuend
Subtrahend

STEPS 1 and 2.
$$\begin{array}{r} 4\,14 \\ \not5\,\not4\,8 \\ -2\,6\,3 \\ \hline 2\,8\,5 \end{array}$$
Borrowings
Minuend
Subtrahend
Difference

Observe that there was only one borrowing; ten was added to 4 to make 14, and 1 was subtracted from 5 to make 4. In actual practice, we do not cross out the 4 and write 14, but simply place a 1 in front of the 4:

$$\begin{array}{r} 4 \\ \not5^{1}4\,8 \\ -2\,6\,3 \\ \hline 2\,8\,5 \end{array}$$

The process becomes more complex when one needs to borrow from a digit which is 0. Then two or more borrows must be made toward the left. Specifically, we borrow from the first nonzero digit to the left, whereupon the intervening 0 digits become $10 - 1 = 9$.

EXAMPLE 1.11   Consider the difference

$$\begin{array}{r} 84\,003 \\ -21\,625 \end{array}$$

First we have

$$\begin{array}{r} 3\,9\,9 \\ 8\,\not4\,\not0\,\not0^{1}3 \\ -2\,1\,6\,2\,5 \end{array}$$

Here the 4, the first nonzero digit to the left of 3, becomes 3; the two intervening 0s become 9s; and the 3 becomes 13. Then we obtain the required difference as

$$\begin{array}{r} 3\,9\,9 \\ 8\,\not4\,\not0\,\not0^{1}3 \\ -2\,1\,6\,2\,5 \\ \hline 6\,2\,3\,7\,8 \end{array}$$

The subtraction facts required for the two-step algorithm consist in the differences of each digit from larger digits and the differences of each digit from smaller digits plus ten. Again, such facts are early memorized.

Binary subtraction also uses the two-step algorithm. Moreover, the only subtraction facts needed for binary subtraction are the four listed in Table 1-5. The first three entries are translations of the addition facts

$$0 + 0 = 0 \qquad 1 + 0 = 1 \qquad 0 + 1 = 1$$

(Recall that subtraction is the inverse operation to addition.) The last entry comes from

$$1 + 1 = 10 \qquad \text{and so} \qquad 10 - 1 = 1$$

That is, the difference $0 - 1$ requires borrowing, which then yields $10 - 1 = 1$.

Table 1-5.　Binary Subtraction Facts

$$0 - 0 = 0$$
$$1 - 0 = 1$$
$$1 - 1 = 0$$
$$0 - 1 = 1, \text{ with a borrow of 1 from the next column}$$

EXAMPLE 1.12　To evaluate the binary difference $11101 - 1011$, we apply the subtraction facts in Table 1-5 and obtain

$$
\begin{array}{r}
0 \\
11\cancel{1}01 \\
-\ \ 1011 \\
\hline
10010
\end{array}
$$

Observe that we borrowed 1 from the third column because of the difference $0 - 1$ in the second column.

As with decimal subtraction, binary subtraction becomes more complex when we need to borrow from a digit which is 0.　Again, we borrow from the first nonzero digit to the left, but now each intervening 0 becomes $10 - 1 = 1$.

EXAMPLE 1.13　Consider the difference

$$
\begin{array}{r}
11000 \\
-\ 10011
\end{array}
$$

We obtain

$$
\begin{array}{r}
0\,1\,1 \\
1\cancel{1}\cancel{0}\cancel{0}0 \\
-\ 10011 \\
\hline
101
\end{array}
$$

Here a difference $0 - 1$ occurs in the first column; hence we borrow from the fourth column, where the first nonzero digit to the left appears, and the two intervening 0s become 1s.

EXAMPLE 1.14

(a)　Calculate the difference $1100101001 - 110110110$.　We have

$$
\begin{array}{r}
0\,0\,1\,1\,0\ \ 0\,1 \\
\cancel{1}\cancel{1}\cancel{0}\cancel{0}\cancel{1}0\cancel{1}\cancel{0}01 \\
-\ \ \ 110110110 \\
\hline
101110011
\end{array}
$$

(b)　Calculate the difference $1101.101 - 11.10111$.　As with decimal fractions, we must first vertically align the binary points before subtracting.

$$
\begin{array}{r}
0\,1\,0\,0\,1\,0\,1 \\
1\cancel{1}\cancel{0}\cancel{1}.\cancel{1}\cancel{0}\cancel{1}\cancel{0}0 \\
-\ \ \ \ 11.10111 \\
\hline
1001.11101
\end{array}
$$

Binary Division

Recall that the division of decimal numbers can be reduced to multiplying the divisor by individual digits of the dividend and subtraction.

EXAMPLE 1.15 Calculate $42\,558 \div 123$. Here 123 is the divisor. Our algorithm for division yields:

$$
\begin{array}{r}
346 \\
\hline
123)\overline{42558} \\
369 \\
\hline
565 \\
492 \\
\hline
738 \\
738 \\
\hline
0
\end{array}
$$

That is, we multiply 123 by 3 and subtract the product, 369, from 425; then we multiply 123 by 4 and subtract the product, 492, from 565; lastly we multiply 123 by 6 and subtract the product, 738, from 738, to obtain a 0 remainder. [Because of the geometry of the scheme, what these steps actually accomplish is first to subtract 3×10^2 times the divisor from the dividend, then 4×10 times the divisor from what is left, and then 6 times the divisor from what is left. At that point the dividend is exhausted, showing that the dividend originally contained the divisor

$$3 \times 10^2 + 4 \times 10 + 6 = 346$$

times.]

The above algorithm also works for binary division. In fact, multiplying the divisor by the only nonzero digit, 1, does not change the number; hence the algorithm for division reduces to repeated subtraction of the divisor (times a power of 2).

EXAMPLE 1.16 Evaluate $1010001 \div 11$. We have

$$
\begin{array}{r}
11011 \\
\hline
11)\overline{1010001} \\
11 \\
\hline
100 \\
11 \\
\hline
100 \\
11 \\
\hline
11 \\
11 \\
\hline
0
\end{array}
$$

Thus the quotient is 11011.

As in decimal division of integers, a remainder is possible when one binary integer is divided by another. Also, the division of binary fractions is handled the same way as the division of decimal fractions; that is, one converts the divisor to an integer by moving the binary point in both the divisor and the dividend the same number of places.

EXAMPLE 1.17

(a) Evaluate $1110111 \div 1001$. Applying the usual division algorithm, we obtain

$$
\begin{array}{r}
1101 \\
\hline
1001)\overline{1110111} \\
1001 \\
\hline
1011 \\
1001 \\
\hline
1011 \\
1001 \\
\hline
10
\end{array}
$$

The quotient is 1101, with a remainder 10.

(b) Evaluate $111.0000 \div 1.01$. First move the binary point in both the divisor and dividend two places to convert the divisor, 1.01, into an integer, 101; then divide as usual:

$$
\begin{array}{r}
1\ 01.101 \\
1.01.)\overline{111.00.001} \\
101 \\
\hline
10\ 00 \\
1\ 01 \\
\hline
11\ 0 \\
10\ 1 \\
\hline
101 \\
\underline{101}
\end{array}
$$

The quotient is 101.101.

1.6 COMPLEMENTS

Arithmetic complements appear in two separate but related situations. First of all, complements come up in storing numbers in the computer. While human beings use the signs $+$ and $-$ to denote positive and negative numbers, the computer can process data only in terms of bits. Although it is possible to reserve a bit to denote the sign of a number (say, 0 for $+$ and 1 for $-$), many computers store negative numbers in the form of their arithmetic complements.

Complements also arise in the operation of subtraction. In fact, complements can be used to reduce subtraction to addition. This is especially useful as it avoids the possibility of repeated borrowing from one column to another.

There are two types of complements, the *radix-minus-one complement* and the *radix complement*. (The term *complement* by itself usually signifies the radix complement.) First we discuss these complements in the familiar decimal system, where they are called the *nines complement* and *tens complement*, respectively. Then we will discuss them in the binary system, where they are called the *ones complement* and *twos complement*, respectively.

Decimal Complements

Let A be a decimal number. The nines complement of A is obtained by subtracting each digit of A from 9; and the tens complement of A is its nines complement plus one. For example,

Decimal number	4308	123 123	9672	751 620
Nines complement	5691	876 876	0327	248 379
Tens complement	5692	876 877	0328	248 380

To illustrate the use of complements in subtraction, let A and B be two decimal integers with the same number of digits—say, four—and suppose that A is less than B. We can rewrite the difference $Y = B - A$ as

$$
\begin{aligned}
Y &= B - A + (9999 + 1 - 10\,000) \\
&= B + (9999 - A + 1) - 10\,000 \\
&= B + [(9999 - A) + 1] - 10\,000
\end{aligned}
$$

In other words, we can calculate Y either by adding the tens complement of A to B, or by adding the nines complement of A to B and then adding 1. In either case, we must subtract 10 000; but since A and B have only four digits, subtracting 10 000 will simply delete the leading 1 from the sum. Although the evaluation of Y using complements requires four calculations, it has the important advantage that no borrowing is necessary in finding the nines complement.

EXAMPLE 1.18   Evaluate the difference $Y = B - A$, where $A = 4816$ and $B = 6142$.

(*a*)   First we use ordinary subtraction:

$$\begin{array}{r} 6142 \\ -\ 4816 \\ \hline 1326 \end{array}$$

Observe that we had to borrow twice.

(*b*)   Here we add the nines complement of A, which is 5183:

$$\begin{array}{r} 6142 \\ +\ 5183 \\ \hline ⓵\ 1325 \\ \longrightarrow 1 \\ \hline 1326 \end{array}$$

We delete the leading 1 (equivalent to subtracting 10 000), and then add the 1 to the sum.   (This method is called *end-around carry*.)

(*c*)   Here we add the tens complement of A, which is $5183 + 1 = 5184$:

$$\begin{array}{r} 6142 \\ +\ 5184 \\ \hline ⓵1326 \end{array}$$

Now we simply delete the leading 1 to obtain the solution.

The requirement that A and B have the same number of digits represents no real restriction, since one can always introduce 0s at the beginning of a number, if necessary.   In fact, many calculating devices, including many computers, use registers and storage locations of which the capacity is a fixed number of digits.   In such devices, the leading 1, which is to be deleted, is lost automatically.

EXAMPLE 1.19   Consider a mechanical desk calculator whose registers hold decimal numbers of exactly eight digits.   We assume that the calculator subtracts a number by adding its tens complement.   Furthermore, if an addition operation results in more than eight digits, then the most significant digits, i.e. the leading digits, are lost.   Given $A = 216$ and $B = 563$, we want to evaluate the difference $Y = B - A$.   The numbers A and B will appear in registers as follows:

0	0	0	0	0	5	6	3	*B*
0	0	0	0	0	2	1	6	*A*

During the process of subtraction, register contents will appear as follows:

	0	0	0	0	0	5	6	3	*B*
+	9	9	9	9	9	7	8	4	**Complement of A**
=	0	0	0	0	0	3	4	7	**Difference**

The leading 1 has been automatically lost.

Remark 1:   *Overflow* is that part of the result of an operation which is lost because the resulting value exceeds the capacity of the intended storage location.   In subtraction, overflow always occurs when A is less than B and we are adding the complement of A to B.

Remark 2:   Suppose now that A is greater than B, say $A = 5872$ and $B = 2148$.   Then the difference $Y = B - A$ is negative:

$$\begin{array}{r} 2148 \\ -\,5872 \\ \hline -\,3724 \end{array}$$

Adding the complement of A to B, we obtain

$$\begin{array}{r} 2148 \\ +\,4128 \\ \hline 6276 \end{array}$$

Observe that now there is no leading 1 to be deleted (such is always the case when A is greater than B).   Hence we need to subtract 10 000 from 6276 to obtain Y.   In other words, the *negative of the tens complement* of the sum 6276 is our required difference, $Y = -3724$.

Binary Complements

The terminology and principle of complements in the decimal system can be easily translated into the binary system.   Specifically, if A is a binary number, the ones complement of A is obtained by subtracting each digit of A from 1, and the twos complement of A is its ones complement plus 1. For example,

Binary number	111100001111	110011001100	111000111000
Ones complement	000011110000	001100110011	000111000111
Twos complement	000011110001	001100110100	000111001000

Observe that taking the ones complement simply inverts each digit, i.e. 0 is replaced by 1 and 1 is replaced by 0.

As in the decimal system, we perform binary subtraction by adding the radix-minus-one (ones) complement plus one or by adding the radix (twos) complement.

EXAMPLE 1.20   Evaluate the difference $Y = B - A$, where $A = 10001110$ and $B = 11110000$.

(*a*)   First we perform ordinary binary subtraction:

$$\begin{array}{rl} 011 & \\ 11110000 & \quad B \\ -\,10001110 & \quad A \\ \hline 01100010 & \quad Y \end{array}$$

Observe that the borrowing was propagated to the third digit to the left.

(*b*)   The ones complement of A is 01110001.   We add this to B and then add 1:

$$\begin{array}{rl} 11110000 & \quad B \\ +\,01110001 & \quad \textbf{Ones complement of } A \\ \hline (1)01100001 & \\ \quad\quad\longrightarrow 1 & \\ \hline 01100010 & \end{array}$$

(This method is also given the name *end-around carry*.)

(*c*)   The twos complement of A is 01110010.   We add this to B:

$$\begin{array}{rl} 11110000 & \quad B \\ +\,01110010 & \quad \textbf{Twos complement of } A \\ \hline (1)01100010 & \end{array}$$

Deleting the 1 (which would be an overflow in an 8-bit register) gives us the difference Y.

EXAMPLE 1.21   Consider the difference $Y = B - A$, where $A = 110011$ and $B = 101010$. The (twos) complement of A is 001101. Adding this to B, we obtain

$$
\begin{array}{ll}
101010 & \mathbf{B} \\
+\,001101 & \textbf{Complement of } \mathbf{A} \\
\hline
110111 & \mathbf{Z}
\end{array}
$$

Observe that there is no 1 in the seventh place to delete. Hence, Z is not the difference Y. The reason for this is that A is greater than B. To obtain Y we must now subtract 1000000 from Z. In other words, the *negative of the twos complement* of Z, or -001001, is the required difference Y.

Solved Problems

DECIMAL SYSTEM

1.1   Evaluate: (a) 10^3, (b) 10^{-2}, (c) 10^0, (d) 2^4, (e) 2^{-1}, (f) 2^0.

For any positive integer n, we have $a^n = a \times a \times \cdots \times a$ (n factors); $a^{-n} = 1/a^n$; and $a^0 = 1$.

(a) $10^3 = 10 \times 10 \times 10 = 1000$ (d) $2^4 = 2 \times 2 \times 2 \times 2 = 16$

(b) $10^{-2} = \dfrac{1}{10^2} = \dfrac{1}{100} = 0.01$ (e) $2^{-1} = \dfrac{1}{2^1} = \dfrac{1}{2}$

(c) $10^0 = 1$ (f) $2^0 = 1$

1.2   Find the face value and the place value of the digit 4 in (a) 7425, (b) 146 723, (c) 305.54, (d) 0.012 345.

The face value of 4 is always 4 itself.   The place values are:

(a) $10^2 = 100$ (b) $10^4 = 10\,000$ (c) $10^{-2} = 0.01$ (d) $10^{-5} = 0.000\,01$

1.3   Rewrite in expanded notation (a) 2468, (b) 54.321.

Express each number as the sum of each digit times its place value.

(a) $2468 = 2 \times 10^3 + 4 \times 10^2 + 6 \times 10 + 8 \times 1$

(b) $54.321 = 5 \times 10 + 4 \times 1 + 3 \times 10^{-1} + 2 \times 10^{-2} + 1 \times 10^{-3}$

1.4   Evaluate $7.32 + 33.3 + 24.678$.

Write the numbers with their decimal points vertically aligned, attach trailing 0s so that all the numbers have the same number of decimal places, and then perform the addition.

$$
\begin{array}{r}
7.320 \\
33.300 \\
+\,24.678 \\
\hline
65.298
\end{array}
$$

1.5   Calculate (a) $62.47 - 35.7$, (b) $8 - 3.246$.

Vertically align the decimal points, and attach trailing 0s so that the numbers have the same number of decimal places.

$$
\begin{array}{ll}
(a) & \begin{array}{r}
5\;1 \\
6\!\!\!/2.47 \\
-\,35.70 \\
\hline
2\,6.77
\end{array}
\qquad
(b)\;
\begin{array}{r}
7\;9\;9 \\
8\!\!\!/.0\!\!\!/0\!\!\!/0 \\
-\,3.246 \\
\hline
4.7\,54
\end{array}
\end{array}
$$

1.6   Calculate (a) 62.04×2.5, (b) 0.054×0.0023.

Multiply as though integers.   Then place the decimal point in the product by counting the number of decimal places in the factors.

(a)
$$
\begin{array}{r}
6\,2.0\,4 \quad \text{(2 places)} \\
\times\quad 2.5 \quad \text{(1 place)} \\
\hline
3\,1\,0\,2\,0 \\
1\,2\,4\,0\,8 \\
\hline
1\,5\,5.1\,0\,0 \quad \text{(}2+1=3\;\text{places)}
\end{array}
$$

(b)
$$
\begin{array}{r}
0.054 \quad \text{(3 places)} \\
\times\,0.0\,023 \quad \text{(4 places)} \\
\hline
162 \\
1\,08 \\
\hline
0.000\,1\,242 \quad \text{(}3+4=7\;\text{places)}
\end{array}
$$

1.7   Evaluate $64.698 \div 1.23$.

Shift the decimal point in the divisor, 1.23, to the right, making the divisor an integer; then shift the decimal point in the dividend, 64.698, the same number of places to the right.   (In this case, we have to shift the decimal point 2 places to the right.)   Then perform the division.

$$
\begin{array}{r}
52.6 \\
1{.}23\,\overline{)\,64{.}69{.}8} \\
\underline{61\;5} \\
3\;19 \\
\underline{2\;46} \\
73\;8 \\
\underline{73\;8}
\end{array}
$$

Thus the quotient is 52.6 and there is no remainder.

BINARY SYSTEM

1.8   How many binary digits are there, which symbols are used for them, and what are they usually called?

There are two binary digits, symbolized 0 and 1.   They are called bits.

1.9   Give the face value and the place value of each underlined bit: (a) 101_10, (b) 1_011001, (c) 101.1101_01, (d) 11_0.00101.

The face value is the bit itself.   The place values are:

$$
\begin{array}{ll}
(a)\quad 2^2 = 4 & (c)\quad 2^{-4} = 1/16 \\
(b)\quad 2^5 = 32 & (d)\quad 2^0 = 1
\end{array}
$$

1.10   Rewrite in expanded notation (a) 110110, (b) 11.01101.

Express each binary number as the sum of each bit times its place value.

(a) $110110 = 1 \times 2^5 + 1 \times 2^4 + 0 \times 2^3 + 1 \times 2^2 + 1 \times 2 + 0 \times 1$

(b) $11.01101 = 1 \times 2 + 1 \times 1 + 0 \times 2^{-1} + 1 \times 2^{-2} + 1 \times 2^{-3} + 0 \times 2^{-4} + 1 \times 2^{-5}$

BINARY-DECIMAL INTERCONVERSION

1.11   Convert each binary number to its decimal form: (a) 1110011, (b) 110.1011.

Write the place value over each bit, and then add up those place values whose bit is 1.

(a)

$$
\begin{array}{ccccccc}
2^6 & 2^5 & 2^4 & 2^3 & 2^2 & 2 & 1 \\
1 & 1 & 1 & 0 & 0 & 1 & 1
\end{array}
$$

$$
\begin{aligned}
& 1 \\
& 2 \\
& 16 \\
& 32 \\
& \underline{64} \\
& 115
\end{aligned}
$$

Hence, 115 is the decimal equivalent.

(b)   Use Table 1-1 for the decimal values of negative powers of 2.

$$
\begin{array}{ccccccc}
2^2 & 2 & 1 & 2^{-1} & 2^{-2} & 2^{-3} & 2^{-4} \\
1 & 1 & 0 & .1 & 0 & 1 & 1
\end{array}
$$

$$
\begin{aligned}
& 0.0625 \\
& 0.1250 \\
& 0.5000 \\
& 2 \\
& \underline{4} \\
& 6.6875
\end{aligned}
$$

Hence, 6.6875 is the decimal equivalent.

1.12   Convert the binary number 110101 to its decimal form.

Since 110101 is an integer, it can be converted by use of the following algorithm rather than the method of Problem 1.11(a).

Conversion of Binary Integers: Double the first (leftmost) digit and add it to the next digit to the right. Double the sum and add it to the next digit. Repeat this process until the last (rightmost) digit is added. The final sum is the required decimal equivalent.

Applying the algorithm to 110101, we double the first bit, 1, to get 2, write 2 under the second bit, and add:

$$
\begin{array}{cccccc}
1 & 1 & 0 & 1 & 0 & 1 \\
  & 2 \\ \hline
  & 3
\end{array}
$$

We now double the sum, 3, to get 6, write 6 under the next bit, and add:

$$
\begin{array}{cccccc}
1 & 1 & 0 & 1 & 0 & 1 \\
  & 2 & 6 \\ \hline
  & 3 & 6
\end{array}
$$

We double the last sum, 6, to get 12, write 12 under the next bit, 1, and add to get 13:

$$
\begin{array}{cccccc}
1 & 1 & 0 & 1 & 0 & 1 \\
  & 2 & 6 & 12 & & \\
\hline
  & 3 & 6 & 13 & &
\end{array}
$$

Doubling the last sum, 13, yields 26, and adding 26 to the next bit, 0, gives the sum 26:

$$
\begin{array}{cccccc}
1 & 1 & 0 & 1 & 0 & 1 \\
  & 2 & 6 & 12 & 26 & \\
\hline
  & 3 & 6 & 13 & 26 &
\end{array}
$$

Doubling the last sum, 26, gives 52, and adding 52 to the last bit, 1, gives 53:

$$
\begin{array}{cccccc}
1 & 1 & 0 & 1 & 0 & 1 \\
  & 2 & 6 & 12 & 26 & 52 \\
\hline
  & 3 & 6 & 13 & 26 & \circledcirc{53}
\end{array}
$$

$$
\begin{array}{r}
1 \\
\times 2 \\
\hline
2 \\
+ 1 \\
\hline
3 \\
\times 2 \\
\hline
6 \\
+ 0 \\
\hline
6 \\
\times 2 \\
\hline
12 \\
+ 1 \\
\hline
13 \\
\times 2 \\
\hline
26 \\
+ 0 \\
\hline
26 \\
\times 2 \\
\hline
52 \\
+ 1 \\
\hline
\circledcirc{53}
\end{array}
$$

Fig. 1-1

We have circled the last sum, which is the required decimal equivalent.

This algorithm is called *double summing* or *Horner's method*. The calculations may also be carried out in the format of Fig. 1-1. Observe that one alternates between multiplying by the base $b = 2$ and adding the next digit.

1.13   Convert the decimal number $N = 437.406\,25$ to its binary equivalent.

The integral part of N is $N_I = 437$, and the fractional part of N is $N_F = 0.406\,25$. We convert each part separately.

Divide $N_I = 437$ and each succeeding quotient by 2, noting the remainders:

$$
\begin{array}{rl}
 & \textbf{Remainders} \\
2\overline{)437} & \\
2\overline{)218} & 1 \uparrow \\
2\overline{)109} & 0 \\
2\overline{)54} & 1 \\
2\overline{)27} & 0 \\
2\overline{)13} & 1 \\
2\overline{)6} & 1 \\
2\overline{)3} & 0 \\
1 & 1
\end{array}
$$

The remainders in reverse order, as indicated by the arrow, 1 1011 0101, give the binary equivalent of $N_I = 437$.

Multiply $N_F = 0.406\,25$ and each succeeding fractional part by 2, noting the integral part of each product:

$$
\begin{array}{r}
0.40625 \\
\times 2 \\
\hline
0.81250 \\
2 \\
\hline
1.6250 \\
2 \\
\hline
1.250 \\
2 \\
\hline
0.50 \\
2 \\
\hline
1.0
\end{array}
$$

The sequence of integral parts, in the order indicated by the arrow, yields the required binary equivalent, .01101.

Putting the equivalents of N_I and N_F together gives us the binary equivalent of N:

$$N = 1\,1011\,0101.0110\,1_2$$

1.14   Convert the decimal number 91 into its binary form.

We convert 91 to its binary form by use of the following algorithm rather than the division method of Problem 1.13.

    Subtraction Algorithm: Beginning with the given decimal number, subtract the largest power of base 2. Repeat the process of subtracting the largest power of base 2 from each difference until zero is obtained. The binary number with bit 1 in those places whose place values were subtracted, and with bit 0 elsewhere, is the required binary equivalent.

Using Table 1-1, which lists powers of base 2, we obtain:

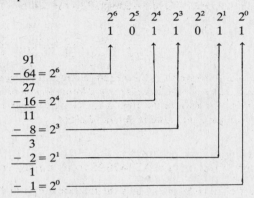

Thus, $91 = 1011011_2$.

1.15   Using the formula for the sum of a geometric series,

$$1 + r + r^2 + r^3 + \cdots = \frac{1}{1-r} \qquad (-1 < r < +1)$$

verify the conversion $0.6 = 0.1001\ 1001\ 1001 \cdots_2$.

We have

$$0.1001\ 1001\ 1001 \cdots_2 = (1 \times 2^{-1} + 1 \times 2^{-4}) + (1 \times 2^{-5} + 1 \times 2^{-8}) + (1 \times 2^{-9} + 1 \times 2^{-12}) + \cdots$$
$$= (1 \times 2^{-1} + 1 \times 2^{-4})[1 + 2^{-4} + 2^{-8} + \cdots]$$

The bracketed expression is a geometric series with $r = 2^{-4}$, and so

$$0.1001\ 1001\ 1001 \cdots_2 = (1 \times 2^{-1} + 1 \times 2^{-4})\left[\frac{1}{1 - 2^{-4}}\right]$$
$$= \left(\frac{1}{2} + \frac{1}{16}\right)\left[\frac{16}{15}\right] = \frac{8}{15} + \frac{1}{15} = \frac{9}{15} = \frac{6}{10}$$

BINARY ARITHMETIC

1.16   Evaluate the binary sums (a) $11011 + 1010$, (b) $110.1101 + 1011.011$.

Use Table 1-4.

(a)

$$
\begin{array}{r}
1\ 1\\
11011\\
+\quad 1010\\
\hline
100101
\end{array}
$$

(*b*) Vertically align the binary points and attach 0s so that both numbers have the same number of binary places:

$$
\begin{array}{r}
1111\ 1 \\
110.1101 \\
+\ 1011.0110 \\
\hline
10010.0011
\end{array}
$$

1.17 Evaluate the binary sum $11011 + 111001 + 1001 + 11001$.

Add the numbers one at a time to a running total, beginning with the first number:

$$
\begin{array}{rl}
11011 & \textbf{First number} \\
+\ \ 111001 & \\
\hline
1010100 & \textbf{First sum} \\
+\ \ \ \ \ 1001 & \\
\hline
1011101 & \textbf{Second sum} \\
+\ \ \ \ 11001 & \\
\hline
1110110 & \textbf{Final sum}
\end{array}
$$

1.18 Evaluate the following binary products: (*a*) 110110×101, (*b*) 111.001×1.11.

It is essential to keep proper alignment.

(*a*)

$$
\begin{array}{r}
110110 \\
\times \ \ \ \ 101 \\
\hline
110110 \\
110110\ \ \ \\
\hline
100001110
\end{array}
$$

Thus 1 0000 1110 is the required product.   As usual, the product by the 0 digit of the multiplier is omitted.

(*b*) Omitting binary points, add the digitwise products one at a time to a running total, beginning with the first product:

$$
\begin{array}{rl}
111001 & \\
\times \ \ \ \ \ 111 & \\
\hline
111001 & \textbf{First product} \\
+\ 111001 & \textbf{Second product} \\
\hline
10101011 & \\
+\ 111001 & \textbf{Third product} \\
\hline
110001111 &
\end{array}
$$

The original factors contained 3 and 2 binary places, respectively.   Hence the product should contain $3 + 2 = 5$ binary places.   Inserting a binary point 5 places from the right, we obtain the required product, 1100.01111.

1.19 Evaluate the binary differences (*a*) $1110001 - 111011$, (*b*) $1101.0011 - 110.11011$.

Use Table 1-5.

(*a*)

$$
\begin{array}{r}
0\ 0\ 1\ 1 \\
1\cancel{1}\cancel{1}\cancel{0}\cancel{0}0\ 1 \\
-\ 1\ 1\ 1\ 0\ 1\ 1 \\
\hline
1\ 1\ 0\ 1\ 1\ 0
\end{array}
$$

Observe that we had to borrow for the second difference, but the first available 1 was in the fifth column (from the right); hence the two intermediate column digits became $10 - 1 = 1$.

(b) Vertically align the binary points, and append 0s so that both numbers have the same number of binary places.

$$
\begin{array}{r}
00\ \ 01\ \ 00 \\
\cancel{1}\cancel{1}0\cancel{1}.\cancel{0}0\cancel{1}\cancel{1}0 \\
-\ 110.11011 \\
\hline
110.01011
\end{array}
$$

1.20 Evaluate $111001 \div 1001$ to two binary places.

Binary division reduces to repeated subtraction of the divisor, since the only nonzero digit is 1.

$$
\begin{array}{r}
110.01 \\
1001\overline{)111001.00} \\
\underline{1001} \\
1010 \\
\underline{1001} \\
11\ 00 \\
\underline{10\ 01} \\
11
\end{array}
$$

The quotient to two binary places is 110.01.

COMPLEMENTS

1.21 Determine the nines and tens complements of the decimal numbers (a) 3268, (b) 479 200, (c) 99 132 756, (d) 2 233 778 899.

For each number, subtract each digit from 9 to get the nines complement.   The nines complement plus 1 gives the tens complement.

	(a)	(b)	(c)	(d)
Nines complement	6731	520 799	00 867 243	7 766 221 100
Tens complement	6732	520 800	00 867 244	7 766 221 101

1.22 Consider the decimal numbers of Problem 1.21 as input to an 8-place decimal calculating machine.   Find the nines and tens complements of the numbers in the machine.

First introduce 0s at the left so that each number has 8 digits; then find complements.

	(a)	(b)	(c)
Number	00 003 268	00 479 200	99 132 756
Nines complement	99 996 731	99 520 799	00 867 243
Tens complement	99 996 732	99 520 800	00 867 244

(d) If the machine is built to omit the most significant (leading) digits when any number with more than 8 digits is inputted, then the number will be registered as 33 778 899.   In this case, we have

Nines complement	66 221 100
Tens complement	66 221 101

1.23 Find the following differences using (tens) complements:

(a)	53 726	(b)	215 743	(c)	71 566 220	(d)	2658
	− 14 503		− 56 100		− 44 332 211		− 4321

Add the tens complement, and then delete the overflow 1 (shown circled):

(a)	53 726	(b)	215 743	(c)	71 566 220
	+ 85 497		+ 943 900		+ 55 667 789
	①39 223		①159 643		①27 234 009

(d) Here there is no overflow:

$$\begin{array}{r} 2658 \\ + 5679 \\ \hline 8337 \end{array}$$

which means the difference is a negative number.   The tens complement of the sum is 1663; hence the difference is −1663, the negative of the tens complement.

1.24 Find the maximum positive integer and the minimum negative integer that can be registered in a 6-place, decimal, desk calculator, if (a) the first place is reserved for the sign of the integer, (b) negative numbers are registered as their (tens) complement. Which way can accommodate more numbers?

(a)

Maximum positive integer	+99 999
Minimum negative integer	−99 999

(b) Assume that a stored integer is positive if its first digit is 0 to 4, and negative if its first digit is 5 to 9. Then

Maximum positive integer	499 999
Minimum negative integer	−500 000

where −500 000 is registered as its tens complement, 500 000.   Under this system, five times as many numbers can be stored, as compared with the system of part (a).

1.25 Determine the ones and twos complements of the binary numbers (a) 110110, (b) 11100111, (c) 110011001100, (d) 11110000111000.

Replace each 0 bit by 1 and each 1 bit by 0 to get the ones complement. (This is the same as subtracting each bit from 1.)   The ones complement plus 1 yields the twos complement.

	(a)	(b)	(c)	(d)
Ones complement	001001	00011000	001100110011	00001111000111
Twos complement	001010	00011001	001100110100	00001111001000

1.26 Consider the binary numbers of Problem 1.25 as input to a 12-place binary calculating machine. Find the ones and twos complements of the numbers in the machine.

First introduce 0s at the left so that each number has 12 bits.   Then find complements.

	(a)	(b)	(c)
Number	000000110110	000011100111	110011001100
Ones complement	111111001001	111100011000	001100110011
Twos complement	111111001010	111100011001	001100110100

(d) If the calculator is built to omit the leading bits when numbers with more than 12 bits are inputted, then the number is registered as 110000111000.   Hence,

Ones complement	001111000111
Twos complement	001111001000

1.27 Find the following differences using (twos) complements:

(a)	111000	(b)	11001100	(c)	111100001111	(d)	11000011
	− 110011		− 101110		− 110011110011		− 11101000

Add the twos complement and then delete the overflow 1 bit (shown circled).

(a)   111000   (b) 11001100   (c) 111100001111
     + 001100   + 11010001   + 001100001100
     ①000100   ①10011101   ①001000011011

(d) Here there is no overflow:

$$11000011$$
$$+ 00010111$$
$$11011010$$

This means the difference is a negative number, the negative of the twos complement of the sum, or -00100110.

1.28 Solve Problem 1.24 for an 8-place binary calculator.

(a)   **Maximum positive integer** $+1111111$
                       Minimum negative integer -1111111

(b) Assume that a stored number is positive if the first bit is 0 and negative if the first bit is 1.

                       Maximum positive integer   01111111
                       Minimum negative integer -10000000

where -10000000 is registered as its twos complement, 10000000. This second method can accommodate only one more number than can the first method.

Supplementary Problems

DECIMAL SYSTEM

1.29 Evaluate (a) 10^4, (b) 10^{-3}, (c) 10^{-1}, (d) 2^6, (e) 2^{-3}, (f) 2^0, (g) 2^3.

1.30 Evaluate (a) 8^3, (b) 8^0, (c) 8^{-2}, (d) 16^2, (e) 16^{-1}, (f) 16^{-3}.

1.31 Give the place value of each underlined digit: (a) 44 333, (b) 22 555.66, (c) 444.555, (d) 22.334 455.

1.32 Write in expanded notation (a) 13 579, (b) 321.789.

1.33 Find each sum: (a) $835.24 + 70.456$, (b) $55.5 + 6.66 + 0.777$.

1.34 Find each difference: (a) $456.7 - 35.79$, (b) $12 - 4.888$.

1.35 Find each product: (a) 38.24×3.7, (b) 0.0345×1.02.

1.36 Evaluate to two decimal places: (a) $36 \div 11$, (b) $83.472 \div 2.4$.

BINARY SYSTEM

1.37 Give the place value of each underlined bit: (a) 111000, (b) 11001100, (c) 111.000111, (d) 11.00110011.

1.38 Rewrite in expanded notation (a) 11001100, (b) 111.000111.

1.39 Convert each binary number to its decimal equivalent: (a) 110110, (b) 111000111.

1.40   Convert to its decimal equivalent (*a*) 110.11, (*b*) 1010.10101.

1.41   Convert each decimal number to its binary equivalent: (*a*) 285, (*b*) 473, (*c*) 694.

1.42   Convert to its binary equivalent (*a*) 0.390 625, (*b*) 24.625, (*c*) 0.8, (*d*) 0.3.

1.43   Show that the decimal equivalent of a terminating binary fraction also terminates (in a 5).

BINARY ARITHMETIC

1.44   Find the binary sums (*a*) 1101 + 111, (*b*) 110011 + 11101, (*c*) 11100111 + 11000011, (*d*) 110.1101 + 1011.101.

1.45   Find the binary sums (*a*) 11001 + 11100 + 1011 + 110011, (*b*) 11.101 + 110.01 + 111.101 + 1101.1.

1.46   Find the binary products (*a*) 11100111 × 11, (*b*) 111011 × 1011, (*c*) 11.101 × 11.01.

1.47   Find the binary differences (*a*) 1100011 − 110111, (*b*) 10101010 − 110011, (*c*) 110.001 − 11.111.

1.48   Find the binary quotients (*a*) 1011011 ÷ 111, (*b*) 100.0001 ÷ 10.1, (*c*) 1011 ÷ 11.

COMPLEMENTS

1.49   Determine the nines and tens complements of the decimal numbers (*a*) 3201, (*b*) 453 800, (*c*) 78 923 019, (*d*) 3 334 455 566.

1.50   If the numbers of Problem 1.49 are input to an 8-place decimal calculator, find their nines and tens complements in the machine.

1.51   Find the ones and twos complements of the binary numbers (*a*) 110011, (*b*) 10001000, (*c*) 10111011101, (*d*) 111000000111.

1.52   Find the following binary differences using complements: (*a*) 1101 − 110, (*b*) 11100111 − 11001100, (*c*) 11000010 − 10111001, (*d*) 10101 − 11011.

Answers to Supplementary Problems

1.29   (*a*) 10 000,   (*b*) 0.001,   (*c*) 0.1,   (*d*) 64,   (*e*) 1/8,   (*f*) 1,   (*g*) 8

1.30   (*a*) 512,   (*b*) 1,   (*c*) 1/64,   (*d*) 256,   (*e*) 1/16,   (*f*) 1/4096

1.31   (*a*) $10^3 = 1000$,   (*b*) $10^2 = 100$,   (*c*) $10^{-2} = 0.01$,   (*d*) $10^{-4} = 0.0001$

1.32   (*a*) $1 \times 10^4 + 3 \times 10^3 + 5 \times 10^2 + 7 \times 10^1 + 9 \times 10^0$
         (*b*) $3 \times 10^2 + 2 \times 10^1 + 1 \times 10^0 + 7 \times 10^{-1} + 8 \times 10^{-2} + 9 \times 10^{-3}$

1.33   (*a*) 905.696,   (*b*) 62.937

1.34   (*a*) 420.91,   (*b*) 7.112

1.35   (*a*) 141.488,   (*b*) 0.035 190

1.36   (*a*) 3.27,   (*b*) 34.78

1.37   (*a*) 2^3,   (*b*) 2^5,   (*c*) 2^{-2},   (*d*) 2^{-5}

1.38   (*a*) $1 \times 2^7 + 1 \times 2^6 + 0 \times 2^5 + 0 \times 2^4 + 1 \times 2^3 + 1 \times 2^2 + 0 \times 2^1 + 0 \times 2^0$
    (*b*) $1 \times 2^2 + 1 \times 2^1 + 1 \times 2^0 + 0 \times 2^{-1} + 0 \times 2^{-2} + 0 \times 2^{-3} + 1 \times 2^{-4} + 1 \times 2^{-5} + 1 \times 2^{-6}$

1.39   (*a*) 54,   (*b*) 455

1.40   (*a*) 6.75,   (*b*) 10.656 25

1.41   (*a*) 1 0001 1101,   (*b*) 1 1101 1001,   (*c*) 10 1011 0110

1.42   (*a*) 0.011001,   (*b*) 11000.101,   (*c*) 0.1100 1100 1100 · · ·, (*d*)   0.0 1001 1001 1001 · · ·

1.43   (*Hint*:   Consider the decimal representation of a negative power of 2.)

1.44   (*a*) 1 0100,   (*b*) 101 0000,   (*c*) 1 1010 1010,   (*d*) 1 0010.0111

1.45   (*a*) 111 0011,   (*b*) 1 1111.000

1.46   (*a*) 10 1011 0101,   (*b*) 10 1000 1001,   (*c*) 1011.1100 1

1.47   (*a*) 10 1100,   (*b*) 111 0111,   (*c*) 10.010

1.48   (*a*) 1101,   (*b*) 1.101,   (*c*) 11.101010 · · ·

1.49   (*a*) 6798, 6799   (*c*) 21 076 980, 21 076 981
    (*b*) 546 199, 546 200   (*d*) 6 665 544 433, 6 665 544 434

1.50   (*a*) 99 996 798, 99 996 799   (*c*) 21 076 980, 21 076 981
    (*b*) 99 546 199, 99 546 200   (*d*) 65 544 433, 65 544 434

1.51   (*a*) 001100, 001101   (*c*) 01000100010, 01000100011
    (*b*) 01110111, 01111000   (*d*) 000111111000, 000111111001

1.52   (*a*) 0111,   (*b*) 00011011,   (*c*) 00001001,   (*d*) −00110

Chapter 2

Computer Codes

2.1 INTRODUCTION

To be processed by the computer, data must be encoded in some fashion into sequences of bits, i.e. into binary numbers. However, people find such strings as

$$110100110110 \qquad 110101100110 \qquad 110110110110$$

very difficult to remember and/or distinguish between; for us, the more compact decimal notation—3382, 3430, 3510—is far superior. To bridge the gap between the binary and decimal systems, other number systems are used in connection with computers—principally, the *octal* (base $b = 8$) and the *hexadecimal* (base $b = 16$) systems. On the one hand, because 8 and 16 are powers of 2, there is almost instant interconversion between the octal and hexadecimal systems and the binary system. On the other hand, the octal and hexadecimal systems are comparable in compactness with the decimal system.

2.2 NUMBER SYSTEMS

Any positive integer $b > 1$ can be chosen as the base for a positional number system similar to the decimal system ($b = 10$) or the binary system ($b = 2$). Such a system uses b symbols for the integers

$$0, \quad 1, \quad 2, \quad 3, \quad \ldots, \quad b - 1$$

These symbols are called the *digits* of the system.

Any integer N is represented in the system by a sequence of base-b digits:

$$N = a_n a_{n-1} \cdots a_1 a_0$$

Then b^k is the place value of a_k, and

$$N = a_n \times b^n + a_{n-1} \times b^{n-1} + \cdots + a_2 \times b^2 + a_1 \times b^1 + a_0 \times b^0$$

which is called the *expanded form* or *expanded notation* for N.

More generally, any number M is represented in the system by a sequence of digits with possibly an embedded base-b point. (We also include a subscript b at the end of the number, if there is any ambiguity about the base.) As with the decimal and binary systems, the place values of the digits to the right of the base-b point are the negative powers of b.

EXAMPLE 2.1 Consider the base $b = 5$ (*quintal*) number system, whose five digits are 0, 1, 2, 3, 4.

(*a*) Given $N = 41323_5$, then

$$N = 4 \times 5^4 + 1 \times 5^3 + 3 \times 5^2 + 2 \times 5 + 3 \times 1$$

is the expanded form for N. Calculating, we find that

$$N = 4 \times 625 + 1 \times 125 + 3 \times 25 + 2 \times 5 + 3 \times 1 = 2500 + 125 + 75 + 10 + 3 = 2713$$

That is, 2713 is the decimal representation of N.

(*b*) Given $M = 32.304_5$, then

$$M = 3 \times 5 + 2 \times 1 + 3 \times 5^{-1} + 0 \times 5^{-2} + 4 \times 5^{-3}$$

is the expanded form for M. Observe that negative powers of 5 correspond to the digits to the right of the quintal point.   Again calculating, we find that

$$M = 3 \times 5 + 2 \times 1 + 3 \times 0.2 + 0 \times 0.04 + 4 \times 0.008 = 15 + 2 + 0.6 + 0 + 0.032 = 17.632$$

In other words, 17.632 is the decimal representation of M.

Base-b-to-Decimal Conversion

One can convert a base-b number, N_b, into its decimal representation by writing N_b in expanded notation and calculating by decimal arithmetic, as in Example 2.1.   This conversion can also be accomplished by the following algorithm, which distinguishes between the integral part and the fractional part of the number.

Conversion from Base-b to Decimal Representation:

(i) **Integral part.** Multiply the leftmost digit by the base b and add the next digit to the right.   Multiply the sum by the base b and add the next digit.   Repeat the process until the rightmost digit is added.   The final sum is the required decimal equivalent.

(ii) **Fractional part.** Multiply the rightmost digit by $1/b$ and add the next digit to the left.   Multiply the sum by $1/b$ and add the next digit.   Repeat the process until the leftmost digit is added and the sum multiplied by $1/b$. The final product is the required decimal equivalent.

Part (ii) presupposes that the fractional part of the base-b number terminates.   In (i) the process ends when the last digit is added, but in (ii) the process ends when the last digit is added and the sum multiplied by $1/b$.

EXAMPLE 2.2   Consider the quintal number $N = 2401.2314_5$.   Then $N_I = 2401$ is the integral part of the number and $N_F = 0.2314$ is the fractional part.

(a)   We convert N_I into its decimal form using (i) of the above algorithm.

$$
\begin{array}{r}
2 \\
\times\ 5 \\
\hline
10 \\
+\ 4 \\
\hline
14 \\
\times\ 5 \\
\hline
70 \\
+\ 0 \\
\hline
70 \\
\times\ 5 \\
\hline
350 \\
+\ 1 \\
\hline
351 \quad \textbf{Final sum}
\end{array}
$$

Thus, 351 is the decimal form for N_I.

(b)   We convert N_F into its decimal form using (ii) of the above algorithm.   We first note that

$$\frac{1}{b} = \frac{1}{5} = 0.2$$

Whenever we multiply by $1/b$ we shall obtain a number less than one (see Problem 2.6).   Hence, we can add the next digit by simply writing the digit in front of the product.   That is,

$$
\begin{array}{r}
4 \\
\times .2 \\
\hline
\textcircled{1}.8 \\
\times .2 \\
\hline
\textcircled{3}.36 \\
\times .2 \\
\hline
\textcircled{2}.672 \\
\times .2 \\
\hline
.5344 \qquad \textbf{Final product}
\end{array}
$$

We have circled the digits which are added. The final product, 0.5344, is the decimal form for N_F.

The decimal representation of N is obtained by putting the decimal representations of N_I and N_F together:

$$N = N_I + N_F = 351.5344$$

Decimal-to-Base-b Conversion

Let N be a decimal number with integral part N_I and (terminating) fractional part N_F. We can convert N to its base-b representation using the following algorithm, which also distinguishes between the integral part of the number and its fractional part.

Conversion from Decimal to Base-b Representation:

(i)   **Integral part.**   Divide N_I and each succeeding quotient by b until a zero quotient is obtained. The sequence of remainders, in reverse order, yields the base-b representation of N_I.

(ii) **Fractional part.**   Multiply N_F, and the fractional part of each succeeding product, by b, until a zero fractional part or a duplicate fractional part is obtained. Then the finite sequence or infinite repeating sequence of integral parts of the products gives the base-b representation of N_F.

We emphasize that each remainder in (i) is less than b and hence is a base-b digit, and that each integral part in (ii) is also less than b and hence is a base-b digit.

EXAMPLE 2.3

(a)   To convert the decimal number $A = 684$ to its quintal (base-5) representation we divide A, and each subsequent quotient, by $b = 5$, noting all remainders:

Divisions	Quotients	Remainders			Remainders
$684 \div 5$	136	4 ↑		5)684	↑
$136 \div 5$	27	1	*or*	5)136	4
$27 \div 5$	5	2		5)27	1
$5 \div 5$	1	0		5)5	2
$1 \div 5$	0	1		1	0

The sequence of remainders in reverse order, as indicated by the arrows, gives the quintal form for A. In other words, $A = 10214_5$.

(b)   To convert the decimal number $B = 0.4704$ to its quintal (base-5) representation we multiply B, and each subsequent fractional part, by $b = 5$, noting the integral part of each product:

Multiplications	Integral parts		
$0.4704 \times 5 = \underline{2}.3520$	2		0.4704
$0.352 \times 5 = \underline{1}.760$	1		$\times 5$
$0.76 \times 5 = \underline{3}.80$	3	*or*	2.3520
$0.8 \times 5 = \underline{4}.0$	4 ↓		5
			1.760
			5
			3.80
			5
			4.0

The sequence of integral parts, as indicated by the arrows, yields the required quintal form for B. That is, $B = 0.2134_5$.

(c) We convert the decimal number $N = 684.4704$ to its quintal form by adding the quintal representations found in (a) and (b):

$$N = A + B = 10214.2134_5$$

(d) Proceeding as in (b), we convert the decimal number $C = 0.4703$ to base-5 as follows:

$$
\begin{array}{r}
0.4703 \\
\times 5 \\
\hline
\underline{2}.3515 \\
\times 5 \\
\hline
\underline{1}.7575 \\
\times 5 \\
\hline
\underline{3}.7875 \\
\times 5 \\
\hline
\underline{3}.9375^* \\
\times 5 \\
\hline
\underline{4}.6875 \\
\times 5 \\
\hline
\underline{3}.4375 \\
\times 5 \\
\hline
\underline{2}.1875 \\
\times 5 \\
\hline
\underline{0}.9375^*
\end{array}
$$

The eighth product has the same fractional part as the fourth product. Hence the block 4320 of integral parts will be obtained over and over again, so that

$$C = 0.2133\ 4320\ 4320\ 4320\ \cdots_5$$

2.3 OCTAL SYSTEM

The octal number system is the system having base $b = 8$. The eight octal digits are 0, 1, 2, 3, 4, 5, 6, and 7. Since $8 = 2^3$, each octal digit has a unique 3-bit binary representation, given in Table 2-1. The place values in the octal system are powers of 8; some of these powers appear in Table 2-2.

Table 2-1

Octal Digits	Binary Equivalents
0	000
1	001
2	010
3	011
4	100
5	101
6	110
7	111

Table 2-2

Octal Place Values	Decimal Values
8^{-3}	$1/512 = 0.001\ 953\ 125$
8^{-2}	$1/64 = 0.015\ 625$
8^{-1}	$1/8 = 0.125$
8^0	1
8^1	8
8^2	64
8^3	512
8^4	4 096
8^5	32 768

Octal-Decimal Interconversion

Conversion between the octal and decimal systems is accomplished by means of the two algorithms of Section 2.2, with $b = 8$. Octal-to-decimal conversion can also be effected by decimal evaluation of the expanded octal form.

Octal-Binary Interconversion

One may view each octal digit as simply a shorthand notation for the equivalent 3-bit value appearing in Table 2-1. Accordingly, convert an octal number to its binary form by replacing each octal digit by its binary equivalent. Conversely, convert a binary number to its octal form by partitioning the number into 3-bit blocks (emanating from the binary point, with zeros added if necessary) and replacing each block by its equivalent octal digit.

EXAMPLE 2.4

(a) To convert 4206_8 to its binary equivalent, replace each octal digit by its 3-bit equivalent, as follows:

$$4206$$
$$100 \quad 010 \quad 000 \quad 110$$

Thus, 100010000110 is the binary equivalent of 4206_8.

(b) To convert 10101011111_2 to its octal equivalent, partition the binary number into 3-bit blocks, beginning at the rightmost bit, and then replace each block by its equivalent octal digit:

$$010 \quad 101 \quad 011 \quad 111$$
$$2537$$

Thus, 2537 is the octal equivalent of 10101011111_2. (Observe that a 0 bit was prefixed to the binary number so that the number of bits becomes a multiple of three.)

(c) To convert 11011.0101_2 to its octal form, partition the binary number into 3-bit blocks, beginning to the left and to the right of the binary point, and then replace each block by its equivalent octal digit:

$$011 \quad 011 \quad . \quad 010 \quad 100$$
$$33 . 24$$

Thus, 33.24 is the octal form of the given binary number.

Octal Arithmetic

Here we cover only octal addition and octal subtraction, the latter using complements. (Octal multiplication and octal division are beyond the scope of this book.)

The sum of two octal numbers can be reduced by the usual addition algorithm to the repeated addition of two octal digits (with possibly a carry of 1). This entails repeated use of Table 2-3, which would be rather tedious.

Table 2-3. Addition of Octal Digits

+	0	1	2	3	4	5	6	7
0	0	1	2	3	4	5	6	7
1	1	2	3	4	5	6	7	10
2	2	3	4	5	6	7	10	11
3	3	4	5	6	7	10	11	12
4	4	5	6	7	10	11	12	13
5	5	6	7	10	11	12	13	14
6	6	7	10	11	12	13	14	15
7	7	10	11	12	13	14	15	16

On the other hand, we have the following

Remark: The sum of two base-b digits, or of two base-b digits plus one, is less than $2b$.

This means that dividing such a sum by b can only yield a quotient of 0 or 1.   Accordingly, we have the following useful result:

Octal Addition: The sum of two octal digits, or the sum of two octal digits plus 1, can be obtained by (i) finding their decimal sum and (ii) modifying the decimal sum, if it exceeds 7, by subtracting 8 and carrying 1 to the next column.

EXAMPLE 2.5   Evaluate $(a)\, 5_8 + 4_8, (b)\, 6_8 + 7_8, (c)\, 3_8 + 2_8, (d)\, 7_8 + 4_8, (e)\, 1_8 + 4_8 + 2_8, (f)\, 1_8 + 6_8 + 3_8, (g)\, 1_8 + 5_8 + 6_8.$

	(a)	(b)	(c)	(d)	(e)	(f)	(g)
					1	1	1
	5	6	3	7	4	6	5
	+4	+7	+2	+4	+2	+3	+6
Decimal sum	9	13	5	11	7	10	12
Modification	−8	−8	−0	−8	−0	−8	−8
Octal sum	11	15	5	13	7	12	14

The carried 1 associated with a subtraction of 8 is shown as the 8s digit, since we are adding only one column of digits.

EXAMPLE 2.6

(a) To evaluate the octal sum $7346_8 + 5263_8$, align the two numbers in the usual way and apply separately to each column the rule for addition of octal digits.   Notice how the carries that arise when 8 is subtracted in the modification step travel from the very bottom of the column to the very top of the next column to the left.

	1		1	1		
		7	3	4	6	
	+ 5		2	6	3	
		12	6	11	9	Decimal sums
		−8	−0	−8	−8	Modifications
	1	4	6	3	1	Octal sum

Thus, 14631_8 is the desired octal sum.

(b) To evaluate the octal difference $Y = B - A$, where $A = 3142_8$ and $B = 7526_8$, first find the radix-minus-one (sevens) complement of A by subtracting each digit of A from 7.   Then add 1 to obtain the (radix) complement of A:

Sevens complement of A	4635
Complement of A	4636

Now add the complement of A to B:

	1	1		1		
		7	5	2	6	B
	+ 4	6		3	6	Complement of A
		12	11	6	12	Decimal sums
		−8	−8	−0	−8	Modifications
(1)		4	3	6	4	Octal sum

Deleting the circled 1, we finally get the required difference, $Y = 4364_8$.

2.4   HEXADECIMAL SYSTEM

The number system to the base $b = 16$ is called the hexadecimal system (sometimes abbreviated *hex*).   The system requires 16 digits, for which the symbols are the 10 decimal digits together with the first 6 letters of the alphabet (see Table 2-4).   Since $16 = 2^4$, each hexadecimal digit has a unique

Table 2-4

Hexadecimal Digits	Decimal Values	Binary Equivalents
0	0	0000
1	1	0001
2	2	0010
3	3	0011
4	4	0100
5	5	0101
6	6	0110
7	7	0111
8	8	1000
9	9	1001
A	10	1010
B	11	1011
C	12	1100
D	13	1101
E	14	1110
F	15	1111

Table 2-5

Hexadecimal Place Values	Decimal Values
16^{-3}	$1/4096 = 0.000\ 244\ 140\ 625$
16^{-2}	$1/256 = 0.003\ 906\ 25$
16^{-1}	$1/16 = 0.062\ 5$
16^{0}	1
16^{1}	16
16^{2}	256
16^{3}	4 096
16^{4}	65 536
16^{5}	1 048 576

4-bit representation, which is also shown in Table 2-4. The place values in the hexadecimal system are the powers of 16, some of which are listed, along with their decimal values, in Table 2-5.

Hexadecimal-Decimal Interconversion

Conversion between the hexadecimal and decimal systems is accomplished via the two algorithms of Section 2.2, with $b = 16$. There is an added difficulty in that one has to know how to handle the hexadecimal digits A, B, C, D, E, and F. One can also convert from hexadecimal to decimal by decimal evaluation of the expanded hexadecimal form.

EXAMPLE 2.7

(a) To convert $73D5_{16}$ to its decimal equivalent, express the number in expanded notation, changing D to 13, and then calculate using decimal arithmetic.

$$73D5_{16} = 7 \times 16^3 + 3 \times 16^2 + 13 \times 16^1 + 5 \times 16^0 = 7 \times 4096 + 3 \times 256 + 13 \times 16 + 5 \times 1$$
$$= 28\ 672 + 768 + 208 + 5 = 29\ 653$$

Alternatively, one can apply the conversion algorithm as follows:

$$
\begin{array}{r}
7 \\
\times 16 \\
\hline
112 \\
+3 \\
\hline
115 \\
\times 16 \\
\hline
1840 \\
+13 \\
\hline
1853 \\
\times 16 \\
\hline
29648 \\
+5 \\
\hline
29653 = 73D5_{16}
\end{array}
$$

(b) Convert $39.B8_{16}$ to its decimal equivalent as follows:

$$39.B8_{16} = 3 \times 16^1 + 9 \times 16^0 + 11 \times 16^{-1} + 8 \times 16^{-2}$$
$$= 3 \times 16 + 9 \times 1 + 11 \times 0.0625 + 8 \times 0.003\,906\,25$$
$$= 48 + 9 + 0.6875 + 0.031\,25 = 57.718\,75$$

(c) To convert the decimal number $P = 9719$ to its hexadecimal equivalent, divide P, and each successive quotient, by the base $b = 16$, noting the remainders, as follows:

Divisions	Quotients	Remainders
$9719 \div 16$	607	7
$607 \div 16$	37	15
$37 \div 16$	2	5
$2 \div 16$	0	2

The sequence of remainders, in which we replace the decimal remainder 15 by the hexadecimal digit F, in reverse order, gives the hexadecimal form for P; i.e. $P = 25F7_{16}$. (Since division by 16 generally involves long rather than short division, we do not have a compact form for obtaining the remainders as we had for the bases 2, 5, and 8.)

(d) To convert the decimal fraction $Q = 0.781\,25$ to its hexadecimal equivalent, apply the integral-part algorithm, with $b = 16$, as follows:

Multiplications	Integral parts
$0.781\,25 \times 16 = 12.500\,00$	12
$0.500\,00 \times 16 = 8.000\,00$	8

In this case a zero fractional part is reached. The sequence of integral parts, in which we replace the decimal 12 by the hexadecimal digit C, gives the required hexadecimal form for Q; i.e. $Q = 0.C8_{16}$.

(e) To convert the decimal number $N = 9719.781\,25$ to its hexadecimal form, add the representations found in (c) and (d):

$$N = P + Q = 25F7.C8_{16}$$

Hexadecimal-Binary Interconversion

This is accomplished exactly as octal-binary interconversion, except that 4-bit equivalents are now involved.

Table 2-6.   Addition of Hexadecimal Digits

+	0	1	2	3	4	5	6	7	8	9	A	B	C	D	E	F
0	0	1	2	3	4	5	6	7	8	9	A	B	C	D	E	F
1	1	2	3	4	5	6	7	8	9	A	B	C	D	E	F	10
2	2	3	4	5	6	7	8	9	A	B	C	D	E	F	10	11
3	3	4	5	6	7	8	9	A	B	C	D	E	F	10	11	12
4	4	5	6	7	8	9	A	B	C	D	E	F	10	11	12	13
5	5	6	7	8	9	A	B	C	D	E	F	10	11	12	13	14
6	6	7	8	9	A	B	C	D	E	F	10	11	12	13	14	15
7	7	8	9	A	B	C	D	E	F	10	11	12	13	14	15	16
8	8	9	A	B	C	D	E	F	10	11	12	13	14	15	16	17
9	9	A	B	C	D	E	F	10	11	12	13	14	15	16	17	18
A	A	B	C	D	E	F	10	11	12	13	14	15	16	17	18	19
B	B	C	D	E	F	10	11	12	13	14	15	16	17	18	19	1A
C	C	D	E	F	10	11	12	13	14	15	16	17	18	19	1A	1B
D	D	E	F	10	11	12	13	14	15	16	17	18	19	1A	1B	1C
E	E	F	10	11	12	13	14	15	16	17	18	19	1A	1B	1C	1D
F	F	10	11	12	13	14	15	16	17	18	19	1A	1B	1C	1D	1E

Hexadecimal Arithmetic

As with the octal system, we cover only hexadecimal addition, and hexadecimal subtraction using complements.

The sum of two hexadecimal numbers can be reduced to the repeated use of Table 2-6, the addition of hexadecimal digits. As with octal addition, we have an alternate method, based on the following analogous result:

Hexadecimal Addition: The sum of two hexadecimal digits, or the sum of two hexadecimal digits plus 1, can be obtained by (i) finding their decimal sum and (ii) modifying the decimal sum, if it exceeds 15, by subtracting 16 and carrying 1 to the next column.

Since the base here exceeds ten, we need mentally to change each hexadecimal letter digit to its decimal form when finding the decimal sum, and each decimal difference greater than nine to its hexadecimal form when modifying the decimal sum. That is, we must memorize the equivalences

$$A = 10 \qquad B = 11 \qquad C = 12 \qquad D = 13 \qquad E = 14 \qquad F = 15$$

EXAMPLE 2.8 Evaluate the hexadecimal sums (a) $8+9$, (b) $3+5$, (c) $6+7$, (d) $A+9$, (e) $C+D$, (f) $3+B$, (g) $F+D$, (h) $1+4+6$, (i) $1+5+C$, (j) $1+E+6$.

	(a)	(b)	(c)	(d)	(e)	(f)	(g)	(h)	(i)	(j)
								1	1	1
	8	3	6	A	C	3	F	4	5	E
	+9	+5	+7	+9	+D	+B	+D	+6	+C	+6
Decimal sum	17	8	13	19	25	14	28	11	18	21
Modification	−16	−0	−0	−16	−16	−0	−16	−0	−16	−16
Hexadecimal sum	11	8	D	13	19	E	1C	B	12	15

Again the carried 1 is written in the next column of the sum.

EXAMPLE 2.9

(a) To evaluate the hexadecimal sum $C868 + 72D9$, apply the addition rule column by column, carrying, if necessary, from the very bottom of a column to the very top of the next column to the left.

1		1	1		
	C	8	6	8	
	+7	2	D	9	
	19	11	20	17	Decimal sums
	−16	−0	−16	−16	Modifications
1	3	B	4	1	Hexadecimal sum

Thus, 13B41 is the desired hexadecimal sum.

(b) To evaluate the hexadecimal difference $Y = L - M$, where $L = 72A4$ and $M = 4E86$, first find the radix-minus-one (fifteens) complement of M by subtracting each digit of M from 15, and then add 1 to obtain the (radix) complement of M.

$$
\begin{array}{ll}
\textbf{Fifteens complement of } M & \text{B179} \\
\textbf{Complement of } M & \text{B17A}
\end{array}
$$

Now add the complement of M to L:

1		1			
	7	2	A	4	*L*
	+B	1	7	A	Complement of *M*
	18	4	17	14	Decimal sums
	−16	−0	−16	−0	Modifications
①	2	4	1	E	Hexadecimal sum

Deleting the circled 1, we finally obtain the required difference, $Y = 241E$.

2.5   4-BIT BCD CODES

There are many ways of representing numerical data in binary form.   One way is simply to write the numbers to the base 2.   This is called *straight binary* coding and is discussed in Chapter 3. Another way is to encode decimal numbers digit by digit.   These codes, which require at least 4 bits for each decimal digit, are called *BCD* (*binary-coded decimal*) codes.   This section will discuss two of the more popular 4-bit BCD codes; BCD codes using 6-bit or 8-bit representations of decimal digits are discussed in Section 2.6.

Table 2-7

Decimal Digits	BCD Codes	
	8-4-2-1	XS-3
0	0000	0011
1	0001	0100
2	0010	0101
3	0011	0110
4	0100	0111
5	0101	1000
6	0110	1001
7	0111	1010
8	1000	1011
9	1001	1100

Weighted 8-4-2-1 BCD Code

Two 4-bit BCD codes are shown in Table 2-7.   The first one is a weighted code, in which the bits are given, from left to right, the weights 8, 4, 2, and 1, respectively.   Since these weights are just the place values in the binary system, a decimal digit is encoded as its binary representation.

EXAMPLE 2.10   The 8-4-2-1 BCD representation of $N = 469$ is

4	**6**	**9**
0100	0110	1001

On the other hand, the straight binary representation of N is

$$N = 111010101_2$$

which involves 3 fewer bits.

Other weighted 4-bit codes are sometimes used, and they are all decoded the same way.   Thus, for the weighted 4-2-2-1 BCD code,

$$1000 \rightarrow 4$$
$$0111 \rightarrow 2 + 2 + 1 = 5$$
$$1110 \rightarrow 4 + 2 + 2 = 8$$

(Notice that *encoding* is not unique in this system; e.g. $2 \rightarrow 0010$ and $2 \rightarrow 0100$.   The same is true for all weighted 4-bit codes other than the 8-4-2-1.)   By a 4-bit BCD code we shall understand the 8-4-2-1, unless otherwise specifically stated.

Nonweighted XS-3 BCD Code

The second BCD code in Table 2-7 is the *excess-three* (XS-3) BCD code. This nonweighted code is related to the weighted 8-4-2-1 BCD as follows: the XS-3 code for a decimal digit d is obtained by adding $3 = 0011_2$ to the 8-4-2-1 BCD code for d.

EXAMPLE 2.11   One codes the decimal integer $N = 469$ in XS-3 as follows:

4	6	9	Decimal digits
0100	0110	1001	8-4-2-1 BCD code
+ 0011	0011	0011	Addition of three
0111	1001	1100	XS-3 BCD code

For the decimal integer $L = 530$, the coding is, similarly,

5	3	0	
0101	0011	0000	
+ 0011	0011	0011	
1000	0110	0011	

The XS-3 has an important arithmetic property: *it encodes a pair of nines complements as a pair of ones complements.* (This is verified in Example 2.11 for the nines complements 469 and 530.) Because of this feature, arithmetic in the XS-3 BCD code is simpler in many respects than arithmetic in the 8-4-2-1 BCD code.

Comparison with Straight Binary Coding

BCD codes have various advantages and disadvantages relative to straight binary coding. On the positive side, conversion between decimal and BCD is much simpler than between decimal and straight binary. Also, there is no roundoff error in BCD coding, but there may be in straight binary coding. For example, the straight binary coding of the decimal fraction 0.6 is (see Problem 1.15)

$$0.1001\ 1001\ 1001 \cdots$$

so there will necessarily be some error in representing 0.6 in the computer. On the other hand, straight binary coding usually requires fewer bits to represent a number than do BCD codes; and performing arithmetic with straight binary coding is easier than with BCD coding.

Remark: Any 4-bit code allows $2^4 = 16$ combinations. Because the 4-bit BCD codes need only 10 of the combinations to represent the decimal digits, 6 combinations remain available for other uses (e.g. to encode the plus and minus signs).

2.6   6-BIT BCD CODE

The computer processes both numerical data and nonnumerical data. (Nonnumerical data may include numbers, as in street addresses, which are not, however, calculated with.) Nonnumerical data are expressed in a set of characters consisting of 10 digits, 26 letters, and a dozen or more *special characters*. Normally, a character set includes the 48 characters shown in Fig. 2-1 (in which the character _ stands for a blank space).

Digits	0	1	2	3	4	5	6	7	8	9
Alphabetic characters	A	B	C	D	E	F	G	H	I	J
	K	L	M	N	O	P	Q	R	S	T
	U	V	W	X	Y	Z				
Special characters	+	−	*	/	,	.	'	=	$	(	)	_

Fig. 2-1

Some character sets contain other special characters, such as the inequality symbols ($<$ and $>$), the question mark (?), etc.   Data consisting of both numeric and nonnumeric items are called *alphameric* data.   Clearly, a 4-bit code, with its 16 combinations, is insufficient to represent alphameric data; but a 6-bit code can do so, with capacity for $2^6 - 36 = 28$ special characters.

The 6-bit BCD code adds two bits, called *zone bits* and labeled *position B* and *position A*, to the four 8-4-2-1 *numeric bits*, as shown:

zone bits		numeric bits			
B	A	8	4	2	1

Digits are coded with 0s for both zone bits and their 8-4-2-1 BCD code for the numeric bits (except the digit 0, which is coded as if it were ten).   Alphabetical characters and special characters are encoded by combinations of both zone bits and numeric bits.   The 6-bit BCD coding for all the letters and digits, and some of the special characters, appears in Table 2-8.   Although a 6-bit coding is partitioned into two zone bits and four numeric bits, it is sometimes specified in terms of two octal digits.   Observe that the letters are divided into three groups: the first nine, A through I, have zone bits 11; the next nine, J through R, have zone bits 10; and the remainder have zone bits 01.

Actually, a 6-bit code appears within the computer in a 7-bit form.   This seventh bit, called a *check bit* or *parity bit*, is included as shown:

check bit	zone bits		numeric bits			
C	B	A	8	4	2	1

Table 2-8.   6-Bit BCD Code

Char.	Zone	Numeric	Octal	Char.	Zone	Numeric	Octal
A	11	0001	61	1	00	0001	01
B		0010	62	2		0010	02
C		0011	63	3		0011	03
D		0100	64	4		0100	04
E		0101	65	5		0101	05
F		0110	66	6		0110	06
G		0111	67	7		0111	07
H		1000	70	8		1000	10
I	11	1001	71	9		1001	11
J	10	0001	41	0	00	1010	12
K		0010	42				
L		0011	43	Char.	Zone	Numeric	Octal
M		0100	44				
N		0101	45	+	11	0000	60
O		0110	46	−	10	0000	40
P		0111	47	*	10	1100	54
Q		1000	50	/	01	0001	21
R	10	1001	51	=	00	1011	13
S	01	0010	22	(	01	1100	34
T		0011	23	)	11	1100	74
U		0100	24	.	11	1011	73
V		0101	25	;	10	1110	56
W		0110	26	$	10	1011	53
X		0111	27	*blank*	00	0000	00
Y		1000	30				
Z	01	1001	31				

For each character, the value of the check bit (0 or 1) is such as to make the sum of the bits, including the check bit, odd or even, according as the machine operates on odd or even parity.

EXAMPLE 2.12 If the computer uses odd parity, the characters 7, 9, P, and W are stored as follows:

	Check	Zone	Numeric
7	0	0 0	0 1 1 1
9	1	0 0	1 0 0 1
P	1	1 0	0 1 1 1
W	0	0 1	1 1 0 0

That is, the check bit for 7 is 0 because the sum of the bits in the 6-bit code for 7 is three, which is already odd. On the other hand, the check bit for P is 1 because the sum of the bits in the 6-bit code for P is four, which is even.

The purpose of the check bit is to ensure that no bit is lost or gained when data are transmitted internally in a computer. After a character is transmitted, the computer adds up the bits in the character. If a single error occurred, the sum of the bits will not have the same parity as the parity of the computer. The computer would then retransmit the data. Clearly, the computer cannot use this type of checking to see if two errors occurred; but such an occurrence is very unlikely.

Remark: An extra, check bit is also used with 4-bit BCD codes.

2.7 8-BIT BCD CODES

Modern data processing frequently requires more than the 28 special characters possible under any 6-bit BCD code. (In fact, some data processing equipment may even want both lowercase and uppercase letters.) Accordingly, various 8-bit codes have been developed. Each coded character, or *byte*, is normally divided into four *zone bits* and four 8-4-2-1 *numeric bits*, as shown;

```
      zone bits    numeric bits
      ┌─────────┐  ┌─────────┐
      │ Z │ Z │ Z │ Z │ 8 │ 4 │ 2 │ 1 │
      └─────────────────────────────┘
```

(More generally, the word "byte" is used to denote any group of eight bits.) It is seen that a byte may be represented by two hexadecimal digits, the first corresponding to the zone bits and the second to the numeric bits. As with the 4- and 6-bit BCD codes, an extra, check bit is utilized in the computer.

There are two 8-bit BCD codes predominant in the computer industry today:

EBCDIC, pronounced "ebb-see-dick" and short for Extended Binary-Coded Decimal Interchange Code. This code was developed by IBM as an extension of the 6-bit BCD code; it is used mainly by IBM and IBM-compatible computer systems. See Table 2-9.

ASCII-8, pronounced "ass-key" and short for American Standard Code for Information Interchange. This code was originally developed as a 7-bit standardization of various special codes, and was then extended to an 8-bit code. It is used mainly by non-IBM computer systems. See Table 2-10.

Table 2-9. EBCDIC

Char.	Zone	Numeric	Hex	Char.	Zone	Numeric	Hex	Char.	Zone	Numeric	Hex
				S	1110	0010	E2	blank	0100	0000	40
A	1100	0001	C1	T		0011	E3	.		1011	4B
B		0010	C2	U		0100	E4	<		1100	4C
C		0011	C3	V		0101	E5	(		1101	4D
D		0100	C4	W		0110	E6	+	0100	1110	4E
E		0101	C5	X		0111	E7	&	0101	0000	50
F		0110	C6	Y		1000	E8	$		1011	5B
G		0111	C7	Z	1110	1001	E9	*		1100	5C
H		1000	C8					)		1101	5D
I	1100	1001	C9	**Char.**	**Zone**	**Numeric**	**Hex**	;	0101	1110	5E
J	1101	0001	D1					−	0110	0000	60
K		0010	D2	0	1111	0000	F0	/		0001	61
L		0011	D3	1		0001	F1	,		1011	6B
M		0100	D4	2		0010	F2	%		1100	6C
N		0101	D5	3		0011	F3	>		1110	6E
O		0110	D6	4		0100	F4	?	0110	1111	6F
P		0111	D7	5		0101	F5	:	0111	1010	7A
Q		1000	D8	6		0110	F6	#		1011	7B
R	1101	1001	D9	7		0111	F7	@		1100	7C
				8		1000	F8	=	0111	1110	7E
				9	1111	1001	F9				

Table 2-10. ASCII-8

Char.	Zone	Numeric	Hex	Char.	Zone	Numeric	Hex	Char.	Zone	Numeric	Hex
0	0101	0000	50	A	1010	0001	A1	P	1011	0000	B0
1		0001	51	B		0010	A2	Q		0001	B1
2		0010	52	C		0011	A3	R		0010	B2
3		0011	53	D		0100	A4	S		0011	B3
4		0100	54	E		0101	A5	T		0100	B4
5		0101	55	F		0110	A6	U		0101	B5
6		0110	56	G		0111	A7	V		0110	B6
7		0111	57	H		1000	A8	W		0111	B7
8		1000	58	I		1001	A9	X		1000	B8
9	0101	1001	59	J		1010	AA	Y		1001	B9
				K		1011	AB	Z	1011	1010	BA
				L		1100	AC				
				M		1101	AD				
				N		1110	AE				
				O	1010	1111	AF				

Observe that in both systems a digit has its binary representation as the numeric portion of its code.   For the zone portion, EBCDIC uses 1111 and ASCII-8 uses 0101.

2.8 ZONED DECIMAL AND PACKED DECIMAL FORMATS

EBCDIC is also used to represent numerical data in the computer. However, the sign of a number is not encoded by an 8-bit group at the beginning of the number, but by a 4-bit group which occupies the zone portion of the rightmost digit. The coding is given in Table 2-11; note that the code for an unsigned (positive) number is the normal digit zone code. For example, the numbers 275, +275, and −275 are encoded as in Fig. 2-2. Numbers encoded in this way are said to be in *zoned decimal format*. Other 8-bit codes use a similar convention.

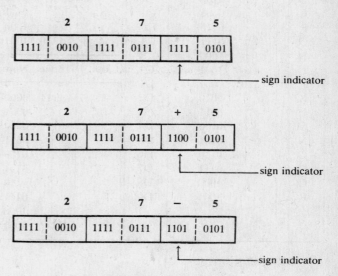

Table 2-11

Sign	EBCDIC
+	1100
−	1101
unsigned	1111

Fig. 2-2. Zoned Decimal Format

Although one can input/output data in zoned decimal format, the arithmetic/logic unit of the computer cannot generally handle data in this form, and the data must be converted to another form before calculation can occur. One acceptable form for calculations is the *packed decimal format*. Here, each digit is encoded in 4-bit BCD, as is the sign, with the code for the sign (see Table 2-11) placed at the end of the number. For example, 275, +275, and −275 are encoded in packed decimal format as in Fig. 2-3.

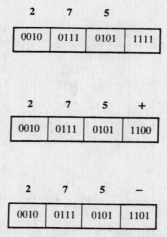

Fig. 2-3. Packed Decimal Format

Conversion from zoned decimal format to packed decimal format is accomplished in two steps. First, the zone and numeric portions of the rightmost byte are interchanged, bringing the sign of the number to the end of the format.   Next, the other zone portions are deleted and the remaining numeric portions "packed" together.   The conversion for -275 is illustrated in Fig. 2-4.

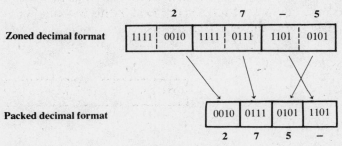

Fig. 2-4

Solved Problems

NUMBER SYSTEMS

2.1 Enumerate and give the symbols for the digits in the number system having base (a) $b = 4$, (b) $b = 9$, (c) $b = 12$, (d) $b = 1$.

The number of digits is the same as the base, and the digits are $0, 1, 2, \ldots, b-1$.
(a)   Four; 0, 1, 2, and 3.
(b)   Nine; 0, 1, 2, 3, 4, 5, 6, 7, 8.
(c)   Twelve; 0, 1, 2, 3, 4, 5, 6, 7, 8, 9, A, B, where A and B symbolize the numbers ten and eleven.
(d)   The base b must be greater than one; hence there is no number system with $b = 1$.

2.2 Write in expanded notation: (a) 3102_4, (b) 416_9, (c) 5713_6.

(a) $$3102_4 = 3 \times 4^3 + 1 \times 4^2 + 0 \times 4 + 2 \times 1$$
(b) $$416_9 = 4 \times 9^2 + 1 \times 9 + 6 \times 1$$
(c) 5713_6 is not a number to base 6, since 7 is not a digit in that system.

2.3 Write 735.426_9 in expanded notation.

The negative powers of $b = 9$ appear to the right of the point:

$$735.426_9 = 7 \times 9^2 + 3 \times 9 + 5 \times 1 + 4 \times 9^{-1} + 2 \times 9^{-2} + 6 \times 9^{-3}$$

2.4 Convert 2153_6 to decimal form.

Method 1.
Write in expanded notation and calculate using decimal arithmetic.

$$2153_6 = 2 \times 6^3 + 1 \times 6^2 + 5 \times 6 + 3 \times 1 = 2 \times 216 + 1 \times 36 + 5 \times 6 + 3 \times 1$$
$$= 432 + 36 + 30 + 3 = 501$$

Method 2.

Use the conversion algorithm.

$$
\begin{array}{r}
2 \\
\times\,6 \\
\hline
12 \\
+\,1 \\
\hline
13 \\
\times\,6 \\
\hline
78 \\
+\,5 \\
\hline
83 \\
\times\,6 \\
\hline
498 \\
+\,3 \\
\hline
501 \quad \textbf{Final sum}
\end{array}
$$

The final sum is the desired decimal form.

2.5   Convert 0.3123_4 to decimal form.

Method 1.

Write in expanded notation and calculate using decimal arithmetic.

$$
\begin{aligned}
0.3123_4 &= 3 \times 4^{-1} + 1 \times 4^{-2} + 2 \times 4^{-3} + 3 \times 4^{-4} \\
&= 3 \times 0.25 + 1 \times 0.0625 + 2 \times 0.015\,625 + 3 \times 0.003\,906\,25 \\
&= 0.75 + 0.0625 + 0.031\,25 + 0.011\,718\,75 = 0.855\,468\,75
\end{aligned}
$$

Method 2.

Use the conversion algorithm, with $1/b = 1/4 = 0.25$.

$$
\begin{array}{r}
3 \\
\times\,0.25 \\
\hline
\textcircled{2}.75 \\
0.25 \\
\hline
\textcircled{1}.6875 \\
0.25 \\
\hline
\textcircled{3}.421875 \\
0.25 \\
\hline
0.85546875 \quad \textbf{Final product}
\end{array}
$$

The final product is the required decimal form.

2.6   In the algorithm for converting a base-b fraction to a decimal fraction, show that every product (not just the final product) is less than one.

The "worst case" is posed by the base-b fraction each digit of which is as large as possible, i.e.

$$
F = 0.\overline{b-1}\ \overline{b-1}\ \cdots\ \overline{b-1}
$$

Applying the algorithm to F,

$$
\textbf{First product} \qquad \frac{1}{b} \times (b-1) = \frac{b-1}{b} < 1
$$

$$
\textbf{Second product} \qquad \frac{1}{b} \times \left(b - 1 + \frac{b-1}{b}\right) = \frac{1}{b} \times \left(\frac{b^2-1}{b}\right) = \frac{b^2-1}{b^2} < 1
$$

$$
\textbf{Third product} \qquad \frac{1}{b} \times \left(b - 1 + \frac{b^2-1}{b^2}\right) = \frac{1}{b} \times \left(\frac{b^3-1}{b^2}\right) = \frac{b^3-1}{b^3} < 1
$$

....

2.7   Convert the decimal number $N = 626.4375$ to its base-4 form.

First, convert $N_I = 626$, the integral part of N, by dividing it, and each successive quotient, by $b = 4$, noting the remainders:

$$
\begin{array}{rl}
 & \textbf{Remainders} \\
4)\overline{626} & \\
4)\overline{156} & 2 \\
4)\overline{39} & 0 \\
4)\overline{9} & 3 \\
2 & 1
\end{array}
$$

The remainders in reverse order yield 21302_4, the required form for N_I.

Next, convert $N_F = 0.4375$, the fractional part of N, by multiplying it, and each successive fractional part, by $b = 4$, noting the integral parts of the products:

$$
\begin{array}{r}
0.4375 \\
\times\,4 \\
\hline
1.7500 \\
\times\,4 \\
\hline
3.0000
\end{array}
$$

The process terminates after two steps, yielding 0.13_4 as the required form for N_F.

Finally, adding the two representations gives $N = 21302.13_4$.

OCTAL SYSTEM

2.8   Convert the decimal integer $A = 1476$ to its octal form.

Divide A, and each subsequent quotient, by the base $b = 8$, noting the remainders:

$$
\begin{array}{rl}
 & \textbf{Remainders} \\
8)\overline{1476} & \\
8)\overline{184} & 4 \\
8)\overline{23} & 0 \\
2 & 7
\end{array}
$$

The sequence of remainders in reverse order, as indicated by the arrow, gives the required octal form, $A = 2704_8$.

2.9   Convert the octal integer $B = 25146_8$ to its decimal form.

Apply the conversion algorithm as follows:

$$
\begin{array}{r}
2 \\
\times\,8 \\
\hline
16 \\
+\,5 \\
\hline
21 \\
\times\,8 \\
\hline
168 \\
+\,1 \\
\hline
169 \\
\times\,8 \\
\hline
1352 \\
+\,4 \\
\hline
1356 \\
\times\,8 \\
\hline
10848 \\
+\,6 \\
\hline
10854 \qquad \textbf{Final sum}
\end{array}
$$

The final sum is the desired decimal form, $A = 10854$.

2.10   Convert to binary form (*a*) 43027_8, (*b*) 21.673_8.

Replace each octal digit by its 3-bit binary equivalent (Table 2-1):

(*a*)

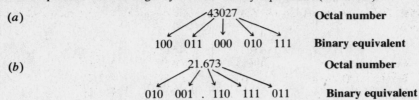

		43027			**Octal number**
100	011	000	010	111	**Binary equivalent**

(*b*)

		21.673				**Octal number**
010	001	.	110	111	011	**Binary equivalent**

2.11   Convert to octal form (*a*) 11101010110_2, (*b*) 1001101.01100001_2.

Partition each binary number into 3-bit blocks extending from the binary point, adding 0s if necessary. Then replace each 3-bit block by its octal equivalent (Table 2-1).

(*a*)

011	101	010	110	**Binary number**
	3526			**Octal equivalent**

(*b*)

001	001	101	.	011	000	010	**Binary number**
		115.302				**Octal equivalent**

2.12   Add the following octal digits:

```
                                 1         1
      3        6        7        2         5        7        3
    + 2      + 4      + 5      + 4       + 6      + 7      + 5
```

Find the decimal sum, subtracting 8 and carrying 1 (which becomes the next digit) whenever the sum exceeds 7:

					1	1		
		3	6	7	2	5	7	3
		+2	+4	+5	+4	+6	+7	+5
Decimal sum		5	10	12	7	12	14	8
Modification		−0	−8	−8	−0	−8	−8	−8
Octal sum		5	12	14	7	14	16	10

2.13   Evaluate the following octal sums:

(*a*) 6254_8 (*b*) 36517_8 (*c*) $465.37_8 + 31.613_8$
       $+ 4176_8$ $+ 64753_8$

Proceed as in Example 2.6.

(*a*)

```
          1    1
     6    2    5    4
   + 4    1    7    6
    10    4   13   10
   - 8   -0   -8   -8
 1   2    4    5    2
```

(*b*)

```
          1    1         1
     3    6    5    1    7
   + 6    4    7    5    3
    10   11   12    7   10
   - 8   -8   -8   -0   -8
 1   2    3    4    7    2
```

(*c*)   As usual, one must line up the octal points.

```
        1         1    1
   4    6    5 .  3    7    0
 +      3    1 .  6    1    3
   5    9    7   10    8    3
 - 0   -8   -0   -8   -8   -0
   5    1    7 .  2    0    3
```

2.14 Find the radix-minus-one (sevens) complement, and the radix (eights) complement of the octal numbers (a) 40613_8, (b) 716520_8, (c) 335500_8.

Subtract each digit from seven to obtain the radix-minus-one complement, and then add one to get the (radix) complement.

	(a)	(b)	(c)
Octal number	40613	716520	335500
Sevens complement	37164	061257	442277
Complement	37165	061260	442300

2.15 Evaluate each octal difference:

$$(a) \quad \begin{array}{r} 6214_8 \\ -\ 3527_8 \end{array} \qquad (b) \quad \begin{array}{r} 4617263_8 \\ -\ 1423736_8 \end{array}$$

Perform the subtraction by adding the complement of the subtrahend to the minuend:

(a)

1				
6	2	1	4	Minuend
+ 4	2	5	1	Complement of subtrahend
10	4	6	5	Decimal sums
− 8	− 0	− 0	− 0	Modifications
① 2	4	6	5	

(b)

1	1		1		1		
4	6	1	7	2	6	3	Minuend
+ 6	3	5	4	0	4	1	Complement of subtrahend
11	9	7	11	3	10	4	Decimal sums
− 8	− 8	− 0	− 8	− 0	− 8	− 0	Modifications
① 3	1	7	3	3	2	4	

Deleting the circled 1 yields the required difference.

2.16 Convert each decimal fraction to its octal form: (a) 0.4375, (b) 0.2.

Multiply the decimal fraction, and the fractional part of each succeeding product, by the base $b = 8$, noting the integral part of each product.

(a)

$$\begin{array}{r} 0.4375 \\ \times\ 8 \\ \hline 3.5000 \\ 8 \\ \hline 4.0000 \end{array}$$

The sequence of integral parts gives the required octal form, $0.4375 = 0.34_8$.

(b)

$$\begin{array}{r} 0.2 \\ \times\ 8 \\ \hline 1.6 \\ 8 \\ \hline 4.8 \\ 8 \\ \hline 6.4 \\ 8 \\ \hline 3.2 \end{array}$$

At the fourth step we again obtain .2 as the fractional part; hence the digits 1463 will repeat, giving

$$0.2 = 0.1463\ 1463\ 1463 \cdots_8$$

HEXADECIMAL SYSTEM

2.17   Convert the decimal number $X = 15\,321$ to its hexadecimal form.

Divide X, and each successive quotient, by $b = 16$, noting the remainders, as follows:

$$
\begin{array}{ccc}
957 & 59 & ③ \\
16)\overline{15321} & 16)\overline{957} & 16)\overline{59} \\
\underline{144} & \underline{80} & \underline{48} \\
92 & 157 & ⑪ \\
\underline{80} & \underline{144} & \\
121 & ⑬ & \\
\underline{112} & & \\
⑨ & &
\end{array}
$$

(The remainders, including the last quotient, 3, which becomes the last remainder, are circled.)   The sequence of remainders in reverse order, as indicated by the arrow, is the hexadecimal form for X. That is, $X = 3BD9_{16}$, where we have replaced the remainders 11 and 13 by their equivalent hexadecimal digits, B and D, respectively.

2.18   Convert the hexadecimal number $Z = 1A74_{16}$ to its decimal form.

Method 1.

Write Z in expanded notation and calculate using decimal arithmetic:

$$Z = 1A74_{16} = 1 \times 16^3 + A \times 16^2 + 7 \times 16 + 4 \times 1 = 1 \times 4096 + 10 \times 256 + 7 \times 16 + 4 \times 1$$

$$= 4096 + 2560 + 112 + 4 = 6772$$

Method 2.

Use the conversion algorithm as follows:

$$
\begin{array}{r}
1 \\
\times 16 \\
\hline
16 \\
+ 10 \\
\hline
26 \\
\times 16 \\
\hline
156 \\
26 \\
\hline
416 \\
+ 7 \\
\hline
423 \\
\times 16 \\
\hline
2538 \\
423 \\
\hline
6768 \\
+ 4 \\
\hline
6772 \quad \textbf{Final sum}
\end{array}
$$

The final sum is the required decimal form: $Z = 6772$.

2.19   Convert to binary form (*a*) $3D59_{16}$, (*b*) $27.A3C_{16}$.

Replace each hexadecimal digit by its 4-bit representation (Table 2-4).

(*a*)   3 D 5 9

0011   1101   0101   1001

Hence, $3D59_{16} = 11110101011001_2$.

(b)
$$27 . A 3 C$$

0010   0111 . 1010   0011   1100

Hence, $27.A3C_{16} = 100111.1010001111_2$.

2.20 Convert to hexadecimal form (a) 10110100101110_2, (b) 11100.1011011011_2.

Partition each binary number into 4-bit blocks to the left and right of the binary point, adding 0s if necessary.   Then replace each 4-bit block by its equivalent hexadecimal digit (Table 2-4).

(a)
0010   1101   0010   1110

2 D 2 E

Hence, $2D2E_{16}$ is the required hexadecimal form.

(b)
0001   1100 . 1011   0110   1100

1 C . B 6 C

Hence, $1C.B6C_{16}$ is the required hexadecimal form.

2.21 Add the following hexadecimal digits:

$$
\begin{array}{ccccccc}
 & & & & & & 1 \\
4 & 8 & 7 & D & 3 & 5 & E \\
+3 & +9 & +5 & +8 & +B & +1 & +F
\end{array}
$$

First, find the usual decimal sum (using $A = 10$, $B = 11, \ldots$, $F = 15$, when necessary). Then, subtract 16 and carry 1 whenever the decimal sum exceeds 15.

	4	8	7	D	3	5	1 E
	+3	+9	+5	+8	+B	+1	+F
Decimal sum	7	17	12	21	14	6	30
Modification	−0	−16	−0	−16	−0	−0	−16
Hexadecimal sum	7	11	C	15	E	6	1E

2.22 Evaluate

    (a) $82C5_{16} + 9D86_{16}$ (b) $83A7F4_{16} + B5B63_{16}$ (c) $4C.3E_{16} + 2.5D8_{16}$

Proceed as in Example 2.9.

(a)
$$
\begin{array}{ccccc}
1 & 1 & 1 & \\
8 & 2 & C & 5 \\
+9 & D & 8 & 6 \\
\hline
18 & 16 & 20 & 11 \\
-16 & -16 & -16 & -0 \\
\hline
1 \;\; 2 & 0 & 4 & B
\end{array}
$$

(b)
$$
\begin{array}{cccccc}
 & 1 & 1 & 1 & & \\
8 & 3 & A & 7 & F & 4 \\
+ & B & 5 & B & 6 & 3 \\
\hline
8 & 15 & 16 & 19 & 21 & 7 \\
-0 & -0 & -16 & -16 & -16 & -0 \\
\hline
8 & F & 0 & 3 & 5 & 7
\end{array}
$$

(c) As usual, line up the hexadecimal points:

$$
\begin{array}{ccccc}
 & & 1 & & \\
4 & C . & 3 & E & 0 \\
+ & 2 . & 5 & D & 8 \\
\hline
4 & 14 . & 9 & 27 & 8 \\
-0 & -0 & -0 & -16 & -0 \\
\hline
4 & E . & 9 & B & 8
\end{array}
$$

2.23 Find the radix-minus-one (15s) complement and the radix (16s) complement of (a) $74B9_{16}$, (b) $5C0F8_{16}$, (c) $2A7600_{16}$.

Subtract each digit from 15 to obtain the 15s complement, and then add 1 to get the 16s complement.

	(a)	(b)	(c)
Hexadecimal number	74B9	5C0F8	2A7600
15s complement	8B46	A3F07	D589FF
Complement	8B47	A3F08	D58A00

2.24 Evaluate $74B64_{16} - 42AF1_{16}$ and $9C4D819_{16} - 23C0482_{16}$.

Perform the subtraction by adding the complement of the subtrahend to the minuend:

	1	1		1		1				1		1		1		
Minuend		7	4	B	6	4			9	C	4	D	8	1	9
Complement		+B	D	5	0	F			+D	C	3	F	B	7	E
Decimal sums		19	18	16	7	19			23	24	8	29	19	9	23
Modifications		−16	−16	−16	−0	−16			−16	−16	−0	−16	−16	−0	−16
	①	3	2	0	7	3		①	7	8	8	D	3	9	7

Deleting the circled 1 gives us the required difference.

BINARY CODING

2.25 What does "BCD" mean, and what is the major difference between BCD coding and straight binary coding?

BCD is short for Binary-Coded Decimal. Given a decimal number A, BCD coding encodes each decimal digit of A, whereas straight binary coding encodes the number A as a whole.

2.26 How many combinations occur in a (a) 4-bit code? (b) 6-bit code? (c) 7-bit code? (d) 8-bit code? (e) 12-bit code?

An n-bit code permits 2^n combinations. Hence: (a) $2^4 = 16$, (b) $2^6 = 64$, (c) $2^7 = 128$, (d) $2^8 = 256$, (e) $2^{12} = 4096$.

2.27 How many zone bits and how many numeric bits appear in a (a) 4-bit BCD code? (b) 6-bit BCD code? (c) 8-bit BCD code?

All three codes have four numeric bits.

(a) A 4-bit BCD code has no zone bits.

(b) A 6-bit BCD code has two zone bits, usually on the left of the numeric bits:

B	A	8	4	2	1

(c) An 8-bit BCD code has four zone bits, also usually on the left of the numeric bits:

Z	Z	Z	Z	8	4	2	1

2.28 Name and describe the three types of characters contained in the character set of a computer.

A character set contains 10 *digits* (0, 1, ..., 9) and 26 *letters* (A, B, ..., Z). All other characters are called *special characters*. Some special characters are

$$(\quad) \quad . \quad , \quad ; \quad ! \quad + \quad = \quad - \quad /$$

The blank space is also a special character.

2.29 How many special characters can be accommodated by a (*a*) 6-bit code?   (*b*) 7-bit code?
(*c*) 8-bit code?

From Problem 2.26, an *n*-bit code can accommodate $2^n - 36$ special characters.   Hence:

(*a*) $2^6 - 36 = 64 - 36 = 28$ (*b*) $2^7 - 36 = 128 - 36 = 92$ (*c*) $2^8 - 36 = 256 - 36 = 220$

In case the computer uses both uppercase and lowercase letters, there are 26 fewer possible special
characters.

2.30 Determine the number of the characters in the data

256 FIRST AVENUE

There are 16 characters, including two blank spaces.

4-BIT BCD CODES

2.31 Decode each numeric, expressed in the weighted 8-4-2-1 BCD code:

(*a*)   0111 0011 0000 1001   (*b*)   0101 1000 0010   (*c*)   0100 1010 0110

Add up the weights of the 1 bits (i.e. evaluate the expanded binary notation).

(*a*) $0111 \to 4 + 2 + 1 = 7$ $0000 \to 0$
                         $0011 \to 2 + 1 = 3$ $1001 \to 8 + 1 = 9$

Hence, 7309 is the number encoded.

(*b*) $0101 \to 4 + 1 = 5$ $1000 \to 8$ $0010 \to 2$

Hence, 582 is encoded.

(*c*) $0100 \to 4$ $1010 \to 8 + 2 = 10$ $0110 \to 4 + 2 = 6$

Since 10 is no digit, there must be an error in the encoding.

2.32 Decode each numeric, expressed in the XS-3 BCD code:

(*a*)   0110 0011 1011 0111   (*b*)   1010 0101 1100 1000 0100

The XS-3 code for a decimal digit *d* is the weighted 8-4-2-1 code for $d + 3$.   **Therefore, decode**
each 4-bit block using the weighted 8-4-2-1 code, and then subtract 3.

(*a*) $0110 \to 4 + 2 = 6$ and $6 - 3 = 3$
                         $0011 \to 2 + 1 = 3$ and $3 - 3 = 0$
                         $1011 \to 8 + 2 + 1 = 11$ and $11 - 3 = 8$
                         $0111 \to 4 + 2 + 1 = 7$ and $7 - 3 = 4$

Hence, 3084 is the number encoded.

(b)

$1010 \rightarrow 8 + 2 = 10$	and	$10 - 3 = 7$
$0101 \rightarrow 4 + 1 = 5$	and	$5 - 3 = 2$
$1100 \rightarrow 8 + 4 = 12$	and	$12 - 3 = 9$
$1000 \rightarrow 8$	and	$8 - 3 = 5$
$0100 \rightarrow 4$	and	$4 - 3 = 1$

Hence, 72951 is the number encoded.

2.33 Decode the following numeric, which is in the weighted 5-4-2-1 BCD code:

$$0100\ 1100\ 0010\ 1010.$$

Here, the weights of the bits are, from left to right, 5, 4, 2 and 1.

$$0100 \rightarrow 4$$
$$1100 \rightarrow 5 + 4 = 9$$
$$0010 \rightarrow 2$$
$$1010 \rightarrow 5 + 2 = 7$$

Hence, 4927 is the number encoded.

2.34 Encode the decimal number 37 926 using (a) the weighted 8-4-2-1 BCD code, (b) the XS-3 BCD code.

(a) Express each digit d in 4-bit binary form (Table 1-2):

$$3 \rightarrow 0011$$
$$7 \rightarrow 0111$$
$$9 \rightarrow 1001$$
$$2 \rightarrow 0010$$
$$6 \rightarrow 0110$$

Accordingly, the code for 37 926 is 0011 0111 1001 0010 0110.

(b) The XS-3 code for a decimal digit d is the weighted 8-4-2-1 code for $d + 3$. Hence, add 3 to each digit and proceed as in (a).

$$3 + 3 = 6 \rightarrow 0110$$
$$7 + 3 = 10 \rightarrow 1010$$
$$9 + 3 = 12 \rightarrow 1100$$
$$2 + 3 = 5 \rightarrow 0101$$
$$6 + 3 = 9 \rightarrow 1001$$

Thus, the XS-3 code for 37 926 is 0110 1010 1100 0101 1001.

2.35 Given that the XS-3 code for 2185 is 0101 0100 1011 1000, find the XS-3 code for 7814.

Observe that 7814 is the 9s complement of 2185. Hence, the 1s complement of the above code for 2185, 1010 1011 0100 0111, is the XS-3 code for 7814.

2.36 How many special characters can be encoded in a 4-bit BCD code?

There are no special characters in a 4-bit code. In fact, since a 4-bit code allows only 16 combinations, it cannot even accommodate the 26 letters.

6-BIT BCD CODE

2.37 Consider the 6-bit BCD code (Table 2-8). (a) Which are the zone bits, and how are they labeled? (b) How are the decimal digits encoded? (c) What is the code for $A = 6927$?

(a) The first two bits are the zone bits, and they are labeled B and A, as shown:

B	A	8	4	2	1

(b) For each digit d, both zone bits are 0, and the four numeric bits are the weighted 8-4-2-1 code for d.

(c)

$$6 \rightarrow 00\,0110 \qquad 2 \rightarrow 00\,0010$$
$$9 \rightarrow 00\,1001 \qquad 7 \rightarrow 00\,0111$$

Hence, A is coded as 000110 001001 000010 000111.

2.38 Using the 6-bit BCD code, how does one encode the zone bits for (a) the 26 letters? (b) the special characters?

(a) The two zone bits are 11 for the letters A through I, 10 for the letters J through R, and 01 for the letters S through Z.

(b) There is no uniform pattern for the special characters. Some have zone bits 11, some 01, some 10, and some 00.

2.39 What abbreviated form is used for 6-bit codes?

Each 6-bit block is represented as two octal digits, whose binary forms are the first three bits and the second three bits, respectively.

40 The following is the 6-bit BCD code for the data item AUDREY:

A	U	D	R	E	Y
110001	010100	110100	101001	110101	011000

Suppose the computer uses an odd parity check. How would the computer encode the check bit for each character?

The codes for A, D, and R already contain an odd number of 1s, so their check bits will be 0s. The codes for U, E, and Y contain an even number of 1s, so their check bits will be 1s. With each code preceded by its check bit, the data item will appear as follows:

A	U	D	R	E	Y
0110001	1010100	0110100	0101001	1110101	1011000

2.41 If a computer switches from an odd parity check to an even parity check, what, if anything, happens to the code of a character?

The original 6-bit code of a character does not change. However, the check bit of each character will change, either from 0 to 1 or from 1 to 0.

8-BIT BCD CODES; PACKED DECIMAL FORMAT

2.42 What is the meaning of the acronym (*a*) EBCDIC? (*b*) ASCII-8? Which computer systems use which?

(*a*) Extended Binary-Coded Decimal Interchange Code. It is primarily used by IBM and IBM-compatible systems.

(*b*) American Standard Code for Information Interchange. It is primarily used by non-IBM computer systems.

2.43 How many different zone combinations appear among the 26 letter codes in (*a*) EBCDIC? (*b*) ASCII-8?

(*a*) Three. The four zone bits are coded 1100 for the letters A through I, 1101 for the letters J through R, and 1110 for the letters S through Z.

(*b*) Two. The four zone bits are coded 1010 for the letters A through O, and 1011 for the letters P through Z.

2.44 Why does ASCII-8 carry the number 8 in its name?

The original ASCII, a 7-bit code, was extended to an 8-bit code. As both codes are still in use, we include a distinguishing 8 in the name for the 8-bit code.

2.45 How are the decimal digits encoded in (*a*) EBCDIC? (*b*) ASCII-8?

Both encode a digit d with the 4-bit binary representation of d as the numeric portion. The zone portion of d is encoded 1111 in EBCDIC, but 0101 in ASCII-8.

2.46 Where and how does EBCDIC encode the sign of a number? What is the name for such a coding?

EBCDIC encodes the sign of a number in the zone portion of the rightmost digit (rather than at the beginning of the number); it uses 1100 for +, 1101 for −, and 1111 for an unsigned number. This coding is called *zoned decimal format*.

2.47 Encode the numbers +619, −619, and 619 in EBCDIC zoned decimal format.

The numbers are encoded as follows:

6		1		+	9
1111	0110	1111	0001	1100	1001

6		1		−	9
1111	0110	1111	0001	1101	1001

6		1			9
1111	0110	1111	0001	1111	1001

Observe that the zone portion of the rightmost digit, 9, incorporates the sign of the number. All other zone portions contain 1111.

2.48 Numbers stored in zoned decimal format are normally processed in *packed decimal format*. (*a*) Describe packed decimal format coding. (*b*) How are the numbers +619, −619, and 619 encoded in EBCDIC packed decimal format? (*c*) How does the computer change a number from zoned decimal format into packed decimal format?

(*a*) Each digit is encoded by its 4-bit binary form, and the 4-bit code of the sign of the number appears at the end of the number.

(*b*)

6	1	9	+
0110	0001	1001	1100

6	1	9	−
0110	0001	1001	1101

6	1	9
0110	0001	1001	1111

(*c*) The computer (i) interchanges the zone and numeric portions of the rightmost digit, (ii) deletes the zone portions of the other digits, and (iii) "packs" together all the bits. Thus, if $Z \equiv$ the zone, $D \equiv$ the digit, and $S \equiv$ the sign,

Zoned decimal format   ZD ZD ZD ZD SD

Packed decimal format   DDDDDS

Supplementary Problems

NUMBER SYSTEMS

2.49 Write in expanded notation: (*a*) 2043_6, (*b*) 435.621_7.

2.50 Convert to decimal form: (*a*) 4205_6, (*b*) 142032_5.

2.51 Convert to decimal form: (*a*) 24.042_5, (*b*) 2.13_4.

2.52 Rewrite the decimal number 3263 to the base (*a*) 5, (*b*) 4, (*c*) 12 (using A = 10 and B = 11).

2.53 Rewrite the decimal number 1547 to the base (*a*) 6, (*b*) 9, (*c*) 12 (using A = 10 and B = 11).

2.54 Convert the decimal number 274.824 to its base-5 form.

2.55 Convert the decimal number 145.6875 to its base-4 form.

2.56 Convert the decimal number 0.3 to its base-4 form.

OCTAL SYSTEM

2.57 Convert each decimal number to its octal form: (*a*) 12 345, (*b*) 44 444.

2.58 Convert to decimal form: (*a*) 12345_8, (*b*) 44444_8.

2.59 Convert each decimal number to its octal form: (*a*) 0.4375, (*b*) 0.4.

2.60 Convert to binary form: (*a*) 617025_8, (*b*) 43.0276_8.

2.61 Convert to octal form: (*a*) 10101111100_2, (*b*) 1000110111_2, (*c*) 1011.01011_2.

2.62   Add the following octal digits:

$$
\begin{array}{ccccccccc}
4 & 5 & 2 & 6 & 7 & 1 & 3 & 4 & 7 \\
+3 & +6 & +4 & +7 & +4 & +4 & +6 & +6 & +7
\end{array}
$$

2.63   Evaluate (a) $45376_8 + 36274_8$, (b) $2573654_8 + 444777_8$, (c) $333.567_8 + 47.4747_8$.

2.64   Find the radix-minus-one (7s) complement and the (8s) complement of (a) 234705_8, (b) 113355_8, (c) 666000_8.

2.65   Evaluate, using complements: (a) $6157_8 - 4325_8$, (b) $671354_8 - 213604_8$.

HEXADECIMAL SYSTEM

2.66   Convert each decimal number to its hexadecimal form: (a) 967, (b) 2893.

2.67   Convert to decimal form: (a) $3E7_{16}$, (b) $4A5C_{16}$.

2.68   Convert the decimal fraction 0.3 to its hexadecimal form.

2.69   Convert to binary form: (a) $B9E4_{16}$, (b) $50C7F6_{16}$.

2.70   Convert to hexadecimal form: (a) 11101101101100_2, (b) 1110001111110_2, (c) 111110.101111_2.

2.71   Add the following hexadecimal digits:

$$
\begin{array}{ccccccccc}
5 & 9 & B & 7 & 2 & E & 6 & C & 4 \\
+7 & +8 & +2 & +3 & +4 & +E & +A & +6 & +9
\end{array}
$$

2.72   Evaluate (a) $47B6_{16} + 9C75_{16}$, (b) $8D07A5_{16} + 734F6_{16}$, (c) $67.E9_{16} + A.BCDE_{16}$.

2.73   Find the radix-minus-one (15s) complement and the (16s) complement of (a) $5D309_{16}$, (b) $2A4E61_{16}$, (c) $A1B2C300_{16}$.

2.74   Evaluate, using complements: (a) $76B5_{16} - 432C_{16}$, (b) $A57913_{16} - 64EE00_{16}$.

4-BIT BCD CODES

2.75   Decode each numeric, expressed in the 8-4-2-1 BCD code: (a) 0110 1001 0111, (b) 0011 0100 1000 0101.

2.76   Decode each numeric, encoded in the XS-3 BCD code: (a) 0101 1011 1000, (b) 0111 1100 0011 0100 1010.

2.77   Decode each numeric, encoded in the 5-4-2-1 BCD code: (a) 1010 0010 1001, (b) 1011 0001 0100 1100.

2.78   Encode each decimal number in the 8-4-2-1 BCD code: (a) 395, (b) 70 246.

2.79   Encode each decimal number in the XS-3 BCD code: (a) 395, (b) 70 246.

2.80   Given that 0110 0011 1001 1011 is the XS-3 code for the decimal number A, find the XS-3 code for the (10s) complement of A, without decoding A.

6-BIT BCD CODE

2.81   Encode the decimal number 4839 in the 6-bit BCD system (without using a table) in (a) binary form, (b) octal form.

2.82   What is the minimum number of character blocks required to encode the message IN THE BEGINNING?

2.83   Suppose that a computer uses the 6-bit BCD code, with odd parity.   How would the computer store the name (*a*) MARC?   (*b*) ERIK?

2.84   Suppose that a computer uses even parity and that the data item HAMLET is stored in the computer as follows:

$$1111000\ 1110001\ 0100101\ 1100011\ 0110101\ 1010011$$

Without using a table, find which letters, if any, contain an error.

8-BIT BCD CODES

2.85   Encode the name AUDREY in (*a*) binary EBCDIC, (*b*) hexadecimal EBCDIC.   (*c*) How would the binary code appear in the computer if the computer uses odd parity checking?

2.86   Repeat Problem 2.85 using ASCII-8 instead of EBCDIC.

2.87   Using zoned decimal format, write the EBCDIC codes for (*a*) +3759, (*b*) −3759, (*c*) 3759.

2.88   Using packed decimal format, write the codes for (*a*) +3759, (*b*) −3759, (*c*) 3759.

Answers to Supplementary Problems

2.49   (*a*) $2 \times 6^3 + 0 \times 6^2 + 4 \times 6 + 3 \times 1$, (*b*) $4 \times 7^2 + 3 \times 7 + 5 \times 1 + 6 \times 7^{-1} + 2 \times 7^{-2} + 1 \times 7^{-3}$

2.50   (*a*) 941, (*b*) 5892

2.51   (*a*) 14.176, (*b*) 2.4375

2.52   (*a*) 101023_5, (*b*) 302333_4, (*c*) $1A7B_{12}$

2.53   (*a*) 11055_6, (*b*) 2108_9, (*c*) $A8B_{12}$

2.54 2044.403_5

2.55 2101.23_4

2.56 $0.1030303\cdots$

2.57   (*a*) 30071_8, (*b*) 126634_8

2.58   (*a*) 5349, (*b*) 18 724

2.59   (*a*) 0.34_8, (*b*) $0.3146\,3146\cdots_8$

2.60   (*a*) 1100011110000010101_2, (*b*) 100011.00001011111_2

2.61   (*a*) 2574_8, (*b*) 1067_8, (*c*) 13.26_8

2.62   7, 13, 6, 15, 13, 5, 11, 12, 16

2.63   (*a*) 103672_8, (*b*) 3240653_8, (*c*) 403.2637_8

2.64   (*a*) 543072_8, 543073_8; (*b*) 664422_8, 664423_8; (*c*) 111777_8, 112000_8

2.65   (a) 1632_8, (b) 455550_8

2.66   (a) $3C7_{16}$, (b) $B4D_{16}$

2.67   (a) 999, (b) 19 036

2.68 $0.4CCCC\cdots$

2.69   (a) 1011100111100100, (b) 101000011000111111110110

2.70   (a) $3B6C_{16}$, (b) $1C7E_{16}$, (c) $3E.BC_{16}$

2.71   C, 11, D, A, 6, 1C, 10, 12, D

2.72   (a) $E42B_{16}$, (b) $943C9B_{16}$, (c) $72.A5DE_{16}$

2.73   (a) $A2CF6_{16}$, $A2CF7_{16}$; (b) $D5B19E_{16}$, $D5B19F_{16}$; (c) $5E4D3CFF_{16}$, $5E4D3D00_{16}$

2.74   (a) 3389_{16}, (b) $408B13_{16}$

2.75   (a) 697, (b) 3485

2.76   (a) 285, (b) 49 017

2.77   (a) 726, (b) 8149

2.78   (a) 0011 1001 0101, (b) 0111 0000 0010 0100 0110

2.79   (a) 0110 1100 1000, (b) 1010 0011 0101 0111 1001

2.80   When a decimal digit is increased by one, its XS-3 code is increased by one. Hence:

$$
\begin{array}{l}
1001 \quad 1100 \quad 0110 \quad 0100 \to 9s \text{ complement of } A \\
\underline{+0001 = +1 } \\
1001 \quad 1100 \quad 0110 \quad 0101 \to 10s \text{ complement of } A
\end{array}
$$

2.81   (a) 000100 001000 000011 001001, (b) 04 10 03 11

2.82   Sixteen; there are two blank characters.

2.83   (a) 1100100 0110001 0101001 1110011, (b) 1110101 0101001 1111001 1100010

2.84   M, since 0100101 contains an odd number of 1s.

2.85   (a)   11000001 11100100 11000100 11011001 11000101 11101000
          (b)   C1 E4 C4 D9 C5 E8
          (c)   011000001 111100100 011000100 011011001 111000101 111101000

2.86   (a)   10100001 10110101 10100100 10110010 10100101 10111001
          (b)   A1 B5 A4 B2 A5 B9
          (c)   010100001 010110101 010100100 110110010 110100101 010111001

2.87   (a)   11110011 11110111 11110101 11001001
          (b)   11110011 11110111 11110101 11011001
          (c)   11110011 11110111 11110101 11111001

2.88   (a) 0011 0111 0101 1001 1100, (b) 0011 0111 0101 1001 1101, (c) 0011 0111 0101 1001 1111

Chapter 3

Computer Arithmetic

3.1 MATHEMATICAL PRELIMINARIES

This section will discuss a number of mathematical concepts needed in the study of computer arithmetic.

Approximate Numbers; Significant Digits

A measuring or calculating device, such as a desk calculator, a micrometer, or even a modern electronic computer, can handle only a finite number of digits at any given moment. Thus a recorded number may represent a quantity only approximately. For example, the height of a student might be recorded as 187 centimeters, whereas the actual height might be half a unit more, i.e. 187.5 cm. Approximate numbers also arise when terminating decimal fractions are used to represent irrational numbers, e.g.

$$\sqrt{2} \approx 1.414 \qquad \pi \approx 3.1416$$

(One sometimes uses $\approx$ for *approximately equals*.)

The accuracy of an approximate number A is frequently measured by the number of *significant digits* in A. We need the adjective "significant" since some numbers use 0s simply for placing the decimal point. For example, one ounce may be approximated in the metric system as

$$28 \text{ grams} \qquad \text{or as} \qquad 0.028 \text{ kilograms}$$

In either case, the approximation uses two significant digits; the 0s in 0.028 are not significant, but merely locate the decimal point. The formal rules for significant digits follow.

Rule 1: A nonzero digit is always significant.

Rule 2: The digit 0 is significant if it lies between other significant digits.

Rule 3: The digit 0 is never significant when it precedes all the nonzero digits.

Consider any nonzero approximate number A. The *most significant digit* of A is the first (leftmost) significant digit; it will always be the first nonzero digit in A (by Rule 3). The *least significant digit* of A is the last (rightmost) significant digit. Except in an ambiguous case, discussed in Example 3.1(*d*), the least significant digit of A will be the last digit in A, zero or not. The significant digits of A are all digits between and including the most and least significant digits.

EXAMPLE 3.1

(*a*) Consider the numbers 3.14, 1234, 56.607, 880.077. All the digits are significant; hence the numbers contain 3, 4, 5, and 6 significant digits, respectively.

(*b*) Consider the number 0.000 345. By Rule 3, the number contains only three significant digits, the 3, the 4, and the 5.

(*c*) Final 0s are all significant if the approximate number has an embedded decimal point. Thus, 7.7700, 7770.0, and 0.000 777 00 each contain five significant digits, the three 7s and the two final 0s. The initial 0s in the third number are not significant.

(*d*) The number $A = 56\,700\,000$ has final 0s but no embedded decimal point. The 7 or one of the final 0s may be the least significant digit; but without additional information about A, we cannot tell which. Such ambiguity is avoided by writing numbers in scientific notation or in normalized exponential form; these forms are discussed in Section 3.2.

Remark: Of all decimal numbers that can be stored (in coded form) in a single memory location of a computer, let M have the greatest number of significant digits.   Then the number of significant digits in M is called the *precision* of the computer.

Rounding Numbers

Frequently we want to approximate a numerical value by another number, having fewer decimal digits or having a given number of significant digits.   This is usually accomplished by dropping one or more of the least significant digits and then *rounding* the remaining number.   The rules for rounding, where "test digit" refers to the first (leftmost) digit to be dropped, are as follows:

Rounding down.   If the test digit is smaller than 5, the preceding digits are unchanged.

Rounding up.   If the test digit is greater than 5 or is 5 followed by at least one nonzero digit, the preceding digit is increased by 1 (with a carry of 1 if the preceding digit is 9).

Odd-add rule.   If the test digit is 5 with only 0s following, the preceding digit is unchanged if even but increased by 1 if odd.

Under these rules, the maximum *roundoff error* will be one-half the place value of the last retained digit.

EXAMPLE 3.2

(*a*)   Round each number to 2 decimal places: 1.734 82, 3.1416, 0.0037, 0.5677, 258.678, 0.009 11.   The test digit (underlined) is the digit in the third decimal place.   The first three numbers are rounded down, since the test digit is less than 5; the last three numbers are rounded up, since the test digit exceeds 5.   Thus the rounded numbers are: 1.73, 3.14, 0.00, 0.57, 258.68, 0.01.

(*b*)   Round each number to 1 decimal place: 1.152, 22.250 070, 3.35, 36.6500, 7.85, 9.9500.   The test digit, appearing in the second decimal place, is 5 in all the numbers.   The first two numbers are rounded up, since each test digit 5 is followed by some nonzero digit.   The odd-add rule applies to the last four numbers, since each test digit 5 is followed only by 0s.   The third and last numbers are rounded up, since the preceding digits, 3 and 9, are odd; but the fourth and fifth numbers are rounded down, since the preceding digits, 6 and 8, are even.   Thus the rounded numbers are 1.2, 22.3, 3.4, 36.6, 7.8, 10.0.

(*c*)   Round each number to 3 significant digits: 0.777 77, 5.4321, 66.6503, 888.5, 333.5, 111 500.   The test digit (underlined) is the fourth significant digit.   The rounded numbers are: 0.778, 5.43, 66.7, 888, 334, 112 000.   The odd-add rule was applied to the last three numbers.

Truncating

Many arithmetic calculations in the computer result in more digits than can be stored in memory locations.   Rather than rounding such numbers, most computers are programmed simply to drop the least significant digits.   This operation is called *truncating* or *chopping*.   For example, each of the following numbers has been truncated to 3 significant digits:

Number	88.77	−7.8989	999.111	−0.012 345
Truncated value	88.7	−7.89	999	−0.012 3

Observe that positive numbers always decrease in value when truncated, whereas negative numbers always increase in value.

The *truncation error* (or *chopoff error*) can be almost equal to the full place value of the last retained digit.   For example, if we truncate \$24.99 to its dollar amount, we get \$24, an error of 99¢.   Thus, the maximum chopoff error is double the maximum roundoff error (relative to the same decimal place).   Moreover, if a large set of assorted positive numbers are all chopped to the same number of decimals, the *average* error should be one-half the place value of the last retained digit; whereas the average roundoff error would be zero (the negative errors from rounding down tending to balance the positive errors from rounding up).

Absolute Value

The *absolute value* of a number may be viewed intuitively as its magnitude without regard to sign. We denote the absolute value of a number a by $|a|$. Formally we define $|a|$ to be the greater of a and $-a$; that is,

$$|a| = \begin{cases} a & (a > 0) \\ 0 & (a = 0) \\ -a & (a < 0) \end{cases}$$

We note that $|a| = |-a| \geq 0$ for every number a, and that $|a|$ is positive whenever a is not zero.

EXAMPLE 3.3

(a) Find the absolute value of 15, -8, 3.25, 0, -2.22, -0.075. Simply write the "numerical part" of the number: $|15| = 15$, $|-8| = 8$, $|3.25| = 3.25$, $|0| = 0$, $|-2.22| = 2.22$, $|-0.075| = 0.075$.

(b) Evaluate: (i) $|3 - 8|$, (ii) $|7 - 2|$, (iii) $|3| - |8|$, (iv) $-|-5|$.

　　　　　(i)　$|3 - 8| = |-5| = 5$　　　　(iii)　$|3| - |8| = 3 - 8 = -5$
　　　　　(ii)　$|7 - 2| = |5| = 5$　　　　(iv)　$-|-5| = -(5) = -5$

3.2 EXPONENTIAL FORM

For brevity, a very large number or a very small number may sometimes be written as a number times a power of 10. For example, 28 000 000 000 may be written 28 billion or 28×10^9, and 0.000 000 033 44 may be written 3.344×10^{-8}. Actually, every number can be written as a number times a power of ten, called an *exponential form*. Thus,

$$567 = 5.67 \times 10^2 \qquad 0.005 = 5.00 \times 10^{-3} \qquad 25 = 0.25 \times 10^2$$

Such a form is not unique, e.g.

$$567 = 0.0567 \times 10^4 = 0.567 \times 10^3 = 56.7 \times 10^1 = 56\,700 \times 10^{-2}$$

We even have $567 = 567 \times 10^0$ (where $10^0 = 1$). Observe that the only difference between equivalent exponential forms is in the position of the decimal point and the exponent of ten. This comes from the fact that multiplying a decimal number by a power of 10 merely shifts the decimal point in the number.

Consider now any nonzero decimal number A. We can write A *uniquely* as a number M times a power of ten, $A = M \times 10^n$, where the decimal point appears directly in front of the first nonzero digit in M. This is called the *normalized exponential form* for A. The number M is called the *mantissa* of A, and the exponent n is called the *exponent* of A. We note that $.1 \leq M < 1$ for A positive, and $-1 < M \leq -.1$ for A negative.

EXAMPLE 3.4 Several numbers are written in normalized exponential form in Table 3-1, which explicitly lists the mantissa and exponent of each number.

Observe first that both the mantissa and exponent may be positive or negative. Observe also that we usually write a 0 in front of the decimal point in the normalized form, for easier reading and to prevent possible loss of the decimal point when the number is copied.

Remark 1: Another exponential form frequently used is called *scientific notation*. In this exponential form, the decimal point appears directly *after* the first nonzero digit. For example:

Decimal number	999.111	0.006 66	0.75	22.33
Scientific notation	$9.991\,11 \times 10^2$	6.66×10^{-3}	7.5×10^{-1}	2.233×10^1

The principal merit of scientific notation, other than its brevity, is that there is never any ambiguity about the significant digits in a number; *every digit of the mantissa is significant*. (This is also true for numbers in normalized exponential form.)

Table 3-1

Decimal Number	Normalized Exponential Form	Mantissa	Exponent
222.2	0.2222×10^3	0.2222	3
0.0033	0.33×10^{-2}	0.33	−2
−44.44	-0.4444×10^2	−0.4444	2
0.55	0.55×10^0	0.55	0
−0.000 06	-0.6×10^{-4}	−0.6	−4

Remark 2:   Numerical input to the computer must be represented as a single line of characters.   Hence the letter E followed by an integer, n, is commonly used to denote multiplication by 10^n.

EXAMPLE 3.5

11.22E3	means	$11.22 \times 10^3 = 11\,220$
3.456E−5	means	$3.456 \times 10^{-5} = 0.000\,034\,56$
0.005 566 77E04	means	$0.005\,566\,77 \times 10^4 = 55.6677$
−0.003 33E+2	means	$-0.003\,33 \times 10^2 = -0.333$

Most computers restrict the exponent to be a signed or unsigned integer of at most two digits.

Binary Exponential Form

Binary numbers, like decimal numbers, can be written in exponential form, where powers of two are used instead of powers of ten.   Thus, each nonzero binary number has a unique normalized exponential form in which the binary point appears before the first 1 bit.   This unique form yields a unique mantissa M, and a unique integer n representing the exponent of two.   Either of these numbers may be positive or negative, and the exponent n may also be zero.   Furthermore, the computer usually stores a mantissa as a fixed number of bits by truncating or by adding 0s to the original number.

EXAMPLE 3.6 Table 3-2 gives some binary numbers in normalized exponential form, each mantissa being written with exactly 5 bits.

Table 3-2

Binary Number	Normalized Exponential Form	Mantissa	Exponent
1010.1	0.10101×2^4	0.10101	4
0.001111	0.11110×2^{-2}	0.11110	−2
−111	-0.11100×2^3	−0.11100	3
0.1	0.10000×2^0	0.10000	0
−0.01010101	-0.10101×2^{-1}	−0.10101	−1

3.3 INTERNAL REPRESENTATION

Here we discuss how numerics are represented inside the computer using straight binary coding, which encodes an entire number as a whole.   (BCD coding, which encodes a number digit by digit, was discussed in Chapter 2.)

Straight binary coding requires that numbers be stored in computer locations as a fixed number of bits.   A list of bits so treated as a unit is called a *word*, and the number of bits is called the *length* of the word.   For definiteness, we shall assume, unless otherwise stated or implied, that our computers use words of fixed length 32.

Integer Representation

Integers, or *fixed-point numbers*, are numbers that have no decimal points.   An integer J is represented in the memory of the computer by its binary form if J is positive, and by its 2s complement (i.e. the 2s complement of its absolute value) if J is negative.

EXAMPLE 3.7   The computer stores $423 = 110100111_2$ in a 32-bit memory location by introducing sufficient 0s at the beginning of the binary form:

423	0	0	0	0	0	$\cdots$	0	0	1	1	0	1	0	0	1	1	1

The computer stores -423 in a memory location by taking the 1s complement of the above representation for 423 and then adding 1:

−423	1	1	1	1	1	$\cdots$	1	1	0	0	1	0	1	1	0	0	1

In the first display the dots represent omitted 0s; in the second, omitted 1s.

The computer can tell whether an integer J in memory is positive or negative by looking at the first bit.   If the first bit is 0, then J is positive; if the first bit is 1, then J is negative.   Accordingly, the largest (positive) integer that can be stored in a 32-bit memory location is

$$0 \underbrace{1\ 1\ 1\ 1\ \cdots\ 1\ 1\ 1\ 1}_{31\ ones}$$

or $2^{31} - 1$, which is approximately 2 billion.   Similarly, the smallest (negative) integer that can be stored in a 32-bit memory location is -2^{31}, or approximately -2 billion.

Floating-Point Representation

Floating-point numbers (also called *real numbers*) have embedded decimal points.   Such numbers are stored and processed in their binary exponential forms.   The memory location is divided into three *fields*, or blocks of bits.   One field, the first bit, is reserved for the sign of the number (usually 0 for + and 1 for −); a second field, for the exponent of the number; and the last field, for the mantissa of the number.   Figure 3-1 shows the usual fields of a 32-bit memory location.   With a 24-bit mantissa field, the precision of the computer (Section 3.1) is 8 (significant decimal digits).

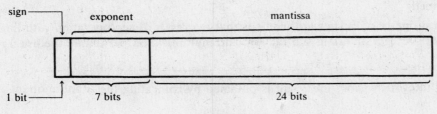

Fig. 3-1

It remains to discuss the way the integer exponent, n, of a floating-point number is represented in its field.   A few computers store n as its binary form when n is positive or zero, and as its 2s complement when n is negative; i.e. the same way that fixed-point integers are stored in memory. However, most computers represent n by its *characteristic*, $n + 2^{t-1}$, where t is the number of bits in the exponent field.   Table 3-3 shows the relationship between the true exponent n and its characteristic when $t = 7$.   Observe that a 7-bit exponent field can represent exponents from -64 to 63, which means that the computer can store floating-point numbers between 2^{-64} and 2^{63}.

Table 3-3

True Exponent	-64	-63	-62	-61	$\cdots$	-1	0	1	$\cdots$	63
Characteristic	0	1	2	3	$\cdots$	63	64	65	$\cdots$	127

EXAMPLE 3.8   Given $A = -419.8125$.   Converting A to binary form yields

$$A = -110100011.1101_2$$

Hence the normalized exponential form of A is

$$A = -0.1101000111101 \times 2^9$$

The true exponent of A being 9, its 7-bit characteristic is

$$9 + 64 = 73 = 1001001_2$$

Thus A will be stored in the 32-bit memory location as follows.

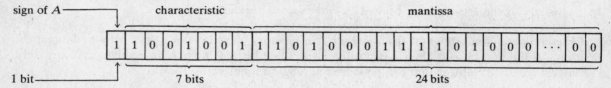

Observe that (i) the first bit is 1, which indicates that A is negative; (ii) the first bit in the characteristic field is 1, which indicates that the exponent of A is nonnegative; and (iii) sufficiently many 0s are attached to the end of the mantissa of A to complete the 24-bit mantissa field.

3.4   COMPUTER ARITHMETIC

Computers normally perform arithmetic calculations with numbers in exponential form; we call this *floating-point arithmetic* or *real arithmetic*.   However, some programming languages, such as FORTRAN, make it possible for computers to perform a separate type of arithmetic for numbers stored as fixed-point integers.   We study this integer type of arithmetic first, and then, in much more detail, study floating-point arithmetic.

Integer Arithmetic

The main property of integer arithmetic is that the result of any operation with integers must be an integer.   For integer addition, subtraction, and multiplication, we obtain the usual results; e.g.

$$12 + 5 = 17 \qquad 12 - 5 = 7 \qquad 12 \times 5 = 60$$

However, for integer division, the result is obtained by truncating the usual quotient to an integer. For example,

$$12 \div 5 = 2 \qquad 7 \div 8 = 0 \qquad -9 \div 2 = -4$$

Thus integer division is different from ordinary division, and, in integer arithmetic, the ordinary rule

$$\frac{a+b}{c} = \frac{a}{c} + \frac{b}{c}$$

does not generally hold.

Floating-Point (Real) Arithmetic

Here all numbers are stored and processed in exponential form. Let P denote the precision of the computer (Section 3.1). The main thing to remember about this arithmetic is that the result of any operation is normalized and the mantissa rounded or truncated to P digits. For illustration purposes, we assume in the examples below that all mantissas are truncated (not rounded) to $P = 4$ decimal digits; we represent the numbers in decimal exponential form rather than the binary exponential form actually employed in the computer.

Real addition. If two numbers to be added have the same exponent, the mantissas are added and the same exponent is used.

$$0.2356 \times 10^4 + 0.4123 \times 10^4 = 0.6479 \times 10^4$$

$$0.5544 \times 10^2 + 0.7777 \times 10^2 = 1.3321 \times 10^2 \approx 0.1332 \times 10^3$$

Here, the last value is obtained by normalizing and truncating the sum of the mantissas. On the other hand, if the two numbers have different exponents, then one of the numbers must be renormalized so that they both do have the same exponent before the addition is performed. Normally, the computer adjusts the smaller exponent. For example,

$$0.1166 \times 10^2 + 0.8811 \times 10^4 = 0.001\,166 \times 10^4 + 0.8811 \times 10^4 = 0.882\,266 \times 10^4 \approx 0.8822 \times 10^4$$

The last value is again obtained by truncation.

Real subtraction is analogous to real addition.

$$0.8844 \times 10^{-2} - 0.3322 \times 10^{-2} = 0.5522 \times 10^{-2}$$

$$0.7777 \times 10^3 - 0.7531 \times 10^3 = 0.0246 \times 10^3 = 0.2460 \times 10^2$$

Note that in the second subtraction a significant digit is lost, so that the final mantissa, .2460, contains a nonsignificant 0.

$$0.6666 \times 10^3 - 0.3333 \times 10^2 = 0.6666 \times 10^3 - 0.033\,33 \times 10^3 = 0.633\,27 \times 10^3 \approx 0.6332 \times 10^3$$

Again truncation has been performed.

Real multiplication. Here we multiply the mantissas and *add* the exponents.

$$(0.3355 \times 10^2) \times (0.4466 \times 10^3) = 0.149\,834\,30 \times 10^5 \approx 0.1498 \times 10^5$$

$$(0.1111 \times 10^2) \times (0.2222 \times 10^4) = 0.024\,686\,42 \times 10^6 \approx 0.2468 \times 10^5$$

$$(0.4444 \times 10^{-5}) \times (0.3579 \times 10^2) = 0.159\,050\,76 \times 10^{-3} \approx 0.1590 \times 10^{-3}$$

As usual, the results of the calculations are renormalized, if necessary, and truncated to $P = 4$ significant digits.

Real division. Here we divide the mantissas and *subtract* the exponents. The division is carried out only to P significant digits.

$$(0.4444 \times 10^7) \div (0.1357 \times 10^4) \approx 3.271 \times 10^3 = 0.3271 \times 10^4$$

$$(0.3333 \times 10^{-4}) \div (0.6543 \times 10^2) \approx 0.5089 \times 10^{-6}$$

$$(0.8877 \times 10^2) \div (0.2121 \times 10^4) \approx 4.185 \times 10^{-2} = 0.4185 \times 10^{-1}$$

3.5 ERRORS

Since the computer retains a limited number of significant digits, most stored numerical values and calculations are only approximations to the true values. Hence it is advantageous to have some notion of absolute error and relative error.

The difference between the true value and the approximate value of a quantity is called the *absolute error* (in the approximation). That is, if $\bar{A}$ is an approximation of the value A, then $e = A - \bar{A}$ is the absolute error. (Some texts define the absolute error to be the approximate value minus the true value.) The ratio of the absolute error to the true value,

$$r = \frac{e}{A} = \frac{A - \bar{A}}{A}$$

is called the *relative error*. Relative errors are frequently written as percents.

EXAMPLE 3.9   If $A = 1.427$, find the absolute error and the relative error when (a) A is rounded to 1.43, (b) A is truncated to 1.42.

(a)   The absolute error e is the difference

$$e = 1.427 - 1.43 = -0.003$$

The relative error r is the ratio

$$r = \frac{e}{A} = \frac{-0.003}{1.427} = -0.0021$$

Thus $|r| = 0.21\%$.

(b)
$$e = 1.427 - 1.42 = 0.007$$

$$r = \frac{e}{A} = \frac{0.007}{1.427} = 0.0049$$

Thus $|r| = 0.49\%$, which is more than twice the relative error in (a).

Consider any numerical value A. Without knowing the true value of A, we can still place a bound on the relative error, r_A, when A is rounded or truncated.

Theorem 3.1:   When A is rounded to P significant decimal digits, then

$$|r_A| < 0.5 \times 10^{-P+1}$$

When A is truncated to P significant decimal digits, then

$$|r_A| < 10^{-P+1}$$

For $P = 4$, the theorem gives $|r_A| < 0.001$ for rounding or truncation of any numeric A.

Besides rounding and truncating, there are other sources of error in the computer. We mention some of these in the next examples.

EXAMPLE 3.10 (Propagation of Errors).   Suppose that a computer truncates all numerical values to $P = 4$ decimal digits. Then $A = 2/3$ would be stored as 0.6666, with a relative error of

$$r = \frac{(2/3) - 0.6666}{2/3} = 0.0001$$

Adding A to itself six times yields

$$
\begin{array}{r}
0.6666 \\
+\,0.6666 \\
\hline
1.333 \\
+\,0.6666 \\
\hline
1.999 \\
+\,0.6666 \\
\hline
2.665 \\
+\,0.6666 \\
\hline
3.331 \\
+\,0.6666 \\
\hline
3.997
\end{array}
$$

Each time the sum is truncated to 4 digits.   The true sum is $6(2/3) = 4$.   Hence the relative error is

$$r = \frac{4 - 3.997}{4} = 0.000\ 75$$

which is more than seven times the original relative error.

EXAMPLE 3.11 (Conversion Errors).   A terminating decimal fraction may convert into a nonterminating binary fraction (see, e.g., Problem 1.15), which must necessarily be chopped to be stored in the computer.   The resultant error, though very small, can be propagated as the number is used repeatedly in calculations.

EXAMPLE 3.12 (Subtractive Cancellation).   Suppose that we want to find the difference, $D = A - B$, of two nearly equal numbers, A and B; say,

$$A = 222.88 \qquad B = 222.11$$

If the computer truncates all numerical values to 4 decimal digits, A and B will be stored as

$$\bar{A} = 0.2228 \times 10^3 \qquad \bar{B} = 0.2221 \times 10^3$$

whence $\bar{D} = 0.0007 \times 10^3 = 0.7000$ (one significant digit!).   The true value is $D = 0.77$.   Therefore, the relative error r is

$$r = \frac{0.7000}{0.77} = 9.1\%$$

This is more than 90 times the 0.1% bound on truncation errors when $P = 4$.   Subtractive cancellation (loss of significant digits when two nearly equal numbers are subtracted) is the source of some of the most serious errors in computer calculations.

Solved Problems

MATHEMATICAL PRELIMINARIES

3.1   Determine the most significant digit, the least significant digit, and the number of significant digits in

$$222.333 \qquad 8.008 \qquad 0.0555 \qquad 0.002\ 200 \qquad 440\ \bar{0}00$$

The most and least significant digits are underlined:

$$\underline{2}22.33\underline{3} \qquad \underline{8}.00\underline{8} \qquad 0.0\underline{5}5\underline{5} \qquad 0.00\underline{2}\ 20\underline{0} \qquad \underline{4}40\ \bar{\underline{0}}00$$

Observe that the most significant digit is the first nonzero digit, and, except in the last number, the least significant digit is the last digit, zero or nonzero.   In the last number, the least significant 0 digit was already indicated by an overbar.   Thus the numbers have 6, 4, 3, 4, and 4 significant digits, respectively.

3.2   Find the number of significant digits in $A = 234\ 000\ 000$.

This is the ambiguous case, where there are final 0s but no embedded decimal point; A can have anywhere from 3 to 9 significant digits.

3.3   Round

$$555.666 \qquad 2222.333 \qquad 333.00 \qquad 44.665 \qquad 9.9950 \qquad 0.005\ 000$$

(*a*) to 2 decimal places, (*b*) to 4 significant digits.

(a)   The digit in the third decimal place is the test digit.   Using the rules for rounding, we obtain:

$$555.67 \qquad 2222.33 \qquad 333.00 \qquad 44.66 \qquad 10.00 \qquad 0.00$$

The odd-add rule was applied to the last three numbers.

(b)   The fifth significant digit is the test digit.   Hence we obtain:

$$555.7 \qquad 2222 \qquad 333.0 \qquad 44.66 \qquad 9.995 \qquad 0.005\,000$$

3.4   How does everyday rounding practice differ from the procedure of this book?

Most people round up when the test digit is 5, without applying the odd-add rule.

3.5   Truncate

$$222.333 \qquad 5.5555 \qquad 0.044\,44 \qquad -7.7777 \qquad -0.009\,999\,99$$

(a) to an integer, (b) to 4 significant digits.

(a)   Simply drop the fractional part of the number, yielding

$$222 \qquad 5 \qquad 0 \qquad -7 \qquad 0$$

(b)   Delete all digits after the fourth significant digit, yielding

$$222.3 \qquad 5.555 \qquad 0.044\,44 \qquad -7.777 \qquad -0.009\,999$$

3.6   Evaluate (a) $|7|, |-7|, |-11|, |0|, |-1|, |13|, |-13|$; (b) $|3-5|, |-3+5|, |-3-5|$.

(a)   The absolute value is the magnitude of the number without regard to sign:

$$|7| = 7 \qquad |-7| = 7 \qquad |-11| = 11 \qquad |0| = 0 \qquad |-1| = 1 \qquad |13| = 13 \qquad |-13| = 13$$

(b)   Evaluate inside the absolute value signs first:

$$|3-5| = |-2| = 2 \qquad |-3+5| = |2| = 2 \qquad |-3-5| = |-8| = 8$$

EXPONENTIAL FORM

3.7   Rewrite the following decimal numbers without exponents:

(a) 2.22×10^4 (c) -444×10^2 (e) 0.0666×10^3
(b) 33.3×10^{-5} (d) $0.005\,55 \times 10^{-3}$ (f) -0.777×10^0

Multiplying a decimal number by 10^n is equivalent to moving the decimal point n places to the right, if n is positive, or $|n|$ places to the left, if n is negative.   Hence:

(a)   22 200   (c) $-44\,400$ (e)   66.6
(b)   0.000 333   (d)   0.000 005 55   (f) -0.777

3.8   Write the following numbers in normalized exponential form: (a) 11.22, (b) -555.666, (c) 0.007 77, (d) $-0.000\,088$, (e) 0.0.   State the mantissa and exponent of each number.

Shift the decimal point to precede the first nonzero digit, producing the mantissa.   Then multiply the mantissa by that power of ten which would restore the decimal point to its original location (see Problem 3.7).   The exponent of that power is the exponent of the form.

	Number	Normalized Form	Mantissa	Exponent
(a)	11.22	0.1122×10^2	0.1122	2
(b)	−555.666	$-0.555\,666 \times 10^3$	−0.555 666	3
(c)	0.007 77	0.777×10^{-2}	0.777	−2
(d)	−0.000 088	-0.88×10^{-4}	−0.88	−4

(e) For the number zero, the exponent is indeterminate, and so a normalized exponential form is not defined for this number.

3.9   Write the numbers of Problem 3.8 in scientific notation.

   Put each number in exponential form, with the decimal point directly after the first nonzero digit. Thus: (a) 1.122×10^1, (b) $-5.556\,66 \times 10^2$, (c) 7.77×10^{-3}, (d) -8.8×10^{-5}.   (e) The number zero has no explicit scientific notation except to leave the number as it stands.   Then one assumes that 0.0 is an approximate number of one significant digit (the second 0).

3.10   Find the value of each decimal number, written in computer E-form:

   (a)   3.33E+03   (c)   0.7E6
   (b)   55.5E−4   (d)   −8.8E−02

   Here E followed by an integer n means multiplication by 10^n.

   (a) $3.33\text{E}+03 = 3.33 \times 10^3 = 3330$ (c) $0.7\text{E}6 = 0.7 \times 10^6 = 700\,000$
   (b) $55.5\text{E}-4 = 55.5 \times 10^{-4} = 0.005\,55$ (d) $-8.8\text{E}-02 = -8.8 \times 10^{-2} = -0.088$

INTERNAL REPRESENTATION

3.11   Find the internal representation of (a) 907, (b) −907, if the computer uses a 32-bit memory location to store each number.

   (a)   First find the binary form for 907:

                                        Remainders
                          2)907
                          2)453   1
                          2)226   1
                          2)113   0
                          2)56   1
                          2)28   0
                          2)14   0
                          2)7   0
                          2)3   1
                          1   1

   Hence, $907 = 1110001011_2$.   The binary form will appear with 0s added on the left to fill out the memory location:

   907 | 0 | 0 | 0 | 0 | 0 | 0 | 0 | 0 | ⋯ | 0 | 0 | 0 | 1 | 1 | 1 | 0 | 0 | 0 | 1 | 0 | 1 | 1 |

(*b*) Replace each 0 by 1 and each 1 by 0 in the 32-bit representation of 907, and then add 1. This gives the representation of −907:

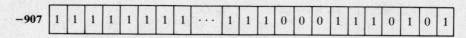

In other words, −907 is stored as the 2s complement of 907.

3.12 How does the computer know whether an integer N in memory is positive or negative?

If the first bit is 0, then N is positive; if the first bit is 1, then N is negative.

3.13 Find the internal representation of $A = 93.625$, assuming a 32-bit memory location.

First convert A to binary form (see Section 1.3).

INTEGRAL PART		FRACTIONAL PART
	Remainders	0.625
2)93		×2
2)46	1	1.250
2)23	0	×2
2)11	1	0.500
2)5	1	×2
2)2	1	1.000
1	0	

Hence, $A = 1011101.101_2$. Now write A in binary normalized exponential form:

$$A = 0.1011101101 \times 2^7$$

Thus the mantissa of A is $M = 0.1011101101$ and the exponent of A is 7. Add $2^6 = 64$ to 7 to give the characteristic, 71. The binary form of 71 is 1000111_2. Therefore, A appears in memory as follows:

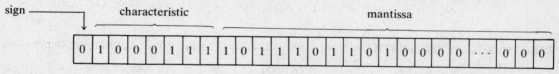

The first bit is 0, since A is positive. Observe that 0s are attached to the end of the mantissa to fill up the memory location.

3.14 How does the computer know whether an exponent n in memory is positive or negative?

If the first bit of the characteristic field is 1, then n is positive (or zero); if the first bit is 0, then n is negative.

COMPUTER ARITHMETIC

3.15 Suppose the computer is programmed to perform fixed-point integer arithmetic. What values does it obtain for (*a*) $4 + 9$, $6 - 11$, 4×5, $-4 + 7$, $-2 - 6$, -7×3? (*b*) 8/3, 24/5, 6/7, −30/7, −4/9, −11/2?

(*a*) Integer addition, subtraction, and multiplication yield the same results as the usual arithmetic operations. Hence:

$$13 \qquad -5 \qquad 20 \qquad 3 \qquad -8 \qquad -21$$

(b)   The result of integer division is the integral part of the usual quotient.   (By the integral part of a negative number, we understand the nonpositive integer a obtained by dropping all digits that follow the decimal point.   In some mathematical contexts, $a - 1$ is taken to be the "integral part" if the given number is nonintegral.)   Hence:

$$2 \qquad 4 \qquad 0 \qquad -4 \qquad 0 \qquad -5$$

3.16   Show how the computer performs the following floating-point additions (where mantissas are truncated to $P = 4$ decimal digits):

(a) $0.5566 \times 10^3 + 0.7777 \times 10^3$ (b) $0.3344 \times 10^2 + 0.8877 \times 10^{-1}$

(a)   The exponents are equal.   Hence the mantissas are added, keeping the common exponent, and then the sum is renormalized, if necessary.

$$0.5566 \times 10^3 + 0.7777 \times 10^3 = 1.3343 \times 10^3 \approx 0.1334 \times 10^4$$

(b)   The exponents are different.   The decimal point in the mantissa corresponding to the smaller exponent is shifted so as to make the two exponents equal.   Now the procedure of (a) is followed.

$$0.3344 \times 10^2 + 0.8877 \times 10^{-1} = 0.3344 \times 10^2 + 0.000\,887\,7 \times 10^2 = 0.335\,287\,7 \times 10^2 \approx 0.3352 \times 10^2$$

3.17   Show how the computer performs the following floating-point subtractions (where mantissas are truncated to $P = 4$ decimal digits):

(a) $0.7744 \times 10^{-2} - 0.6666 \times 10^{-2}$ (b) $0.8844 \times 10^{-2} - 0.2233 \times 10^0$

(a)   The exponents are the same.   Hence the mantissas are subtracted, keeping the common exponent, and then the difference is renormalized, if necessary.

$$0.7744 \times 10^{-2} - 0.6666 \times 10^{-2} = 0.1078 \times 10^{-2}$$

(b)   The exponents are different.   The decimal point in the mantissa corresponding to the smaller exponent is shifted so as to make the two exponents equal.   Now the procedure of (a) is followed.

$$0.8844 \times 10^{-2} - 0.2233 \times 10^0 = 0.008\,844 \times 10^0 - 0.2233 \times 10^0 = -0.214\,456 \times 10^0 \approx -0.2144 \times 10^0$$

3.18   Show how the computer performs the following floating-point calculations (where mantissas are truncated to $P = 4$ decimal digits):

(a) $(0.2255 \times 10^2) \times (0.1234 \times 10^1)$ (b) $(0.8833 \times 10^3) \div (0.5555 \times 10^5)$

(a)   In multiplication, mantissas are multiplied and exponents are added.   Again, results are renormalized, if necessary.

$$(0.2255 \times 10^2) \times (0.1234 \times 10^1) = 0.025\,826\,70 \times 10^3 \approx 0.2582 \times 10^2$$

(b)   In division, mantissas are divided until $P = 4$ significant digits are obtained, and exponents are subtracted.

$$(0.8833 \times 10^3) \div (0.5555 \times 10^5) = 1.590 \times 10^{-2} = 0.1590 \times 10^{-1}$$

ERRORS

3.19   Suppose that 91 637 146 miles, the mean distance between the earth and sun, is rounded to 92 000 000 miles.   Find the absolute error and the relative error in this approximation.

The absolute error, e, is the difference between the true value and the approximate value:

$$e = 91\,637\,146 - 92\,000\,000 = -392\,854 \text{ miles}$$

The relative error, r, is the ratio of the absolute error to the true value:

$$r = \frac{-392\,854 \text{ miles}}{91\,637\,146 \text{ miles}} = -0.00429 = -0.429\%$$

Observe that the relative error is dimensionless.

3.20   Numerical values A and B are stored in the computer as approximations $\bar{A}$ and $\bar{B}$, which are then multiplied together. Neglecting any further truncation or roundoff error, show that the relative error in the product is the sum of the relative errors in the factors.

We have

$$r_A = \frac{A - \bar{A}}{A} = 1 - \frac{\bar{A}}{A} \qquad \text{or} \qquad \bar{A} = A(1 - r_A)$$

Likewise, $\bar{B} = B(1 - r_B)$. It follows that

$$\bar{A}\bar{B} = AB(1 - r_A)(1 - r_B) = AB[1 - (r_A + r_B) + r_A r_B] \approx AB[1 - (r_A + r_B)]$$

since the product of the two small quantities r_A and r_B will be negligible in comparison to their sum. Consequently, $r_{AB} = r_A + r_B$, which is what we were asked to show. One often expresses this result in the statement that *when quantities are multiplied, their percent errors add.*

Supplementary Problems

MATHEMATICAL PRELIMINARIES

3.21   Determine the most significant digit, the least significant digit, and the number of significant digits in (a) 44.44, 30 303, 6.6707, 5.005; (b) 0.222, 0.000 333 3, 00 011, 0.008 008; (c) 2.220 00, 3300.000, 0.004 440 0, 55 500 000.

3.22   Round 22.4444, 1.234 567, 333.777, 0.065 432 1, 0.005 678 (a) to 2 decimal places, (b) to 3 decimal places, (c) to 4 significant digits.

3.23   Round 0.445 00, 7.775, 66.665 000, 8.885 020, 2.3350 (a) to 2 decimal places, (b) to 2 significant digits.

3.24   Truncate 44.44, 30 303, 6.6707, 5.005, −0.044 488 8 (a) to an integer, (b) to 4 significant digits.

3.25   Let CHOP(Q) denote the integral part of the number Q. Solve the equation

$$2 \times \text{CHOP}(N/2) = N$$

3.26   Evaluate:

(a) $|4 - 9|$ (c) $|-4 + 9|$ (e) $|3 - 5| - |6 - 2|$

(b) $|-4 - 9|$ (d) $|-4| - |9|$ (f) $||-6| - |3 - 12||$

EXPONENTIAL FORM

3.27   Rewrite each number without an exponent:

(a) 44.44×10^4 (c) -0.066×10^2 (e) 88.99×10^0
(b) 55.55×10^{-5} (d) $0.000\,77 \times 10^{-3}$ (f) 1.234×10^{-4}

3.28   Write each number in normalized exponential form:

(a) 333.444   (c) 0.000 667 7   (e) 222 000
(b) −1.2345   (d) −0.8899   (f) −0.03

3.29 Find the mantissa and exponent of each number in Problem 3.28.

3.30 Write in scientific notation each number in Problem 3.28.

3.31 Find the value of each number written in computer E-form: (a) 222.2E+2, (b) −3.3E−03, (c) 0.4E4, (d) 0.5E−5.

3.32 Give the binary normalized exponential form of each binary number: (a) 111.000111, (b) 0.0011001100, (c) −1010.1010, (d) 0.1111.

INTERNAL REPRESENTATION

In Problems 3.33 through 3.38, assume that the computer stores each number in a 32-bit memory location.

3.33 Find the internal representation of (a) 118, (b) −118.

3.34 Find the internal representation of (a) 397, (b) −397.

3.35 Find the internal representation of $A = -50.375$.

3.36 Find the internal representation of $B = 0.093\,75$.

3.37 Find the internal representation of $C = 0.2$.

3.38 Suppose that an exponent n is represented in a 7-bit field as follows. The first bit is reserved for the sign, 1 for + and 0 for −. In the remaining 6-bit field, n is represented as its binary form if n is positive, but as its 2s complement if n is negative. Show that this is exactly the same representation as given by storing the 7-bit characteristic, $C = n + 64$, of n.

COMPUTER ARITHMETIC

3.39 If the computer is programmed to perform fixed-point integer arithmetic, what values are obtained for (a) $6 + 10$, $2 - 7$, $3 \times (-5)$, $-4 - 8$, $-4 - (6 - 3)$? (b) 11/4, −15/3, 8/11, 123/4, −26/8, −5/8?

3.40 Under integer arithmetic, the quotient J/K of two integers J and K is less than or equal to the usual quotient. (True or false.)

In Problems 3.41 through 3.44, assume that the computer truncates mantissas to $P = 4$ decimal digits.

3.41 Give the results of the three floating-point additions: (a) $0.2233 \times 10^2 + 0.6688 \times 10^1$, (b) $5.666 + 44.55$, (c) $111.77 + 55.666$.

3.42 Perform the following floating-point subtractions: (a) $0.9922 \times 10^{-3} - 0.4477 \times 10^{-3}$, (b) $33.666 - 2.7777$, (c) $0.8888 \times 10^2 - 0.2222 \times 10^3$.

3.43 Give the results of the following floating-point multiplications: (a) $(0.5432 \times 10^3) \times (0.3333 \times 10^{-5})$, (b) 222.88×1.1177.

3.44 Perform the following floating-point divisions: (a) $(0.2233 \times 10^{-2}) \div (0.6611 \times 10^3)$, (b) $111.99 \div 44.888$.

ERRORS

3.45 Given $A = 66.888$, find the absolute error e and the relative error r if (a) A is rounded to 66.89, (b) A is truncated to 66.88.

3.46 Given $A = 66.888$ and $B = 66.111$. When the computer calculates the difference $D = A - B$ (where mantissas are truncated to $P = 4$ digits), what are the absolute and relative errors?

3.47 Place a bound on the absolute error when two approximate numbers are added.

Answers to Supplementary Problems

3.21 (a) 44.44, four; 30 303, five; 6.6707, five; 5.005, four
    (b) 0.222, three; 0.000 333 3, four; 00 011, two; 0.008 008, four
    (c) 2.220 00, six; 3300.000, seven; 0.004 440 0, five; 55 500 000, six

3.22 (a) 22.44, 1.23, 333.78, 0.06, 0.00
    (b) 22.444, 1.235, 333.777, 0.065, 0.006
    (c) 22.44, 1.235, 333.8, 0.065 43, 0.005 678

3.23 (a) 0.44, 7.78, 66.66, 8.89, 2.34; (b) 0.44, 7.8, 67, 8.9, 2.3

3.24 (a) 44, 30 303, 6, 5, 0; (b) 44.44, 30 3$\overline{0}$0, 6.670, 5.005, −0.044 48

3.25 N/2 = integer, or N = even integer

3.26 (a) 5, (b) 13, (c) 5, (d) −5, (e) −2, (f) 3

3.27 (a) 444 400, (b) 0.000 555 5, (c) −6.6, (d) 0.000 000 77, (e) 88.99, (f) 0.000 123 4

3.28   (a) $0.333\,444 \times 10^3$ (c) -0.6677×10^{-3} (e) $0.222\,000 \times 10^6$
            (b) $-0.123\,45 \times 10^1$ (d) -0.8899×10^0 (f) -0.3×10^{-1}

3.29   (a) $0.333\,444$, $n = 3$ (c) -0.6677, $n = -3$ (e) $0.222\,000$, $n = 6$
            (b) $-0.123\,45$, $n = 1$ (d) -0.8899, $n = 0$ (f) -0.3, $n = -1$

3.30 (a) $3.334\,44 \times 10^2$, (b) -1.2345, (c) 6.677×10^{-4}, (d) -8.899×10^1, (e) $2.220\,00 \times 10^5$, (f) -3×10^{-2}

3.31 (a) 22 220, (b) −0.0033, (c) 4000, (d) 0.000 005

3.32 (a) 0.111000111×2^3, (b) 0.11001100×2^{-2}, (c) -0.10101010×2^4 (d) 0.1111×2^0

3.33

(a) | 0 | 0 | 0 | ··· | 0 | 0 | 1 | 1 | 1 | 0 | 1 | 1 | 0 |

(b) | 1 | 1 | 1 | ··· | 1 | 1 | 0 | 0 | 0 | 1 | 0 | 1 | 0 |

3.34

(a) | 0 | 0 | 0 | ··· | 0 | 0 | 1 | 1 | 0 | 0 | 0 | 1 | 1 | 0 | 1 |

(b) | 1 | 1 | 1 | ··· | 1 | 1 | 0 | 0 | 1 | 1 | 1 | 0 | 0 | 1 | 1 |

3.35 | 1 | 1 | 0 | 0 | 0 | 1 | 1 | 0 | 1 | 1 | 0 | 0 | 1 | 0 | 0 | 1 | 1 | 0 | 0 | 0 | ··· | 0 | 0 | 0 |

3.36 | 0 | 0 | 1 | 1 | 1 | 1 | 0 | 1 | 1 | 1 | 0 | 0 | 0 | 0 | 0 | 0 | ··· | 0 | 0 | 0 |

3.37 | 0 | 0 | 1 | 1 | 1 | 1 | 1 | 0 | 1 | 1 | 0 | 0 | 1 | 1 | 0 | 0 | 1 | 1 | 0 | 0 | ··· | 1 | 1 | 0 | 0 |

3.39   (a) 16, −5, −15, −12, −7; (b) 2, −5, 0, 30, −3, 0

3.40   False; for a negative quotient, truncation generally yields a larger number.

3.41   (a) 0.2901×10^2, (b) 0.5021×10^2, (c) 0.1673×10^3

3.42   (a) 0.5445×10^{-3}, (b) 0.3088×10^2, (c) -0.1333×10^3

3.43   (a) 0.1810×10^{-2}, (b) 0.2488×10^3

3.44   (a) 0.3377×10^{-5}, (b) 0.2493×10^1

3.45   (a) $e = -0.002$, $r = -0.00299\%$; (b) $e = 0.008$, $r = 0.01196\%$

3.46 $D = 0.777$, $\bar{D} = 0.77$, $e = 0.007$, $r = 0.90\%$

3.47 $|e_{A+B}| = |e_A + e_B| \le |e_A| + |e_B|$

Chapter 4

Logic, Truth Tables

4.1 INTRODUCTION

A computer may be programmed to make decisions based on whether certain statements—e.g. "The number just computed exceeds 100"—are true or false. The truth or falsity of a statement is called its *truth value*; a statement is either *true* or *false*, but not both. Some statements are *compound statements*, i.e. are composed of substatements and various connectives.

EXAMPLE 4.1

(*a*) "Roses are red and violets are blue" is a compound statement with substatements "Roses are red" and "Violets are blue".

(*b*) "He is intelligent or studies every night" is, implicitly, a compound statement with substatements "He is intelligent" and "He studies every night".

(*c*) "Where are you going?" is not a statement since it is neither true nor false.

The fundamental property of a compound statement is that its truth value is completely determined by the truth values of its substatements together with the way in which they are connected to form the compound statement. We begin with a study of some of these connectives. In this chapter we will use the letters p, q, r (lowercase or uppercase, with or without subscripts) to denote statements.

4.2 CONJUNCTION, $p \wedge q$

Any two statements can be combined by the word "and" to form a compound statement called the *conjunction* of the original statements. Symbolically,

$$p \wedge q$$

denotes the conjunction of the statements p and q, read "p and q".

The truth value of the compound statement $p \wedge q$ is given by the following table:

p	q	$p \wedge q$
T	T	T
T	F	F
F	T	F
F	F	F

Here, the first line is a short way of saying that if p is true and q is true then $p \wedge q$ is true. The other lines have analogous meaning. We regard this table as defining precisely the truth value of the compound statement $p \wedge q$ as a function of the truth values of p and of q. Observe that $p \wedge q$ is true only in the case that both substatements are true.

EXAMPLE 4.2 Consider the following four statements:

(i)   Paris is in France and $\quad 2+2 = 4$.
(ii) Paris is in France and $\quad 2+2 = 5$.
(iii) Paris is in England and $2+2 = 4$.
(iv) Paris is in England and $2+2 = 5$.

Only the first statement is true.   Each of the other statements is false since at least one of its substatements is false.

4.3   DISJUNCTION, $p \lor q$

Any two statements can be combined by the word "or" (in the sense of "and/or") to form a new statement which is called the *disjunction* of the original two statements.   Symbolically,

$$p \lor q$$

denotes the disjunction of the statements p and q and is read "p or q".

The truth value of $p \lor q$ is given by the following table, which we regard as defining $p \lor q$:

p	q	$p \lor q$
T	T	T
T	F	T
F	T	T
F	F	F

Observe that $p \lor q$ is false only when both substatements are false.

EXAMPLE 4.3   Consider the following four statements:

(i)   Paris is in France or $2 + 2 = 4$.
(ii)   Paris is in France or $2 + 2 = 5$.
(iii) Paris is in England or $2 + 2 = 4$.
(iv)   Paris is in England or $2 + 2 = 5$.

Only (iv) is false.   Each of the other statements is true since at least one of its substatements is true.

Remark: The English word "or" is commonly used in two distinct ways.   Sometimes it is used in the sense of "p or q or both", i.e. at least one of the two alternatives occurs, as above, and sometimes it is used in the sense of "p or q but not both", i.e. exactly one of the two alternatives occurs.   For example, the sentence "He will go to Harvard or to Yale" uses "or" in the latter sense, called the *exclusive disjunction*.   Unless otherwise stated, "or" shall be used in the former sense. This discussion points out the precision we gain from our symbolic language: $p \lor q$ is defined by its truth table and *always* means "p and/or q".

4.4   NEGATION, $\sim p$

Given any statement p, another statement, called the *negation* of p, can be formed by writing "It is false that . . ." before p or, if possible, by inserting in p the word "not".   Symbolically,

$$\sim p$$

denotes the negation of p (read "not p").

The truth value of $\sim p$ is given by the following table:

p	$\sim p$
T	F
F	T

In other words, if p is true then $\sim p$ is false, and if p is false then $\sim p$ is true.   Thus the truth value of the negation of any statement is always the opposite of the truth value of the original statement.

EXAMPLE 4.4   Consider the following statements.

(*a*)   Paris is in France.

(*b*)   It is false that Paris is in France.

(*c*)   Paris is not in France.

(*d*) $2 + 2 = 5$.

(*e*)   It is false that $2 + 2 = 5$.

(*f*) $2 + 2 \neq 5$.

Then (*b*) and (*c*) are each the negation of (*a*); and (*e*) and (*f*) are each the negation of (*d*).   Since (*a*) is true, the statements (*b*) and (*c*) are false; and since (*d*) is false, the statements (*e*) and (*f*) are true.

Remark: The logical notation for the connectives "and", "or", and "not" is not standard.   For example, some texts use

$$p \, \& \, q, p \cdot q \text{ or } pq \quad \text{for} \quad p \wedge q$$

$$p + q \qquad\qquad \text{for} \quad p \vee q$$

$$p', \bar{p} \text{ or } \neg p \quad\quad \text{for} \quad {\sim}p$$

4.5   PROPOSITIONS AND TRUTH TABLES

By repetitive use of the logical connectives ($\wedge$, $\vee$, $\sim$ and others discussed subsequently), we can construct compound statements that are more involved.   In the case where the substatements $p, q, \ldots$ of a compound statement $P(p, q, \ldots)$ are variables, we call the compound statement a *proposition*.

Now the truth value of a proposition depends exclusively upon the truth values of its variables, that is, the truth value of a proposition is known once the truth values of its variables are known.   A simple concise way to show this relationship is through a *truth table*.   The truth table of the proposition ${\sim}(p \wedge {\sim}q)$, for example, is constructed as follows:

p	q	${\sim}q$	$p \wedge {\sim}q$	${\sim}(p \wedge {\sim}q)$
T	T	F	F	T
T	F	T	T	F
F	T	F	F	T
F	F	T	F	T

Observe that the first columns of the table are for the variables $p, q, \ldots$ and that there are enough rows in the table to allow for all possible combinations of T and F for these variables.   (For 2 variables, as above, 4 rows are necessary; for 3 variables, 8 rows are necessary; and, in general, for n variables, 2^n rows are required.)   There is then a column for each "elementary" stage of the construction of the proposition, the truth value at each step being determined from the previous stages by the definitions of the connectives $\wedge$, $\vee$, $\sim$.   Finally we obtain the truth value of the proposition, which appears in the last column.

Remark: The truth table of the above proposition consists precisely of the columns under the variables and the column under the proposition:

p	q	${\sim}(p \wedge {\sim}q)$
T	T	T
T	F	F
F	T	T
F	F	T

The other columns were merely used in the construction of the truth table.

Another way to construct the above truth table for ${\sim}(p \wedge {\sim}q)$ is as follows.   First construct the following table:

p	q	~	(p	∧	~	q)
T	T					
T	F					
F	T					
F	F					
Step

Observe that the proposition is written on the top row to the right of its variables, and that there is a column under each variable or connective in the proposition. Truth values are then entered into the truth table in various steps as follows:

p	q	~	(p	∧	~	q)
T	T		T			T
T	F		T			F
F	T		F			T
F	F		F			F
Step		1			1

(a)

p	q	~	(p	∧	~	q)
T	T		T		F	T
T	F		T		T	F
F	T		F		F	T
F	F		F		T	F
Step		1		2	1

(b)

p	q	~	(p	∧	~	q)
T	T		T	F	F	T
T	F		T	T	T	F
F	T		F	F	F	T
F	F		F	F	T	F
Step		1	3	2	1

(c)

p	q	~	(p	∧	~	q)
T	T	T	T	F	F	T
T	F	F	T	T	T	F
F	T	T	F	F	F	T
F	F	T	F	F	T	F
Step	4	1	3	2	1

(d)

The truth table of the proposition then consists of the original columns under the variables and the last column entered into the table, i.e. the last step.

4.6 TAUTOLOGIES AND CONTRADICTIONS

Some propositions $P(p, q, \ldots)$ contain only T in the last column of their truth tables, i.e. are true for any truth values of their variables. Such propositions are called *tautologies*. Similarly, a proposition $P(p, q, \ldots)$ is called a *contradiction* if it contains only F in the last column of its truth table, i.e. is false for any truth values of its variables. For example, the proposition "p or not p", i.e. $p \vee \sim p$, is a tautology and the proposition "p and not p", i.e. $p \wedge \sim p$, is a contradiction. This is verified by constructing their truth tables.

p	$\sim p$	$p \vee \sim p$
T	F	T
F	T	T

p	$\sim p$	$p \wedge \sim p$
T	F	F
F	T	F

We note that the negation of a tautology is a contradiction since it is always false, and the negation of a contradiction is a tautology since it is always true.

Now let $P(p, q, \ldots)$ be a tautology, and let $P_1(p, q, \ldots)$, $P_2(p, q, \ldots), \ldots$ be any propositions. Since the truth value of $P(p, q, \ldots)$ does not depend upon the particular truth values of its variables $p, q, \ldots$, we can substitute P_1 for p, P_2 for q, $\ldots$ in the tautology $P(p, q, \ldots)$ and still have a tautology. In other words:

Principle of Substitution: If $P(p, q, \ldots)$ is a tautology, then $P(P_1, P_2, \ldots)$ is a tautology for any propositions $P_1, P_2, \ldots$.

EXAMPLE 4.5 By the above truth table, $p \vee \sim p$ is a tautology. Substituting $q \wedge r$ for p we obtain the proposition $(q \wedge r) \vee \sim (q \wedge r)$, which, by the Principle of Substitution, should also be a tautology. This is verified by the following truth table:

q	r	$q \wedge r$	$\sim(q \wedge r)$	$(q \wedge r) \vee \sim(q \wedge r)$
T	T	T	F	T
T	F	F	T	T
F	T	F	T	T
F	F	F	T	T

✳ 4.7   LOGICAL EQUIVALENCE; ALGEBRA OF PROPOSITIONS

Two propositions $P(p, q, \ldots)$ and $Q(p, q, \ldots)$ are said to be *logically equivalent*, or simply *equivalent* or *equal*, denoted by

$$P(p, q, \ldots) \equiv Q(p, q, \ldots)$$

if they have identical truth tables. For example, consider the truth tables of $\sim(p \wedge q)$ and $\sim p \vee \sim q$:

p	q	$p \wedge q$	$\sim(p \wedge q)$
T	T	T	F
T	F	F	T
F	T	F	T
F	F	F	T

p	q	$\sim p$	$\sim q$	$\sim p \vee \sim q$
T	T	F	F	F
T	F	F	T	T
F	T	T	F	T
F	F	T	T	T

Since the truth tables are the same, i.e. both propositions are false in the first case and true in the other three cases, the propositions $\sim(p \wedge q)$ and $\sim p \vee \sim q$ are logically equivalent and we can write:

$$\sim(p \wedge q) \equiv \sim p \vee \sim q$$

EXAMPLE 4.6 Consider the statement

"It is false that roses are red and violets are blue."

This statement can be written in the form $\sim(p \wedge q)$ where p is "roses are red" and q is "violets are blue". However, by the above truth tables, $\sim(p \wedge q)$ is logically equivalent to $\sim p \vee \sim q$. Thus the given statement has the same meaning as the statement

"Roses are not red, or violets are not blue."

Propositions satisfy many logical equivalences, or *laws*, besides the one listed above. Some of the more important laws, with their names, are listed in Table 4-1. In the table, t denotes a tautology and f denotes a contradiction.

4.8   CONDITIONAL AND BICONDITIONAL STATEMENTS

Many statements, particularly in mathematics, are of the form "If p then q". Such statements are called *conditional* statements and are denoted by

$$p \rightarrow q$$

The conditional $p \rightarrow q$ is frequently read "p implies q" or "p only if q".

Another common statement is of the form "p if and only if q". Such statements, denoted by

$$p \leftrightarrow q$$

are called *biconditional* statements.

Table 4-1. Laws of the Algebra of Propositions

Idempotent Laws

1a. $p \lor p \equiv p$ 　　　　　　　　　1b. $p \land p \equiv p$

Associative Laws

2a. $(p \lor q) \lor r \equiv p \lor (q \lor r)$ 　　　　2b. $(p \land q) \land r \equiv p \land (q \land r)$

Commutative Laws

3a. $p \lor q \equiv q \lor p$ 　　　　　　　3b. $p \land q \equiv q \land p$

Distributive Laws

4a. $p \lor (q \land r) \equiv (p \lor q) \land (p \lor r)$ 　　4b. $p \land (q \lor r) \equiv (p \land q) \lor (p \land r)$

Identity Laws

5a. $p \lor f \equiv p$ 　　　　　　　　5b. $p \land t \equiv p$

6a. $p \lor t \equiv t$ 　　　　　　　　6b. $p \land f \equiv f$

Complement Laws

7a. $p \lor \sim p \equiv t$ 　　　　　　　7b. $p \land \sim p \equiv f$

8a. $\sim t \equiv f$ 　　　　　　　　　8b. $\sim f \equiv t$

Involution Law

9. $\sim \sim p \equiv p$

DeMorgan's Laws

10a. $\sim(p \lor q) \equiv \sim p \land \sim q$ 　　　10b. $\sim(p \land q) \equiv \sim p \lor \sim q$

The truth values of $p \to q$ and $p \leftrightarrow q$ are given in the following tables:

p	q	$p \to q$
T	T	T
T	F	F
F	T	T
F	F	T

p	q	$p \leftrightarrow q$
T	T	T
T	F	F
F	T	F
F	F	T

Observe that the conditional $p \to q$ is false only when the first part p is true and the second part q is false. In case p is false, the conditional $p \to q$ is true regardless of the truth value of q. Observe also that $p \leftrightarrow q$ is true when p and q have the same truth values and false otherwise. (For the relationship between biconditional statements and conditional statements, see Problems 4.15 and 4.16.)

Now consider the truth table of the proposition $\sim p \lor q$:

p	q	$\sim p$	$\sim p \lor q$
T	T	F	T
T	F	F	F
F	T	T	T
F	F	T	T

Observe that the above truth table is identical to the truth table of $p \to q$. Hence $p \to q$ is logically equivalent to the proposition $\sim p \lor q$:

$$p \to q \equiv \sim p \lor q$$

In other words, the conditional statement "If p then q" is logically equivalent to the statement "Not p or q" which only involves the connectives $\lor$ and $\sim$ and thus was already a part of our language.

Consider the conditional proposition $p \to q$ and the other simple conditional propositions which contain p and q:

$$q \to p, \quad \sim p \to \sim q \quad \text{and} \quad \sim q \to \sim p$$

These propositions are respectively called the *converse*, *inverse*, and *contrapositive* of the proposition $p \to q$. The truth tables of the four propositions follow:

p	q	Conditional $p \to q$	Converse $q \to p$	Inverse $\sim p \to \sim q$	Contrapositive $\sim q \to \sim p$
T	T	T	T	T	T
T	F	F	T	T	F
F	T	T	F	F	T
F	F	T	T	T	T

Observe that a conditional statement and its converse or inverse are not logically equivalent. On the other hand, a conditional statement and its contrapositive are seen to be logically equivalent. We state this result formally.

Theorem 4.1:   A conditional statement $p \to q$ and its contrapositive $\sim q \to \sim p$ are logically equivalent.

EXAMPLE 4.7

(a)   Consider the following statements about a triangle A:

$p \to q$:   If A is equilateral, then A is isosceles.
$q \to p$:   If A is isosceles, then A is equilateral.

Here $p \to q$ is true, but its converse $q \to p$ is false.

(b)   Let x be an integer.   Prove: $(p \to q)$ If x^2 is odd then x is odd.
   We show that the contrapositive $\sim q \to \sim p$, "If x is even then x^2 is even", is true.   Let x be even; then $x = 2n$ where n is an integer.   Hence

$$x^2 = (2n)(2n) = 2(2n^2)$$

is also even.   Since the contrapositive statement $\sim q \to \sim p$ is true, the original conditional statement $p \to q$ is also true.

4.9   ARGUMENTS

An *argument* is a relationship between a set of propositions $P_1, P_2, \ldots, P_n$, called *premises*, and another proposition Q, called the *conclusion*; we denote an argument by

$$P_1, P_2, \ldots, P_n \vdash Q$$

An argument is said to be *valid* if the premises yield (have as a consequence) the conclusion; more formally, we make the following

Definition: An argument $P_1, P_2, \ldots, P_n \vdash Q$ is *valid* if Q is true whenever all the premises $P_1, P_2, \ldots, P_n$ are true.

An argument that is not valid is called a *fallacy*.

EXAMPLE 4.8

(a)   The following argument is valid:

$$p, p \to q \vdash q \qquad \text{(Law of Detachment)}$$

The proof of this rule follows from the following truth table.

p	q	$p \to q$
T	T	T
T	F	F
F	T	T
F	F	T

For p is true in cases (lines) 1 and 2, and $p \to q$ is true in cases 1, 3 and 4; hence p and $p \to q$ are true simultaneously in case 1. Since in this case q is true, the argument is valid.

(b) The following argument is a fallacy: $p \to q, q \vdash p$. For $p \to q$ and q are both true in case (line) 3 in the above truth table, but in this case p is false.

Now the propositions $P_1, P_2, \ldots, P_n$ are true simultaneously if and only if the proposition $P_1 \wedge P_2 \wedge \cdots \wedge P_n$ is true. Thus the argument $P_1, P_2, \ldots, P_n \vdash Q$ is valid if and only if Q is true whenever $P_1 \wedge P_2 \wedge \cdots \wedge P_n$ is true or, equivalently, if and only if the proposition

$$(P_1 \wedge P_2 \wedge \cdots \wedge P_n) \to Q$$

is a tautology. We state this result formally.

Theorem 4.2: The argument $P_1, P_2, \ldots, P_n \vdash Q$ is valid if and only if $(P_1 \wedge P_2 \wedge \cdots \wedge P_n) \to Q$ is a tautology.

We apply this theorem in the next example.

EXAMPLE 4.9 A fundamental principle of logical reasoning states:

"If p implies q and q implies r, then p implies r."

That is, the following argument is valid:

$$p \to q, q \to r \vdash p \to r \qquad \text{(Law of Syllogism)}$$

This fact is verified by the following truth table which shows that the proposition

$$[(p \to q) \wedge (q \to r)] \to (p \to r)$$

is a tautology:

p	q	r	[(p	$\to$	q)	$\wedge$	(q	$\to$	r)]	$\to$	(p	$\to$	r)
T	T	T	T	T	T	T	T	T	T	T	T	T	T
T	T	F	T	T	T	F	T	F	F	T	T	F	F
T	F	T	T	F	F	F	F	T	T	T	T	T	T
T	F	F	T	F	F	F	F	T	F	T	T	F	F
F	T	T	F	T	T	T	T	T	T	T	F	T	T
F	T	F	F	T	T	F	T	F	F	T	F	T	F
F	F	T	F	T	F	T	F	T	T	T	F	T	T
F	F	F	F	T	F	T	F	T	F	T	F	T	F
Step			1	2	1	3	1	2	1	4	1	2	1

Equivalently, the argument is valid since the premises $p \to q$ and $q \to r$ are true simultaneously only in cases (lines) 1, 5, 7 and 8, and in these cases the conclusion $p \to r$ is also true. (Observe that the truth table required $2^3 = 8$ lines since there are three variables p, q and r.)

We emphasize that the validity of an argument does not depend upon the truth values or the content of the statements appearing in the argument, but only upon the formal structure of the argument. This is illustrated in the following example.

EXAMPLE 4.10   Consider the following argument:

S_1:   If a man is a bachelor, he is unhappy.
S_2:   If a man is unhappy, he dies young.
....
S: Bachelors die young.

Here the statement S below the line denotes the conclusion of the argument, and the statements S_1 and S_2 above the line denote the premises.   We claim that the argument $S_1, S_2 \vdash S$ is valid.   For the argument is of the form

$$p \to q, q \to r \vdash p \to r$$

where p is "He is a bachelor", q is "He is unhappy" and r is "He dies young"; and by Example 4.9 this argument (Law of Syllogism) is valid.

4.10   LOGICAL IMPLICATION

A proposition $P(p, q, \ldots)$ is said to *logically imply* a proposition $Q(p, q, \ldots)$, written

$$P(p, q, \ldots) \Rightarrow Q(p, q, \ldots)$$

if $Q(p, q, \ldots)$ is true whenever $P(p, q, , ..)$ is true.

EXAMPLE 4.11   We claim that p logically implies $p \lor q$.   For consider the truth tables of p and $p \lor q$ in the table below.   Observe that p is true in cases (lines) 1 and 2, and in these cases $p \lor q$ is also true.   In other words, p logically implies $p \lor q$.

p	q	$p \lor q$
T	T	T
T	F	T
F	T	T
F	F	F

Now if $Q(p, q, \ldots)$ is true whenever $P(p, q, \ldots)$ is true, then the argument

$$P(p, q, \ldots) \vdash Q(p, q, \ldots)$$

is valid; and conversely.   Furthermore, the argument $P \vdash Q$ is valid if and only if the conditional statement $P \to Q$ is always true, i.e. a tautology.   We state this result formally.

Theorem 4.3:   For any propositions $P(p, q, \ldots)$ and $Q(p, q, \ldots)$, the following three statements are equivalent:
  (i) $P(p, q, \ldots)$ logically implies $Q(p, q, \ldots)$.
  (ii)   The argument $P(p, q, \ldots) \vdash Q(p, q, \ldots)$ is valid.
  (iii) The proposition $P(p, q, \ldots) \to Q(p, q, \ldots)$ is a tautology.

If $P \Rightarrow Q$ and $Q \Rightarrow P$, then P and Q must have the same truth table, and so $P \equiv Q$.   The converse is also true.   Thus the notion of logical implication is also intimately related to that of logical equivalence.

Solved Problems

STATEMENTS AND COMPOUND STATEMENTS

4.1 Let p be "It is cold" and let q be "It is raining".   Give a simple verbal sentence which describes each of the following statements:

(1) $\sim p$,   (2) $p \wedge q$,   (3) $p \vee q$,   (4) $q \vee \sim p$,   (5) $\sim p \wedge \sim q$,   (6) $\sim \sim q$

In each case, translate $\wedge$, $\vee$ and $\sim$ to read "and", "or" and "It is false that" or "not", respectively, and then simplify the English sentence.

(1) It is not cold.   (4) It is raining or it is not cold.
(2) It is cold and raining.   (5) It is not cold and it is not raining.
(3) It is cold or it is raining.   (6) It is not true that it is not raining.

4.2 Let p be "He is tall" and let q be "He is handsome".   Write each of the following statements in symbolic form using p and q.   (Assume that "He is short" means "He is not tall", i.e. $\sim p$.)

(1) He is tall and handsome.   (4) He is neither tall nor handsome.
(2) He is tall but not handsome.   (5) He is tall, or he is short and handsome.
(3) It is false that he is short or handsome. (6) It is not true that he is short or not handsome.

(1) $p \wedge q$ (3) $\sim(\sim p \vee q)$ (5) $p \vee (\sim p \wedge q)$
(2) $p \wedge \sim q$ (4) $\sim p \wedge \sim q$ (6) $\sim(\sim p \vee \sim q)$

PROPOSITIONS AND THEIR TRUTH TABLES

4.3 Find the truth table of $\sim p \wedge q$.

p	q	$\sim p$	$\sim p \wedge q$
T	T	F	F
T	F	F	F
F	T	T	T
F	F	T	F

Method 1

p	q	$\sim$	p	$\wedge$	q
T	T	F	T	F	T
T	F	F	T	F	F
F	T	T	F	T	T
F	F	T	F	F	F
Step		2	1	3	1

Method 2

4.4 Find the truth table of $\sim(p \vee q)$.

p	q	$p \vee q$	$\sim(p \vee q)$
T	T	T	F
T	F	T	F
F	T	T	F
F	F	F	T

Method 1

p	q	$\sim$	(p	$\vee$	q)
T	T	F	T	T	T
T	F	F	T	T	F
F	T	F	F	T	T
F	F	T	F	F	F
Step		3	1	2	1

Method 2

4.5   Find the truth table of $\sim(p \vee \sim q)$.

p	q	$\sim q$	$p \vee \sim q$	$\sim(p \vee \sim q)$
T	T	F	T	F
T	F	T	T	F
F	T	F	F	T
F	F	T	T	F

Method 1

p	q	$\sim$	$(p$	$\vee$	$\sim$	$q)$
T	T	F	T	T	F	T
T	F	F	T	T	T	F
F	T	T	F	F	F	T
F	F	F	F	T	T	F
Step		4	1	3	2	1

Method 2

(Note that this truth table is identical to that of Problem 4.3.)

TAUTOLOGIES AND CONTRADICTIONS

4.6   Verify that the proposition $p \vee \sim(p \wedge q)$ is a tautology.

Construct the truth table of $p \vee \sim(p \wedge q)$:

p	q	$p \wedge q$	$\sim(p \wedge q)$	$p \vee \sim(p \wedge q)$
T	T	T	F	T
T	F	F	T	T
F	T	F	T	T
F	F	F	T	T

Since the truth value of $p \vee \sim(p \wedge q)$ is T for all truth values of p and q, it is a tautology.

4.7   Verify that the proposition $(p \wedge q) \wedge \sim(p \vee q)$ is a contradiction.

Construct the truth table of $(p \wedge q) \wedge \sim(p \vee q)$:

p	q	$p \wedge q$	$p \vee q$	$\sim(p \vee q)$	$(p \wedge q) \wedge \sim(p \vee q)$
T	T	T	T	F	F
T	F	F	T	F	F
F	T	F	T	F	F
F	F	F	F	T	F

Since the truth value of $(p \wedge q) \wedge \sim(p \vee q)$ is F for all truth values of p and q, it is a contradiction.

LOGICAL EQUIVALENCE

4.8   Prove that disjunction distributes over conjunction; that is, prove the Distributive Law $p \vee (q \wedge r) \equiv (p \vee q) \wedge (p \vee r)$.

Construct the required truth tables.

p	q	r	$q \wedge r$	$p \vee (q \wedge r)$	$p \vee q$	$p \vee r$	$(p \vee q) \wedge (p \vee r)$
T	T	T	T	T	T	T	T
T	T	F	F	T	T	T	T
T	F	T	F	T	T	T	T
T	F	F	F	T	T	T	T
F	T	T	T	T	T	T	T
F	T	F	F	F	T	F	F
F	F	T	F	F	F	T	F
F	F	F	F	F	F	F	F

Since the truth tables are identical, the propositions are equivalent.

4.9 Prove that the operation of disjunction can be written in terms of the operations of conjunction and negation. Specifically, $p \vee q \equiv \sim(\sim p \wedge \sim q)$.

Construct the required truth tables:

p	q	$p \vee q$	$\sim p$	$\sim q$	$\sim p \vee \sim q$	$\sim(\sim p \wedge \sim q)$
T	T	T	F	F	F	T
T	F	T	F	T	F	T
F	T	T	T	F	F	T
F	F	F	T	T	T	F

Since the truth tables are identical, the propositions are equivalent.

4.10 Simplify each proposition by using Table 4-1: (a) $p \vee (p \wedge q)$, (b) $\sim(p \vee q) \vee (\sim p \wedge q)$.

(a)

	Equivalence		Reason
(1)	$p \vee (p \wedge q) \equiv (p \wedge t) \vee (p \wedge q)$	(1)	Identity law
(2)	$\equiv p \wedge (t \vee q)$	(2)	Distributive law
(3)	$\equiv p \wedge t$	(3)	Identity law
(4)	$\equiv p$	(4)	Identity law

(b)

	Equivalence		Reason
(1)	$\sim(p \vee q) \vee (\sim p \wedge q) \equiv (\sim p \wedge \sim q) \vee (\sim p \wedge q)$	(1)	DeMorgan's law
(2)	$\equiv \sim p \wedge (\sim q \vee q)$	(2)	Distributive law
(3)	$\equiv \sim p \wedge t$	(3)	Complement law
(4)	$\equiv \sim p$	(4)	Identity law

NEGATION

4.11 Prove DeMorgan's laws: (a) $\sim(p \wedge q) \equiv \sim p \vee \sim q$; (b) $\sim(p \vee q) \equiv \sim p \wedge \sim q$.

In each case construct the required truth tables.

(a)

p	q	$p \wedge q$	$\sim(p \wedge q)$	$\sim p$	$\sim q$	$\sim p \vee \sim q$
T	T	T	F	F	F	F
T	F	F	T	F	T	T
F	T	F	T	T	F	T
F	F	F	T	T	T	T

(b)

p	q	$p \vee q$	$\sim(p \vee q)$	$\sim p$	$\sim q$	$\sim p \wedge \sim q$
T	T	T	F	F	F	F
T	F	T	F	F	T	F
F	T	T	F	T	F	F
F	F	F	T	T	T	T
			↑			↑

4.12　Verify: $\sim \sim p \equiv p$.

p	$\sim p$	$\sim \sim p$
T	F	T
F	T	F
↑		↑

4.13　Use the results of Problems 4.11 and 4.12 to simplify each of the following statements.
(a)　It is not true that his mother is English or his father is French.
(b)　It is not true that he studies physics but not mathematics.
(c)　It is not true that sales are decreasing and prices are rising.
(d)　It is not true that it is not cold or it is raining.

(a)　Let p denote "His mother is English" and let q denote "His father is French". Then the given statement is $\sim(p \vee q)$. But $\sim(p \vee q) \equiv \sim p \wedge \sim q$. Hence the given statement is logically equivalent to the statement "His mother is not English and his father is not French".
(b)　Let p denote "He studies physics" and let q denote "He studies mathematics". Then the given statement is $\sim(p \wedge \sim q)$. But $\sim(p \wedge \sim q) \equiv \sim p \vee \sim \sim q \equiv \sim p \vee q$. Hence the given statement is logically equivalent to the statement "He does not study physics or he studies mathematics".
(c)　Since $\sim(p \wedge q) \equiv \sim p \vee \sim q$, the given statement is logically equivalent to the statement "Sales are increasing or prices are falling".
(d)　Since $\sim(\sim p \vee q) \equiv p \wedge \sim q$, the given statement is logically equivalent to the statement "It is cold and it is not raining".

CONDITIONALS AND BICONDITIONALS

4.14　Rewrite the following statements without using the conditional.
(a)　If it is cold, he wears a hat.
(b)　If productivity increases, then wages rise.

　　Recall that "If p then q" is equivalent to "Not p or q"; that is, $p \rightarrow q \equiv \sim p \vee q$.
(a)　It is not cold or he wears a hat.
(b)　Productivity does not increase or wages rise.

4.15　(a)　Show that "p implies q and q implies p" is logically equivalent to the biconditional "p if and only if q"; that is, $(p \rightarrow q) \wedge (q \rightarrow p) \equiv p \leftrightarrow q$.
　　(b)　Show that the biconditional $p \leftrightarrow q$ can be written in terms of the original three connectives $\vee$, $\wedge$ and $\sim$.

(a)

p	q	$p \leftrightarrow q$	$p \rightarrow q$	$q \rightarrow p$	$(p \rightarrow q) \wedge (q \rightarrow p)$
T	T	T	T	T	T
T	F	F	F	T	F
F	T	F	T	F	F
F	F	T	T	T	T
		↑			↑

(b)   Now $p \to q \equiv {\sim}p \vee q$ and $q \to p \equiv {\sim}q \vee p$; hence by (a)

$$p \leftrightarrow q \equiv (p \to q) \wedge (q \to p) \equiv ({\sim}p \vee q) \wedge ({\sim}q \vee p)$$

4.16   Show that $p \leftrightarrow q \equiv (p \vee q) \to (p \wedge q)$ (a) by comparing truth tables, (b) by the algebra of propositions.

(a)

p	q	$p \leftrightarrow q$	$p \vee q$	$p \wedge q$	$(p \vee q) \to (p \wedge q)$
T	T	T	T	T	T
T	F	F	T	F	F
F	T	F	T	F	F
F	F	T	F	F	T

(b)   **Equivalence**   **Reason**

(1) $p \leftrightarrow q \equiv ({\sim}p \vee q) \wedge ({\sim}q \vee p)$ (1)   Problem 4.15(b)
(2) $\equiv [({\sim}p \vee q) \wedge {\sim}q] \vee [({\sim}p \vee q) \wedge p]$ (2)   Distributive law
(3) $\equiv [({\sim}q \wedge q) \vee ({\sim}q \wedge {\sim}p)] \vee [(p \wedge q) \vee (p \wedge {\sim}p)]$ (3)   Commutative and Distributive laws
(4) $\equiv [f \vee ({\sim}q \wedge {\sim}p)] \vee [(p \wedge q) \vee f]$ (4)   Complement law
(5) $\equiv ({\sim}q \wedge {\sim}p) \vee (p \wedge q)$ (5)   Identity law
(6) $\equiv [{\sim}(p \vee q)] \vee (p \wedge q)$ (6)   DeMorgan's law
(7) $\equiv (p \vee q) \to (p \wedge q)$ (7)   Section 4.8

4.17   Determine the truth tables of (a) $(p \to q) \to (p \wedge q)$, (b) $(p \to q) \vee {\sim}(p \leftrightarrow {\sim}q)$.

(a)

p	q	$p \to q$	$p \wedge q$	$(p \to q) \to (p \wedge q)$
T	T	T	T	T
T	F	F	F	T
F	T	T	F	F
F	F	T	F	F

(b)

p	q	$(p$	$\to$	$q)$	$\vee$	$\sim$	$(p$	$\leftrightarrow$	$\sim$	$q)$
T	T	T	T	T	T	T	T	F	F	T
T	F	T	F	F	T	F	T	T	T	F
F	T	F	T	T	T	F	F	T	F	T
F	F	F	T	F	T	T	F	F	T	F
Step		1	2	1	5	4	1	3	2	1

4.18   Determine the contrapositive of each statement.
(a)   If John is a poet, then he is poor.
(b)   Only if Marc studies will he pass the test.
(c)   It is necessary to have snow in order for Eric to ski.
(d)   If x is less than zero, then x is not positive.

(a)   The contrapositive of $p \to q$ is ${\sim}q \to {\sim}p$. Hence the contrapositive of (a) is

          If John is not poor, then he is not a poet.

(b)   The given statement is equivalent to "If Marc passes the test, then he studied". Hence the contrapositive of (b) is

          If Marc does not study, then he will not pass the test.

(c)   The given statement is equivalent to "If Eric skis, then it snowed".   Hence the contrapositive of (c) is

If it does not snow, then Eric will not ski.

(d)   The contrapositive of $p \to {\sim}q$ is ${\sim}{\sim}q \to {\sim}p \equiv q \to {\sim}p$.   Hence the contrapositive of (d) is

If x is positive, then x is not less than zero.

ARGUMENTS AND LOGICAL IMPLICATION

4.19   Show that the following argument is valid: $p \leftrightarrow q, q \vdash p$.

Method 1

Construct the truth table on the right.   Now $p \leftrightarrow q$ is true in cases (lines) 1 and 4, and q is true in cases 1 and 3; hence $p \leftrightarrow q$ and q are true simultaneously only in case 1, where p is also true.   Thus the argument $p \leftrightarrow q, q \vdash p$ is valid.

p	q	$p \leftrightarrow q$
T	T	T
T	F	F
F	T	F
F	F	T

Method 2

Construct the truth table of $[(p \leftrightarrow q) \wedge q] \to p$:

p	q	$p \leftrightarrow q$	$(p \leftrightarrow q) \wedge q$	$[(p \leftrightarrow q) \wedge q] \to p$
T	T	T	T	T
T	F	F	F	T
F	T	F	F	T
F	F	T	F	T

Since $[(p \leftrightarrow q) \wedge q] \to p$ is a tautology, the argument is valid.

4.20   Test the validity of the following argument:

If I study, then I will not fail mathematics.
If I do not play basketball, then I study.
But I failed mathematics.
....
Therefore, I played basketball.

First translate the argument into symbolic form.   Let p be "I study", q be "I fail mathematics" and r be "I play basketball".   Then the given argument is as follows:

$$p \to {\sim}q, \; {\sim}r \to p, \; q \vdash r$$

To test the validity of the argument, construct the truth tables of the given propositions $p \to {\sim}q$, ${\sim}r \to p$, q and r:

p	q	r	${\sim}q$	$p \to {\sim}q$	${\sim}r$	${\sim}r \to p$
T	T	T	F	F	F	T
T	T	F	F	F	T	T
T	F	T	T	T	F	T
T	F	F	T	T	T	T
F	T	T	F	T	F	T
F	T	F	F	T	T	F
F	F	T	T	T	F	T
F	F	F	T	T	T	F

Now the premises $p \to \sim q$, $\sim r \to p$ and q are true simultaneously only in case (line) 5, and in that case the conclusion r is also true; hence the argument is valid.

4.21   Show that $p \leftrightarrow q$ logically implies $p \to q$.

Method 1
   Construct the truth tables of $p \leftrightarrow q$ and $p \to q$:

p	q	$p \leftrightarrow q$	$p \to q$
T	T	T	T
T	F	F	F
F	T	F	T
F	F	T	T

Now $p \leftrightarrow q$ is true in lines 1 and 4, and in these cases $p \to q$ is also true.   Hence $p \leftrightarrow q$ logically implies $p \to q$.

Method 2
   Construct the truth table of $(p \leftrightarrow q) \to (p \to q)$.   It will be a tautology; hence, by Theorem 4.3, $p \leftrightarrow q$ logically implies $p \to q$.

Method 3
   Use Problem 4.15(a) and the algebra of propositions.

$$
\begin{aligned}
(p \leftrightarrow q) \to (p \to q) &\equiv [(p \to q) \wedge (q \to p)] \to (p \to q) \\
&\equiv \sim[(p \to q) \wedge (q \to p)] \vee (p \to q) \\
&\equiv [\sim(p \to q) \vee \sim(q \to p)] \vee (p \to q) \\
&\equiv t \vee \sim(q \to p) \\
&\equiv t
\end{aligned}
$$

(This method might also have been applied in Problems 4.19 and 4.20.)

4.22   Show that $p \leftrightarrow \sim q$ does not logically imply $p \to q$.

Method 1
   Construct the truth tables of $p \leftrightarrow \sim q$ and $p \to q$:

p	q	$\sim q$	$p \leftrightarrow \sim q$	$p \to q$
T	T	F	F	T
T	F	T	T	F
F	T	F	T	T
F	F	T	F	T

Recall that $p \leftrightarrow \sim q$ logically implies $p \to q$ if $p \to q$ is true whenever $p \leftrightarrow \sim q$ is true.   But $p \leftrightarrow \sim q$ is true in case (line) 2 in the above table, and in that case $p \to q$ is false.   Hence $p \leftrightarrow \sim q$ does not logically imply $p \to q$.

Method 2
   Construct the truth table of the proposition $(p \leftrightarrow \sim q) \to (p \to q)$.   It will not be a tautology; hence, by Theorem 4.3, $p \leftrightarrow \sim q$ does not logically imply $p \to q$.

Supplementary Problems

STATEMENTS AND COMPOUND STATEMENTS

4.23   Let p be "Marc is rich" and let q be "Marc is happy".   Write each of the following in symbolic form.
  (a)   Marc is poor but happy.
  (b)   Marc is neither rich nor happy.
  (c)   Marc is either rich or unhappy.
  (d)   Marc is poor or else he is both rich and unhappy.

4.24   Let p be "Erik reads *Newsweek*", let q be "Erik reads *The New Yorker*" and let r be "Erik reads *Time*".   Write each of the following in symbolic form.
  (a)   Erik reads *Newsweek* or *The New Yorker*, but not *Time*.
  (b)   Erik reads *Newsweek* and *The New Yorker*, or he does not read *Newsweek* and *Time*.
  (c)   It is not true that Erik reads *Newsweek* but not *Time*.
  (d)   It is not true that Erik reads *Time* or *The New Yorker* but not *Newsweek*.

4.25   Let p be "Audrey speaks French" and let q be "Audrey speaks Danish".   Give a simple verbal sentence which describes each of the following.

  (a) $p \vee q$, (b) $p \wedge q$, (c) $p \wedge {\sim} q$, (d) ${\sim} p \vee {\sim} q$, (e) ${\sim} {\sim} p$, (f) ${\sim}({\sim} p \wedge {\sim} q)$

TRUTH TABLES, LOGICAL EQUIVALENCE

4.26   Find the truth table of each proposition:

  (a) $p \vee {\sim} q$, (b) ${\sim} p \wedge {\sim} q$, (c) ${\sim}({\sim} p \wedge q)$, (d) ${\sim}({\sim} p \vee {\sim} q)$

4.27   Find the truth table of each proposition:

  (a) $(p \wedge {\sim} q) \vee r$, (b) ${\sim} p \vee (q \wedge {\sim} r)$, (c) $(p \vee {\sim} r) \wedge (q \vee {\sim} r)$

4.28   Prove that conjunction distributes over disjunction: $p \wedge (q \vee r) \equiv (p \wedge q) \vee (p \wedge r)$.

4.29   Prove $p \vee (p \wedge q) \equiv p$ by constructing the appropriate truth tables.

4.30   Prove ${\sim}(p \vee q) \vee ({\sim} p \wedge q) \equiv {\sim} p$ by constructing the appropriate truth tables.

4.31   (a)   Express $\vee$ in terms of $\wedge$ and ${\sim}$.
  (b)   Express $\wedge$ in terms of $\vee$ and ${\sim}$.

4.32   Prove the following equivalences by using the laws of algebra of propositions listed in Table 4-1:

  (a) $p \wedge (p \vee q) \equiv p$, (b) $(p \wedge q) \vee {\sim} p \equiv {\sim} p \vee q$, (c) $p \wedge ({\sim} p \vee q) \equiv p \wedge q$

NEGATION

4.33   Simplify: (a) ${\sim}(p \wedge {\sim} q)$, (b) ${\sim}({\sim} p \vee q)$, (c) ${\sim}({\sim} p \wedge {\sim} q)$.

4.34   Write the negation of each of the following statements as simply as possible.
  (a)   He is tall but handsome.
  (b)   He has blond hair or blue eyes.
  (c)   He is neither rich nor happy.
  (d)   He lost his job or he did not go to work today.
  (e)   Neither Marc nor Erik is unhappy.
  (f)   Audrey speaks Spanish or French, but not German.

CONDITIONALS, BICONDITIONALS

4.35　Find the truth table of each proposition:　(a) $(\sim p \vee q) \to p$, (b) $q \leftrightarrow (\sim q \wedge p)$.

4.36　Find the truth table of each proposition:　(a) $(p \leftrightarrow \sim q) \to (\sim p \wedge q)$, (b) $(\sim q \vee p) \leftrightarrow (q \to \sim p)$.

4.37　Prove:　(a) $(p \wedge q) \to r \equiv (p \to r) \vee (q \to r)$, (b) $p \to (q \to r) \equiv (p \wedge \sim r) \to \sim q$.

4.38　Determine the contrapositive of each statement:
　　　　　(a) If he has courage he will win.　(b) Only if he does not tire will he win.

4.39　Find:　(a)　Contrapositive of $p \to \sim q$.　　　　(c)　Contrapositive of the converse of $p \to \sim q$.
　　　　　　　　(b)　Contrapositive of $\sim p \to q$.　　　　(d)　Converse of the contrapositive of $\sim p \to \sim q$.

ARGUMENTS, LOGICAL IMPLICATION

4.40　Test the validity of each argument:　(a) $\sim p \to q, p \vdash \sim q$; (b) $\sim p \to q, q \vdash p$.

4.41　Test the validity of each argument:　(a) $p \to q, r \to \sim q \vdash r \to \sim p$; (b) $p \to \sim q, \sim r \to \sim q \vdash p \to \sim r$.

4.42　Translate into symbolic form and test the validity of the argument:

(a)　If 6 is even, then 2 does not divide 7.
　　　Either 5 is not prime or 2 divides 7.
　　　But 5 is prime.
　　　.......................................
　　　Therefore, 6 is odd (not even).

(b)　Roses are red.
　　　Roses are blue.
　　　...
　　　Therefore, roses are red if and only if they are blue.

(c)　If I work, I cannot study.
　　　Either I work, or I pass mathematics.
　　　I passed mathematics.
　　　.................................
　　　Therefore, I studied.

(d)　If I work, I cannot study.
　　　Either I study, or I pass mathematics.
　　　I worked.
　　　.................................
　　　Therefore, I passed mathematics.

4.43　Show that (a) $p \wedge q$ logically implies p, (b) $p \vee q$ does not logically imply p.

4.44　Show that (a) q logically implies $p \to q$, (b) $\sim p$ logically implies $p \to q$.

4.45　Determine those propositions which logically imply (a) a tautology, (b) a contradiction.

Answers to Supplementary Problems

4.23 $(a) \sim p \wedge q$, $(b) \sim p \wedge \sim q$, $(c) p \vee \sim q$, $(d) \sim p \vee (p \wedge \sim q)$

4.24 $(a) (p \vee q) \wedge \sim r$, $(b) (p \wedge q) \vee \sim (p \wedge r)$, $(c) \sim (p \wedge \sim r)$, $(d) \sim [(r \vee q) \wedge \sim p]$

4.25 (a) Audrey speaks French or Danish.
     (b) Audrey speaks French and Danish.
     (c) Audrey speaks French but not Danish.
     (d) Audrey does not speak French or she does not speak Danish.
     (e) It is not true that Audrey does not speak French.
     (f) It is not true that Audrey speaks neither French nor Danish.

4.26

p	q	$p \vee \sim q$	$\sim p \wedge \sim q$	$\sim(\sim p \wedge q)$	$\sim(\sim p \vee \sim q)$
T	T	T	F	T	T
T	F	T	F	T	F
F	T	F	F	F	F
F	F	T	T	T	F

4.27

p	q	r	(a)	(b)	(c)
T	T	T	T	F	T
T	T	F	F	T	T
T	F	T	T	F	F
T	F	F	T	F	T
F	T	T	T	T	F
F	T	F	F	T	T
F	F	T	T	T	F
F	F	F	F	T	T

4.31 $(a) p \vee q \equiv \sim(\sim p \wedge \sim q)$, $(b) p \wedge q \equiv \sim(\sim p \vee \sim q)$

4.33 $(a) \sim p \vee q$, $(b) p \wedge \sim q$, $(c) p \vee q$

4.34 (c) He is rich or happy. (f) Audrey speaks German, but neither Spanish nor French.

4.35 (a) TTFF, (b) FFFT   **4.36** (a) TFTT, (b) FTFT

4.37 (*Hint*: Construct the appropriate truth tables.)

4.38 (a) If he does not win, then he does not have courage. (b) If he tires, then he will not win.

4.39 $(a) q \rightarrow \sim p$, $(b) \sim q \rightarrow p$, $(c) \sim p \rightarrow q$, $(d) p \rightarrow q$

4.40 (a) fallacy, (b) valid   **4.41** (a) valid, (b) fallacy

4.42 $(a) p \rightarrow \sim q, \sim r \vee q, r \vdash \sim p$; valid. $(c) p \rightarrow \sim q, p \vee r, r \vdash q$; fallacy.
     $(b) p, q \vdash p \leftrightarrow q$; valid. $(d) p \rightarrow \sim q, q \vee r, p \vdash r$; valid.

4.45 (a) Every proposition logically implies a tautology.
     (b) Only a contradiction logically implies a contradiction.

Chapter 5

Algorithms, Flowcharts, Pseudocode Programs

5.1 INTRODUCTION

An *algorithm* is a step-by-step list of instructions for solving a particular problem.

EXAMPLE 5.1 The First National Bank wants to calculate and record the INTEREST on a home mortgage for a certain customer. Assuming that the BALANCE of the mortgage and the RATE of interest appear in the customer's record, a suitable algorithm follows:

STEP 1. . Obtain NAME, BALANCE, RATE from customer's record.

STEP 2.   Calculate: INTEREST = BALANCE × RATE.

STEP 3.   Record customer's NAME and INTEREST in the interest file.

Here a *record* means a collection of related data items, e.g. data on a given customer, and a *file* is a collection of similar records.

There are two ways of presenting an algorithm other than by means of a numbered list of instructions. One way, the most popular, is by a *flowchart*, in which the essential steps of the algorithm are pictured by boxes of various geometrical shapes and the flow of data between steps is indicated by arrows, or *flowlines*. Figure 5-1 is a flowchart of the algorithm of Example 5.1. Although the flowchart may seem redundant in this case, it is frequently an indispensable tool in formulating complex types of algorithms.

Another way of presenting an algorithm is to write it in *pseudocode language*, to be discussed at the end of this chapter.

Suppose now that we want to execute an algorithm using an electronic computer. This can be done by translating the algorithm (from its flowchart or pseudocode form) into a computer program

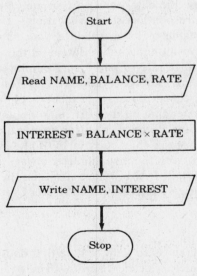

Fig. 5-1

95

written in some *high-level programming language*, such as BASIC, COBOL, or FORTRAN.   We begin the chapter with a general discussion of computer programs.

5.2   COMPUTER PROGRAMS; VARIABLES, CONSTANTS

The computer solves a particular problem by way of a computer *program*, which is a list of statements giving detailed instructions for the computer.   Such instructions must eventually be given to the computer in its own *machine language*.   However, computer programs are usually originally written in some high-level language (FORTRAN, COBOL, BASIC, . . .).   These are called *compiler languages*; a *compiler* is a special program that translates a program written in a high-level language into machine language.   The original program is called the *source program*, and the translated program is called the *object program*.   Figure 5-2 indicates this transition.

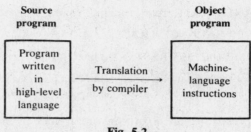

Fig. 5-2

Each programming language has its own character set which includes the 10 digits, the 26 letters, and special characters (see Section 2.6).   This character set, together with the syntax of the language, is used to write a program.   Whatever the language, however, the notions of *variable-names* and *constants* apply.

Variable-Names

The term *variable* in computer terminology normally means (i) a storage location in the memory of the computer reserved for a specific data item.   Alternately, a *variable* may mean (ii) the data item itself that is stored in the corresponding storage location.   Each variable (in either sense) in a computer program is supplied a name or label, called a *data-name* or *variable-name*.   Normally, one chooses names so that they indicate the kind of data item they represent.   For example, RATE could be chosen to denote the hourly rate of pay for an employee in a company.   One may then speak of the "variable RATE," or the "storage location RATE," or the "address RATE."   Note that the *value* of RATE [sense (ii)] may change during the execution of the program.   For example, RATE may be $8.25, $9.00, $9.75, . . . , depending on the particular employee.

Each programming language has its own rules for forming variable-names, e.g. they cannot exceed two characters in BASIC, or six characters in FORTRAN.   Variable-names generally are formed from alphameric characters, with the first character always alphabetic.   Furthermore, certain words (e.g. DATA, STOP, IF, LET in BASIC) have a special meaning in the language and hence cannot be chosen as variable-names.   These are called *reserved words* in the language.

Constants

A *constant* in a computer program is an item of data that does not change during the execution of the program.   Constants are sometimes called *literals*.   There are basically two kinds of constants, which appear in almost every programming language.

Numeric constants.   A numeric constant is a signed or unsigned number, with or without a decimal point, e.g.

$$999 \qquad +2058 \qquad -111 \qquad 333.22 \qquad -0.12345 \qquad +67.89$$

The first three, which do not contain a decimal point, are also called *integers*.   Numeric constants normally include numbers in exponential form, such as

$$1.23E03 \qquad 0.444E{-}2 \qquad 33.77E1$$

(see Section 3.2).   We emphasize that neither commas nor blank spaces can appear in numeric constants.   For example, 2,345,789 and 25 333.6 are inadmissible as numeric constants.   Each programming language also has rules limiting the number of digits that can appear in a constant.

Nonnumeric constants.   A nonnumeric constant is simply a string of characters from the character set of the language; such constants are sometimes called *messages* or *comments*.   One indicates a nonnumeric constant by placing its name between single quotation marks:

'FIND THE AVERAGE'   'RATE'
'088-24-0996'   'TO BE OR NOT TO BE'

Observe that 'RATE' indicates that RATE is a nonnumeric constant, whereas RATE (without quotation marks) indicates that RATE is a variable-name.

Remark: Some programming languages allow some special kinds of constants.   For example, FORTRAN allows the logical constants *true* and *false*, represented by

$$.\text{TRUE.} \qquad \text{and} \qquad .\text{FALSE.}$$

respectively.   COBOL allows figurative constants such as

$$\text{ZERO} \qquad \text{SPACE} \qquad \text{QUOTE}$$

all with their obvious meanings.

5.3   FLOWCHARTS AND THEIR LANGUAGE

A flowchart is a visual representation of an algorithm.   In fact, a flowchart is itself frequently used in the planning, development, and structuring of an algorithm for solving a complex problem.   The flowchart is usually regarded as an essential part of the documentation of any computer program translation of the original algorithm.

Formally, a flowchart is a diagram consisting of symbols (boxes) and arrows, or flowlines, connecting one symbol to another.   The symbols denote the essential steps of the algorithm and the arrows denote the *logic* of the algorithm, or the flow of data from one step of the algorithm to another.   The flowchart is usually drawn so that the flow direction is downward or from left to right.   Under this convention, the heads of the arrows may be omitted, giving flowlines as in Fig. 5-4.

The symbols themselves are of standardized shapes that indicate the type of action taking place at that step of the algorithm.   In fact, each symbol is labeled by its algorithm step, written within the symbol.   These algorithm steps are expressed in a universal flowchart language, which, fortunately, can be learned in less than an hour.

The six most frequently employed symbols in flowcharts appear in Fig. 5-3.   We discuss these symbols individually, along with the statements that can appear in them.

Terminal Symbol

The oval symbol is used to indicate the beginning or end of an algorithm by

$$\boxed{\text{Start}} \qquad \text{or} \qquad \boxed{\text{Stop}}$$

respectively.   Clearly, a flowchart can contain only one Start symbol; however, it can contain more

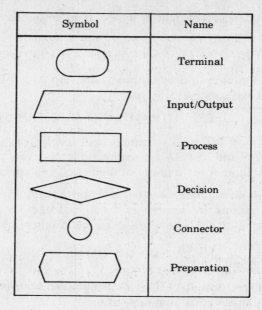

Symbol	Name
	Terminal
	Input/Output
	Process
	Decision
	Connector
	Preparation

Fig. 5-3

than one Stop symbol, since the algorithm may contain alternative branches. Sometimes we omit the Start and/or Stop symbols if it is clear where the algorithm begins and/or ends.

This symbol is also used to indicate a pause in the algorithm; e.g. it may be necessary to change the forms in an output printer. Hence the symbol may sometimes be labeled with "Pause."

Input/Output Symbol

The parallelogram symbol is used to indicate an input or an output operation. Specifically, we write

$$\boxed{\text{Read A, B, C, D}}$$

to indicate that data are to be inputted into the memory locations A, B, C, and D, in that order. Observe that the input statement begins with the word "Read" and is followed by variable-names separated by commas. Similarly,

$$\boxed{\text{Write X, Y, Z}}$$

indicates that the data in the memory locations X, Y, Z are to be outputted. The output statement begins with the word "Write" and also is followed by variable-names separated by commas. We can also output messages by including the message in single quotation marks; e.g.

$$\boxed{\text{Write 'NO SOLUTION'}}$$

indicates that the message "No solution (exists)" is to be outputted.

CHAP. 5]   ALGORITHMS, FLOWCHARTS, PSEUDOCODE PROGRAMS   99

Process Symbol

The rectangle symbol is used to indicate a processing operation. This can be an *assignment statement*, defined below, or it can be a *macroinstruction*, whose programming language translation would otherwise require an entire list of computer statements; e.g.

| Alphabetize names |

| Find mean and standard deviation of the salaries |

are macroinstructions.

The assignment statement has the form

| *variable = arithmetic expression* |

That is, only a variable-name can appear on the left side of the equals sign, =, but any arithmetic expression involving variables, constants, and arithmetic operations can appear on the right side of the equals sign. (Arithmetic expressions include individual variables or constants as special cases.) Another acceptable form for an assignment statement is

| *variable ← arithmetic expression* |

in which the backward arrow ← is used instead of the equals sign.

EXAMPLE 5.2

(a) Suppose that we want to add the numbers (numeric constants) A, B and C. We could not write

| A + B + C |

since the sum must be placed in a definite memory location. We can write

| SUM = A + B + C |

which directs that the result of the additions be stored in the memory location SUM. We could not write

| A + B + C = SUM |

since only a variable-name may appear on the left side of the assignment statement.

(b) Suppose we want to increase the value of the variable K by 1. We might accomplish this by writing

| Increase the value of K by 1 |

We claim that we may also accomplish this by writing

| K = K + 1 |

Note first that this is not a mathematical statement, for K does not in fact equal K + 1. However, as an assignment statement, it tells the computer to add 1 to the number currently in storage location K and then to assign the sum back to location K. The net effect is to increase the value stored at K by 1.

EXAMPLE 5.3 Figure 5-4 is a flowchart for finding the sum S, the average A, and the product P of three numbers X, Y, and Z. Observe that three assignment statements are put into one process symbol. Observe also that the heads of the arrows are omitted, since the direction of the flow is obviously from top to bottom.

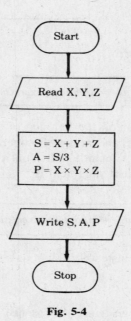

Fig. 5-4

Decision Symbol

One of the powerful properties of a computer is its ability to decide whether or not two field values are equal, and, if they are unequal, to decide which is the larger. That is, the computer can determine the truth value of a statement involving one of the six mathematical relations symbolized in Table 5-1. In practice, the computer is presented not with a true/false statement, but with a question having a "Yes" or "No" answer; e.g.

$$\text{Is } A = B? \qquad \text{Is } K \le 25? \qquad \text{Is SALES} > \$5000?$$

The computer can also be programmed to answer questions which involve one or more of the logical connectives "and," "or," and "not," or questions to which there are more than two possible answers; e.g.

$$\text{Is } X \le 50 \text{ and } Y \le 75? \qquad \text{Is D positive, negative, or zero?}$$

Table 5-1

Symbol	Meaning
$=$	Equals
$\ne$	Not equal
$<$	Less than
$\le$	Less than or equal
$>$	Greater than
$\ge$	Greater than or equal

With each question, the computer can be programmed to take a different course of action depending on the answer.   A step in an algorithm or program that leads to more than one possible continuation is called a *decision*.

The diamond-shaped symbol is used to indicate a decision.   The question is placed inside the symbol, and each alternative answer to the question is used to label the exit arrow which leads to the appropriate next step of the algorithm.   We note that the decision symbol is the only symbol that may have more than one exit.

EXAMPLE 5.4   Figure 5-5 shows a flowchart which reads two numbers, A and B, and prints them in decreasing order, after assigning the larger number to BIG and the smaller number to SMALL.   Observe the two arrows leaving the decision "Is A < B?", one labeled "no" and the other labeled "yes."   For notational convenience, the word "Is" will frequently be omitted from questions.

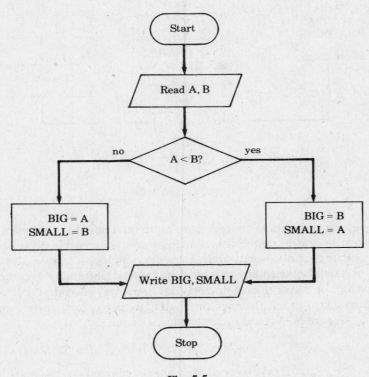

Fig. 5-5

EXAMPLE 5.5 (Quadratic Equation).   Recall that the solutions of the quadratic equation

$$ax^2 + bx + c = 0$$

where $a \neq 0$, are given by the quadratic formula

$$x = \frac{-b \pm \sqrt{b^2 - 4ac}}{2a}$$

The quantity $D = b^2 - 4ac$ is called the *discriminant* of the equation.   If D is negative, then there are no real solutions.   If $D = 0$, then there is only one (double) real solution, $x = -b/2a$.   If D is positive, the formula gives two distinct real solutions.   Figure 5-6 is a flowchart which inputs the coefficients A, B, C of a quadratic equation, and outputs the real solutions, if any.   Observe that there are three exits to the question "Sign of D?", each leading to a different output.

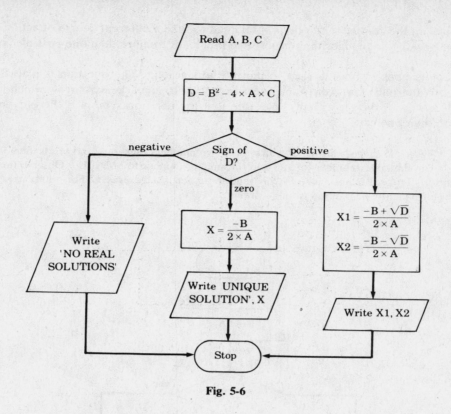

Fig. 5-6

Connector Symbol

A flowchart that is long and complex may require more than one sheet of paper, which means that certain flowlines cannot be drawn; or the flowchart may contain crossing flowlines that could cause confusion. The connector symbol, a small circle, is used to remedy such situations. Specifically, one assumes that a flowline exists between any pair of identically labeled connector symbols such that the flow is out of the flowchart at one of the connectors (*exit connector*) and into the flowchart at the other connector (*entry connector*). For each connector label, there is a unique entry connector, but one or more exit connectors. Figure 5-25 is a flowchart that uses connectors.

Preparation Symbol

This symbol indicates the preparation for some procedure by initializing certain variables. Many programmers indicate a preparation by means of the process symbol. We use the preparation symbol in Section 5.5.

5.4 LOOPS

Suppose that we want to read data from a record, process the data, and output the results. The flowchart of such a procedure appears in Fig. 5-7(*a*). Now suppose that we want to repeat this procedure for every record in the file. Such a repetitive operation is called a *loop*, with the handling of each record constituting one *cycle* or *iteration*. One might picture a loop by drawing an arrow from the last step of the procedure back to the first step, as in Fig. 5-7(*b*), but this would indicate no mechanism for exiting from the loop. A means of controlling a loop will now be discussed, and another way in Section 5.5.

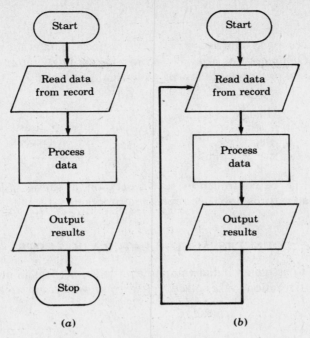

(a)   (b)

Fig. 5-7

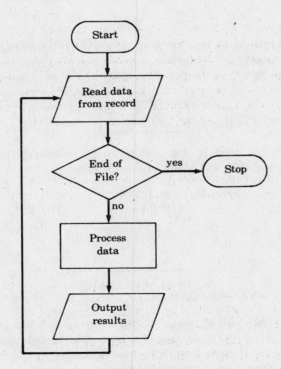

Fig. 5-8

End-of-File Record

Normally a data processing system uses a record with special data at the end of each file.   This is called an *end-of-file* (*EOF*) record or a *trailer* record.   Accordingly, after a record has been read, the computer may be asked one of the following equivalent questions:

The loop of Fig. 5-7(*b*) may be controlled by such a decision, as shown in Fig. 5-8.   It is clear that cycling now ends when all records have been read and processed.

5.5   INITIALIZATION: COUNTERS, ACCUMULATORS, DO LOOPS

Frequently we need to assign an initial value to a variable, perhaps in preparation for some type of procedure.   Such "initialization" takes place, in particular, with counters, accumulators and certain types of loops.

Counters

In order to count the number of records processed by the flowchart of Fig. 5-8, we might initially set a variable, say COUNT, equal to zero, and then increase COUNT by 1 every time a record is processed; we would output the value of COUNT upon leaving the loop.   It is clear from Fig. 5-9 that the final value of COUNT will be exactly equal to the number of records that have been processed.   A variable like COUNT is called a *counter*.   Observe that the counter required an initialization as a preparation for the loop; hence we used the preparation symbol in Fig. 5-9.

Accumulators

Suppose that a company wants to find its total weekly payroll, i.e. the sum of all its employees' salaries as listed in the personnel file.   If there are only three employees, with salaries X, Y, Z, then it is clear that the sum S can be found by use of the assignment statement

$$\boxed{S = X + Y + Z}$$

as in Example 5.3.   On the other hand, with many employees, we find the sum of the salaries as in Fig. 5-10.   That is, we choose a variable, say SUM, and initially set

$$\langle SUM = 0 \rangle$$

Then, using the assignment statement

$$\boxed{SUM = SUM + SALARY}$$

we are able to add the salaries to SUM one after the other as they are inputted from the records. Clearly, the final value of SUM upon leaving the loop will be the sum of all the salaries.   A variable like SUM is called an *accumulator*, for obvious reasons.   Observe that the accumulator SUM required an initialization as a preparation for summing the salaries; hence the use of the preparation symbol.

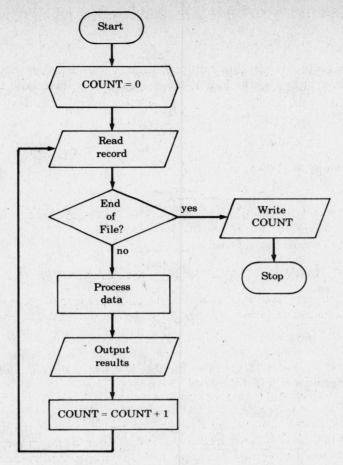

Fig. 5-9

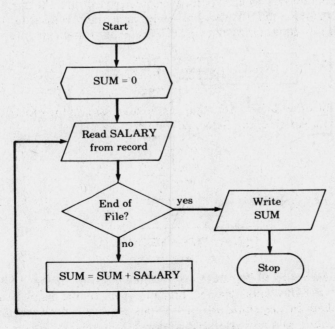

Fig. 5-10

DO Loops

Figure 5-11(a) shows a loop whereby a certain procedure is executed exactly N times, the value of N being a fixed positive integer. The loop involves a variable K, essentially a counter. Initially we set

$$\langle K = 1 \rangle$$

After each cycle, we increment K by 1, using

$$\boxed{K = K + 1}$$

and then test to see if K exceeds the limit value N, by

If the answer is "No," the procedure is repeated. After N cycles an answer of "Yes" is obtained, and we exit from the loop. This type of loop occurs so often that it is given a name, *DO loop*, and a shorthand flowchart notation, Fig. 5-11(b). The variable K is called the *index* of the DO loop, and the procedure which is iterated is called the *body* of the loop.

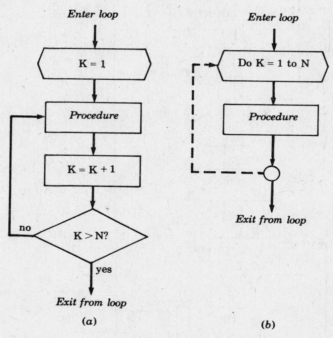

Fig. 5-11

In Fig. 5-11(b), the initial value of the index and its increment are both 1. This is not a necessary restriction. If, instead, a statement such as

$$\langle \text{Do } K = 2 \text{ to } 10 \text{ by } 3 \rangle$$

is used, then 2 will be the initial value, 10 the limit value, and 3 the increment. Thus, we first execute the loop with K = 2. Then we increment K by 3, and execute the loop with K = 5. Again we increment K by 3, and execute the loop with K = 8. However, the next time we increment K by 3, we get 11, which exceeds the limit value 10, and so we exit from the loop. DO loops of this more general type are often required when the index K also figures in the body of the loop.

EXAMPLE 5.6   Find the positive odd integers, along with their squares, that do not exceed N, where N is an integer greater than 1 which is inputted into the program.

Figure 5-12(a) is the flowchart of a suitable program. Observe that the initial value of K is 1, the increment is 2, and the test value is N. The equivalent DO loop appears in Fig. 5-12(b).

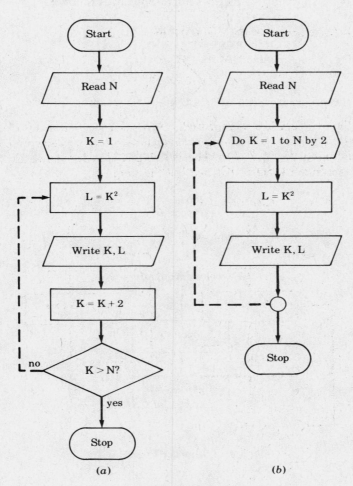

(a)   (b)

Fig. 5-12

5.6   PSEUDOCODE PROGRAMS

For an algorithm that is sufficiently complex to resist simple, top-down structuring, a flowchart may be difficult to read, to modify, or to revise. Consequently, there has been developed a pseudo-code program language, which gives an alternate way of presenting algorithms.

A pseudocode program is composed of a list of statements. Some of these statements are among those used in flowcharts, e.g. Read statements, Write statements, Assignment statements,

Conditions. However, instead of arrows or flowlines to indicate the logic of an algorithm, the pseudocode program uses three types of organization: (i) sequence logic, (ii) selection logic, and (iii) iteration logic.

Sequence Logic

Under this logic, the instructions in a pseudocode program are executed in order, from the top to the bottom (see Fig. 5-13). Although it is not necessary to have a Start statement or Stop statement (since the program begins at the top and ends at the bottom), we frequently signal the completion of a pseudocode program by writing END at the bottom of the program. For example, a pseudocode program equivalent to the flowchart of Fig. 5-1 might appear as follows:

        read NAME, BALANCE, RATE
        INTEREST = BALANCE × RATE
        write NAME, INTEREST
        END

Sometimes we begin the program with a title, e.g. here "Calculating Interest."

Pseudocode Program

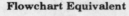

Process A
Process B

Flowchart Equivalent

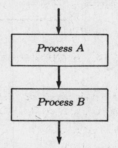

Fig. 5-13.   Sequence Logic

Selection Logic

This logic employs a number of structures, called *IF-structures*, each of which is essentially a selection of one out of several alternatives. Each such structure begins with a statement of the form

IF *condition*

and ends with the statement

ENDIF

The condition in the beginning statement can usually be paraphrased as a question with a "Yes" or "No" answer.

Single alternative.   The appropriate structure is called an *IFTHEN structure*; its logic is illustrated in Fig. 5-14.   That is, IF the condition holds, THEN procedure A, the coding of which may involve one or more statements, is executed; otherwise procedure A is skipped, and control transfers to the first statement following the ENDIF statement.

Pseudocode Program

```
              ⋮

    IF condition
        [Procedure A]
    ENDIF
              ⋮
```

Flowchart Equivalent

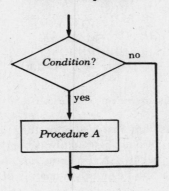

Fig. 5-14.   IFTHEN Structure

EXAMPLE 5.7   The following pseudocode program, which calculates an employee's WAGES given his hourly RATE of pay and the number of HOURS worked, uses an IFTHEN structure:

```
    read NAME, RATE, HOURS
    WAGES = HOURS × RATE
    IF HOURS > 40
        WAGES = WAGES + (HOURS − 40) × 0.5 × RATE
    ENDIF
    write NAME, WAGES
    END
```

In other words, WAGES equals HOURS times RATE, but if the employee has worked overtime (more than 40 hours), then there is an additional payment at half rate for those hours over 40.

Observe that the IF-procedure is indented, a pseudocode convention that makes programs much easier to read.

Double alternative.   An *IFTHENELSE structure* is used for decisions involving two distinct alternatives.   Its logic appears in Fig. 5-15.   Observe that an ELSE statement separates two procedures.   As indicated by the flowchart, IF the condition holds, THEN procedure A (above the ELSE statement) is executed; ELSE procedure B (below the ELSE statement) is executed.

Pseudocode Program   **Flowchart Equivalent**

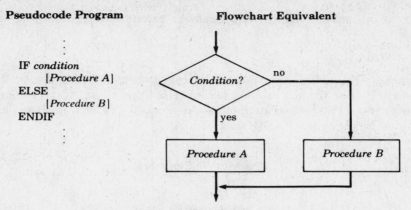

```
    ⋮
IF condition
      [Procedure A]
ELSE
      [Procedure B]
ENDIF
    ⋮
```

Fig. 5-15.   IFTHENELSE Structure

EXAMPLE 5.8   Following is a pseudocode program for Example 5.4 (whose flowchart appears in Fig. 5.5).

```
read A, B
IF A < B
    BIG = B
    SMALL = A
ELSE
    BIG = A
    SMALL = B
ENDIF
write BIG, SMALL
END
```

Again observe how the procedure blocks are indented for easier reading.

Multiple alternatives.   Decisions that involve more than two alternatives may be handled by means of *nested* IF-structures, wherein one IF-structure is contained in the procedure component of another IF-structure.   An example is the *ELSEIF structure*, whose form and logic appear in Fig. 5-16.   These ELSEIF structures use one or more statements of the form

ELSEIF condition

between the IF and ELSE statements.   Observe that only one of the procedures will be executed, the one that follows the first condition which holds.   If none of the conditions holds, then the procedure following the ELSE statement is executed.   If this procedure is empty, then the ELSE statement itself may be omitted.   (See Problem 5.46.)

EXAMPLE 5.9   Following is a pseudocode program for Example 5.5 (flowcharted in Fig. 5-6).

```
read A, B, C
D = B² - 4 × A × C
IF D > 0
    X1 = (-B + √D)/(2 × A)
    X2 = (-B - √D)/(2 × A)
    write X1, X2
ELSEIF D = 0
    X = -B/(2 × A)
    write 'UNIQUE SOLUTION', X
ELSE
    write 'NO REAL SOLUTIONS'
ENDIF
END
```

Again observe the indented procedure blocks.

Pseudocode Program   **Flowchart Equivalent**

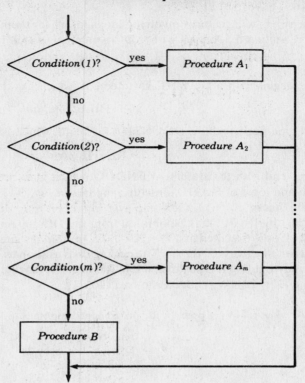

```
        .
        .
IF condition (1)
     [Procedure A₁]
ELSEIF condition(2)
     [Procedure A₂]
        .
        .
ELSEIF condition(m)
     [Procedure Aₘ]
ELSE
     [Procedure B]
ENDIF
        .
        .
```

Fig. 5-16. ELSEIF Structure

Pseudocode Program   **Flowchart Equivalent**

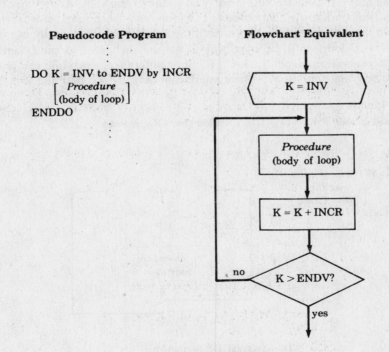

```
        .
        .
DO K = INV to ENDV by INCR
   ⎡   Procedure    ⎤
   ⎣ (body of loop) ⎦
ENDDO
        .
        .
```

Fig. 5-17. DO Structure

Iteration Logic

This logic refers to structures involving loops, of which there are three types. One type, already flowcharted in Section 5.5, begins with a DO statement, which has the form

$$\text{DO } K = 1 \text{ to } N \qquad \text{or} \qquad \text{DO } K = INV \text{ to } ENDV \text{ by } INCR$$

Another type begins with a DOWHILE statement, which has the form

$$\text{DOWHILE } condition$$

and the third type begins with a DOUNTIL statement, which has the form

$$\text{DOUNTIL } condition$$

All three types end with the statement ENDDO, which indicates the end of the loop.

The form and logic of the DO structure appears in Fig. 5-17. Here, K is called the *index* of the loop, INV the *initial value*, ENDV the *end value* or *test value*, and INCR the *increment*. (By convention, the "by INCR" may be omitted from the DO statement when INCR = 1.) Observe that the body of the loop is executed first with K = INV, the initial value. After each cycle, K is incremented by INCR and tested against ENDV. The cycling ends as soon as K exceeds ENDV. The flowchart in Fig. 5-17 assumes that INCR is positive; if INCR is negative, so that K decreases in value, then the cycling ends as soon as K < ENDV. [See Problem 5.16(*d*).]

EXAMPLE 5.10 The following pseudocode program accomplishes the task of Example 5.6:

```
read N
DO K = 1 to N by 2
    L = K²
    write K, L
ENDDO
END
```

The form and logic of the DOWHILE structure appears in Fig. 5-18, and that of the DOUNTIL structure in Fig. 5-19. Since these structures are loops, there must be a statement before the structure that initializes the condition which controls the loop, and there must be a statement in the body of the loop that changes the condition, in order that looping may eventually cease. Looping continues in the DOWHILE loop until the condition is not met, whereas looping continues in the DOUNTIL loop until the condition is met. This is a minor difference, since one can always impose the negation of the underlying condition. More significant is the fact that the condition in the

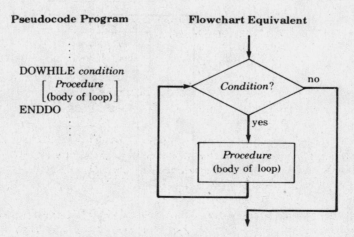

Pseudocode Program

```
    .
    .
    .
DOWHILE condition
  ⎡  Procedure   ⎤
  ⎣ (body of loop)⎦
ENDDO
    .
    .
    .
```

Flowchart Equivalent

Fig. 5-18.   DOWHILE Structure

Pseudocode Program   **Flowchart Equivalent**

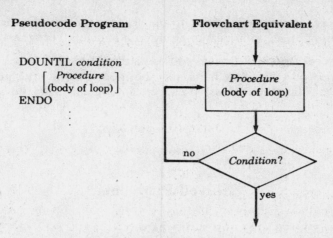

DOUNTIL *condition*
$\begin{bmatrix} Procedure \\ \text{(body of loop)} \end{bmatrix}$
ENDO

Fig. 5-19.   DOUNTIL Structure

DOWHILE loop is tested at the beginning of the loop, whereas the condition in the DOUNTIL loop is tested at the end of the loop.   Therefore, the body of the DOUNTIL loop is always executed at least once, but the body of the DOWHILE loop need not be executed at all.

EXAMPLE 5.11   Either of the following pseudocode programs will accomplish the task of Example 5.6 (see Fig. 5-12).

```
read N                          read N
K = 1                           K = 1
DOWHILE K ≤ N                   DOUNTIL K > N
    L = K²                          L = K²
    write K, L                      write K, L
    K = K + 2                       K = K + 2
ENDDO                           ENDDO
END                             END
```

Observe that the programs are identical, except that the conditions controlling the loops are the negations of each other.

EXAMPLE 5.12   Suppose that a company's personnel file contains the names and ages of all its employees, and that the company wants a list of all its employees who are under 30 years of age.   The following pseudocode program accomplishes this task:

```
read NAME, AGE from record
DOUNTIL End of File
    IF AGE < 30
        write NAME
    ENDIF
    read NAME, AGE from next record
ENDDO
END
```

Observe that the first read statement initializes the condition controlling the loop, and the second read statement changes the condition.   Observe also the double indenting, one for the DO-structure and one for the IF-structure contained inside the loop.   (We assume implicitly that the file contains data for at least one employee; otherwise we would need to use the DOWHILE loop.)

Remark 1:   *Structured programming* refers to a methodology that emphasizes a top-down structure for a computer program rather than a transfer of control back and forth within the program. This methodology has resulted in the introduction of various types of structures into the different programming languages that are very similar to the above-given pseudocode structures. Thus the pseudocode expression of an algorithm fits in nicely with structured programming techniques.

   Remark 2: To "code" normally means to translate an algorithm into a well-defined programming language. *Pseudo*code, like flowcharts, may be used to help write programs in almost any programming language. In other words, pseudocode programs (like flowcharts) are essentially language-independent.

Solved Problems

COMPUTER PROGRAMS; VARIABLES, CONSTANTS

5.1   A compiler is a machine which translates a program written in machine language into a high-level language such as BASIC, COBOL, or FORTRAN. (True or false.)

   False, for two reasons. First of all, a compiler is not a machine, but a program. Second, the compiler translates a program written in a high-level language into machine language.

5.2   What is meant by the *source program* and the *object program* in computer programming?

   The *source program* refers to the original program, written in a high-level language. Translated into machine-language instructions, the source program becomes the *object program*.

5.3   The different programming languages use different character sets. (True or false.)

   True. Although all programming-language character sets include the 10 digits and the 26 letters, they normally include different selections of special characters.

5.4   Which of the following are acceptable variable-names (*a*) in BASIC? (*b*) in FORTRAN?

(i) X	(iii) RATE	(v) 4H	(vii) A	(ix) K*
(ii) X2	(iv) INTEREST	(vi) X/B	(viii) A9	(x) AGE

   (*a*) (i), (ii), (vii), and (viii). Variable-names in BASIC are either a single letter or a letter followed by a digit.
   (*b*) (i), (ii), (iii), (vii), (viii), and (x). A variable-name in FORTRAN consists of one to six alphameric characters, with the first character being a letter.

5.5   Why are the following unacceptable as numeric constants?

   (*a*) 5,000,000   (*b*) $55.50   (*c*) 33 444 555   (*d*) 75¢   (*e*) 4.6E234

   (*a*) Contains commas. (*b*) Contains a dollar sign. (*c*) Contains blank spaces. (*d*) Contains the cents sign. (*e*) The exponent exceeds two digits.

5.6   In a parenthesis-free arithmetic expression, the *strengths* of the operations determine the order in which they are performed. In BASIC (or in FORTRAN, with ** instead of ↑), the symbols for the five fundamental operations and their strengths are as follows:

Strongest:	↑	exponentiation	
Next-strongest:	*	multiplication	/ division
Weakest:	+	addition	− subtraction

Within an equal-strength group, the operations are to be carried out from left to right. The strength order is overridden if parentheses are introduced in the BASIC formula; then, the expression(s) in parentheses is (are) evaluated (following the strength order) before anything else. Give BASIC formulas for

$$(a) \quad x^2 + 2xy + y^2 \qquad\qquad (c) \quad a + \frac{b+c}{2d}$$

$$(b) \quad x^3 - 4x^2 + 7x + 6 \qquad (d) \quad \frac{a}{b} - \frac{c^2}{3a}$$

Note first that only capital letters can be used and that all expressions must be typed on one line.

(a)   X↑2+2∗X∗Y+Y↑2

(b)   X↑3−4∗X↑2+7∗X+6

(c)   A+(B+C)/(2∗D)

(d)   A/B−C↑2/(3∗A)

5.7   Referring to Problem 5.6, evaluate each of the following BASIC expressions:

        (a)   3+2∗6−12/3   (c)   4∗3↑2/2−2↑3+1
        (b)   (3+2)∗(6−12)/3   (d)   (4∗3)↑2/2↑(3+1)

(a)   First, 2∗6 = 12 and 12/3 = 4, giving

$$3 + 12 - 4$$

Then, $3 + 12 = 15$, and, finally, $15 - 4 = 11$.

(b)   First, $3 + 2 = 5$ and $6 - 12 = -6$, giving

$$5*(-6)/3$$

Then, $5*(-6) = -30$, and, finally, $(-30)/3 = -10$.

(c)   First, $3↑2 = 9$ and $2↑3 = 8$, giving

$$4*9/2-8+1$$

Then, $4*9 = 36$ and $36/2 = 18$, giving $18 - 8 + 1$. Next, $18 - 8 = 10$, and, finally, $10 + 1 = 11$.

(d)   First, $4*3 = 12$ and $3 + 1 = 4$, giving

$$12↑2/2↑4$$

Then $12↑2 = 144$ and $2↑4 = 16$, which yields $144/16$. Finally, $144/16 = 9$.

FLOWCHARTS

5.8   A file of incoming freshmen contains, among other data, each student's NAME and SCORE on an entrance examination. Draw a flowchart which will list the names of those students with perfect scores of 100.

      Figure 5-20 is such a flowchart. There are two decisions, one which tests if SCORE = 100 and one which tests for End of File.

5.9   Each record in a student file contains, among other data, the student's WEIGHT. Draw a flowchart which finds the average weight of the students.

      Figure 5-21 is the flowchart. Note the need for an accumulator, SUM, to add the weights of the students, and a counter, N, to enumerate the students. The average weight, MEAN, is the final SUM divided by the final N.

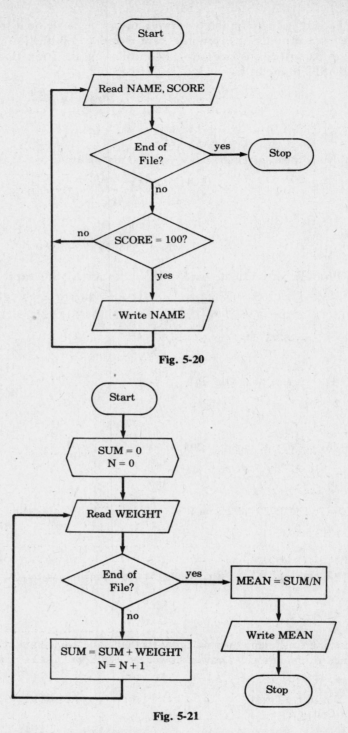

Fig. 5-20

Fig. 5-21

5.10 Each record in a student file contains, among other data, the student's HEIGHT. Draw a flowchart which finds the height of the tallest student.

Figure 5-22 is such a flowchart. Observe that each time we read a new HEIGHT, we compare it to the variable TALL. If the new HEIGHT exceeds TALL, then we reset TALL equal to the new HEIGHT. Thus, at any given moment, TALL contains the height of the tallest student processed up to that point. Accordingly, TALL will contain the height of the tallest student(s) when all the student records have been processed. We also initially assign TALL = 0 to begin the process.

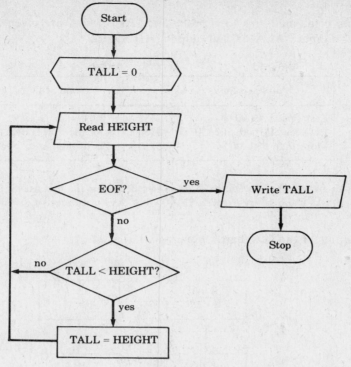

Fig. 5-22

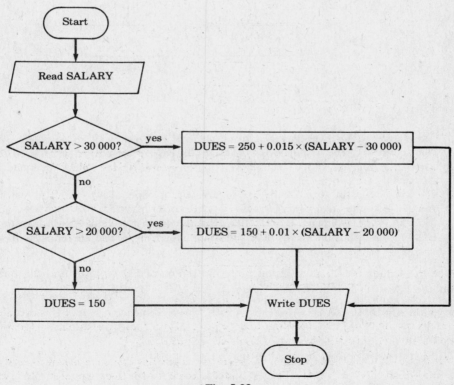

Fig. 5-23

5.11   The amounts of annual dues to a professional association are given in Table 5-2.   Draw a flowchart with input SALARY and output DUES.

Table 5-2

Annual Salary	Annual Dues
Less than $20 000 Between $20 000 and $30 000 More than $30 000	$150 $150 + (1% of the excess over $20 000) $250 + (1.5% of the excess over $30 000)

Figure 5-23 gives such a flowchart.   Of the two decisions, the first asks if the salary exceeds $30 000 and the second asks if the salary exceeds $20 000 (but does not exceed $30 000).

5.12   Find the outputs of the flowchart segments in Fig. 5-24.

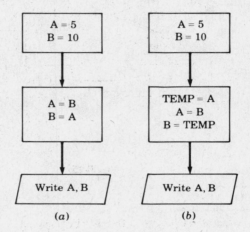

(a)   (b)

Fig. 5-24

(a)   The statement A = B assigns the current value of B, which is 10, to A.   Hence the value of A is changed to 10.   The statement B = A assigns the current value of A, which is 10, to B.   Hence the value of B remains at 10.   Thus the output is 10 and 10; the values of A and B are not interchanged by the two statements.

(b)   Here the values of A and B become interchanged.   Specifically, TEMP = A assigns the original value of A to the "temporary" location TEMP, i.e. TEMP = 5.   Now A = B assigns the value of B to A; so A = 10.   Finally, B = TEMP assigns the original value of A, which is stored in TEMP, to B; so B = 5.   Thus the values of A and B are interchanged, and the output is 10 and 5.

5.13   Trace the values of A and B through the flowchart of Fig. 5-25 to find the outputs given the inputs (a) A = 10, B = 5; (b) A = 3, B = 5; (c) A = 5, B = 10.

(a)   Since A = 10, B = 5, the answer to "A < B?" is "No"; hence A = A^2 + B is executed, giving

$$A = 10^2 + 5 = 100 + 5 = 105$$

Then B = B + 5 × A is executed, giving

$$B = 5 + 5 \times 105 = 5 + 525 = 530$$

Hence the output is A = 105, B = 530.

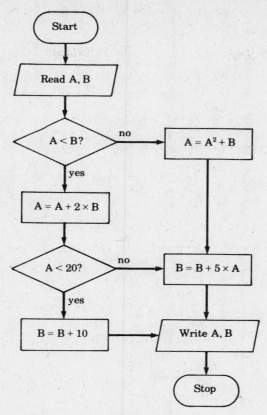

Fig. 5-25

(b)   Since A = 3, B = 5, the answer to "A < B?" is "Yes"; hence A = A + 2 × B is executed, giving

$$A = 3 + 2 \times 5 = 3 + 10 = 13$$

The answer to "A < 20?" is also "Yes"; so B = B + 10 is executed, yielding

$$B = 5 + 10 = 15$$

Hence the output is A = 13, B = 15.

(c)   Since A = 5, B = 10, the answer to "A < B?" is "Yes"; hence A = A + 2 × B is executed, giving

$$A = 5 + 2 \times 10 = 5 + 20 = 25$$

Now the answer to "A < 20?" is "No"; so B = B + 5 × A is executed, yielding

$$B = 10 + 5 \times 25 = 10 + 125 = 135$$

Hence the output is A = 25, B = 135.

5.14   Find the outputs of the flowchart in Fig. 5-26, assuming the inputs (a) X = 2, Y = 5; (b) X = 2, Y = 12.

Note first that there are three connectors labeled A, two labeled B, and three labeled C.   As required, there is only one entrance connector for each label.

(a)   Since X = 2, Y = 5, the answer to "X < 10?" is "Yes"; hence X = X + Y is executed, giving

$$X = 2 + 5 = 7$$

The answer to "Y < 10?" is also "Yes"; hence Y = Y + X is executed, giving

$$Y = 5 + 7 = 12$$

Then $Z = X + Y$ yields

$$Z = 7 + 12 = 19$$

Thus the output is $Z = 19$.

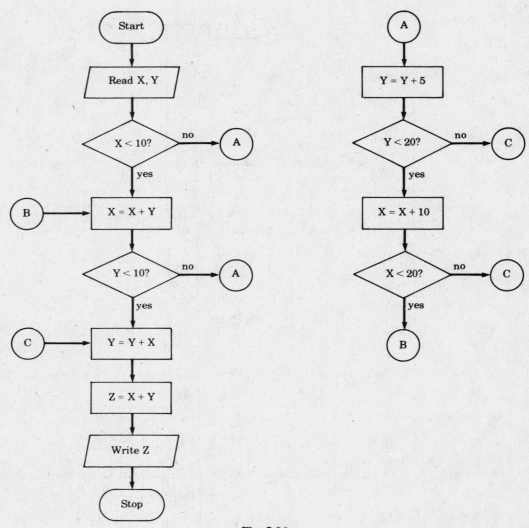

Fig. 5-26

(b) Since $X = 2$, $Y = 12$, the answer to "$X < 10$?" is "Yes"; hence $X = X + Y$ is executed, yielding

$$X = 2 + 12 = 14$$

The answer to "$Y < 10$?" is "No"; hence, by way of connector A, we go to the statement $Y = Y + 5$. This gives

$$Y = 12 + 5 = 17$$

The answer to "$Y < 20$?" is "Yes"; hence $X = X + 10$ is executed, giving

$$X = 14 + 10 = 24$$

The answer to "X < 20?" is "No"; hence, by way of connector C, we go to Y = Y + X.   This gives

$$Y = 17 + 24 = 41$$

Then Z = X + Y yields

$$Z = 24 + 41 = 65$$

Thus the output is Z = 65.

DO LOOPS

5.15 Draw a flowchart having a DO loop, which inputs a positive integer N and outputs

$$(a) \quad 1 + \frac{1}{2} + \frac{1}{3} + \frac{1}{4} + \cdots + \frac{1}{N} \qquad (b) \quad 1 - \frac{1}{2} + \frac{1}{3} - \frac{1}{4} + \cdots \pm \frac{1}{N}$$

(a) Figure 5-27(a) is such a flowchart. Observe that the index K also is involved in the body of the loop. The accumulator SUM is used to add up the 1/K for K = 1 to N.

(b) Figure 5-27(b) is such a flowchart. Here we need to multiply 1/K by $(-1)^{K+1}$ to make sure that the even values of K contribute negative summands.

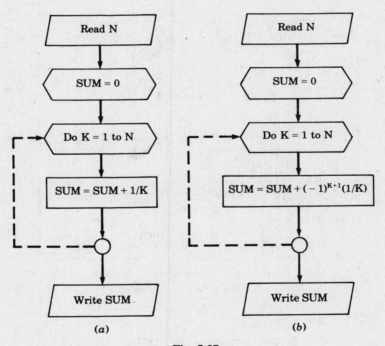

Fig. 5-27

5.16 Find the number of cycles, and the value of K for each cycle, for DO loops headed by:

$$(a) \quad \text{Do K} = 1 \text{ to } 10 \text{ by } 2 \qquad (c) \quad \text{Do K} = 5 \text{ to } 3 \text{ by } 4$$
$$(b) \quad \text{Do K} = 5 \text{ to } 20 \text{ by } 4 \qquad (d) \quad \text{Do K} = 15 \text{ to } 5 \text{ by } -4$$

(a) The initial value for K is K = 1. Then K is increased by 2 each time, giving odd values for K.   But 10 is the end value; hence the loop is executed for

$$K = 1 \qquad K = 3 \qquad K = 5 \qquad K = 7 \qquad K = 9$$

or 5 times [unless there is an abnormal exit, as discussed in Problem 5.17(b)].

(b) The loop is executed for

$$K = 5 \qquad K = 5 + 4 = 9 \qquad K = 9 + 4 = 13 \qquad K = 13 + 4 = 17$$

or 4 times.

(c) Here the initial value, 5, already exceeds the end value, 3. If the test for stopping follows the body of the loop (as in Fig. 5-11), the loop will still be executed once, for the initial value K = 5. Otherwise, the loop will not be executed at all.

(d) The initial value for K is K = 15. Since the increment −4 is negative, K is decreased by 4 after each cycle. Hence the loop is executed for

$$K = 15 \qquad K = 15 - 4 = 11 \qquad K = 11 - 4 = 7$$

but not for K = 7 − 4 = 3, since 3 is less than the test value 5. Thus the loop is executed three times.

5.17 Find the output for each flowchart in Fig. 5-28.

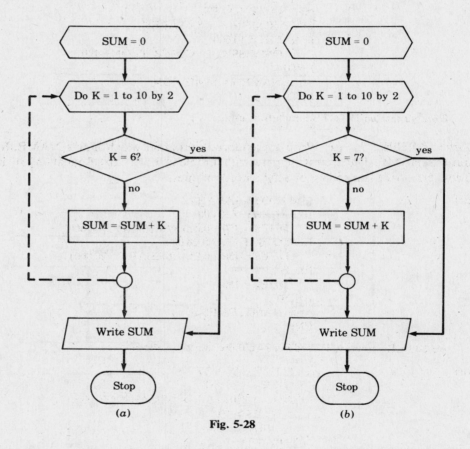

Fig. 5-28

(a) The header statement indicates that the loop is to be executed five times, for K = 1, 3, 5, 7, 9. As the answer to "K = 6?" is "No" for all five values of K, SUM = SUM + K is in fact executed five times. We have:

(1) For K = 1, SUM = 0 + 1 = 1.   (4) For K = 7, SUM = 9 + 7 = 16.
(2) For K = 3, SUM = 1 + 3 = 4.   (5) For K = 9, SUM = 16 + 9 = 25.
(3) For K = 5, SUM = 4 + 5 = 9.

We now exit from the loop to "Write SUM"; hence the output is 25. This is a case of *normal exit*, since all cycles indicated by the header statement were completed.

(b) The header statement again indicates that the loop is to be executed five times, for K = 1, 3, 5, 7, 9. The answer to "K = 7?" is "No" for the first three values of K; hence SUM = SUM + K is executed for these values of K:

(1) For K = 1, SUM = 0 + 1 = 1.
(2) For K = 3, SUM = 1 + 3 = 4.
(3) For K = 5, SUM = 4 + 5 = 9.

But, at the beginning of the fourth cycle of the loop, the answer to "K = 7?" is "Yes"; hence, by Fig. 5-28(b), we exit from the loop to "Write SUM." Thus, the output is 9, the last value of SUM. This is a case of *abnormal exit*, since not all cycles indicated by the header statement were completed.

PSEUDOCODE PROGRAMS

5.18 A salesman's commission is 5% of his total weekly sales, with an extra $100 if the sales exceed $10 000. Write a pseudocode program which identifies a salesman and calculates his COMMISSION.

```
read NAME, SALES
COMMISSION = 0.05 × SALES
IF SALES > 10 000
    COMMISSION = COMMISSION + 100
ENDIF
Write NAME, COMMISSION
END
```

Observe that an IFTHEN structure is used.

5.19 Refer to Table 5-2. (a) Write a pseudocode program which inputs NAME, SALARY and outputs NAME, DUES. (Compare with Problem 5.11.) (b) Repeat (a) if, in addition to Table 5-2, there is a ceiling of $400 placed on dues.

(a)
```
read NAME, SALARY
IF SALARY > 30 000
    DUES = 250 + 0.015 × (SALARY − 30 000)
ELSEIF SALARY > 20 000
    DUES = 150 + 0.01 × (SALARY − 20 000)
ELSE
    DUES = 150
ENDIF
write NAME, DUES
END
```

An ELSEIF structure is used because there are three alternatives.

(b)
```
read NAME, SALARY
IF SALARY > 30 000
    DUES = 250 + 0.015 × (SALARY − 30 000)
    IF DUES > 400
        DUES = 400
    ENDIF
ELSEIF SALARY > 20 000
    DUES = 150 + 0.01 × (SALARY − 20 000)
ELSE
    DUES = 150
ENDIF
write NAME, DUES
END
```

Observe the embedded IFTHEN structure in the first alternative, which limits dues to $400.

5.20   Suppose that y is given as a function of t by

$$y = 2t^3 - 3t^2 - 27t + 28$$

Write a pseudocode program which calculates y for values of t from -5 to 5 in steps of 0.25, and outputs each t and its corresponding y.

```
T = -5
DOWHILE T ≤ 5
    Y = 2 × T³ - 3 × T² - 27 × T + 28
    write T, Y
    T = T + 0.25
ENDDO
END
```

Observe that T is initialized before the DOWHILE structure but is incremented inside the structure.

5.21   Write a pseudocode program for the algorithm in Fig. 5-20, which finds the average weight for a list of students.

```
SUM = 0
N = 0
read WEIGHT
DOUNTIL End of File
    SUM = SUM + WEIGHT
    N = N + 1
    read next WEIGHT
ENDDO
MEAN = SUM/N
write MEAN
END
```

The first read statement, which initializes the condition controlling the DOUNTIL structure, appears before the structure; the second read statement, which changes the condition, appears within the structure.

5.22   Referring to Problem 5.8, write a pseudocode program which lists the names of those students with perfect scores of 100, and also finds the number N of such students.

Again we need two read statements, one initializing the condition controlling the loop and the other changing the condition.

```
N = 0
read NAME, SCORE
DOUNTIL End of File
    IF SCORE = 100
        write NAME
        N = N + 1
    ENDIF
    read next NAME, SCORE
ENDDO
write N
END
```

Observe that N is used as a counter to enumerate the students who score 100. Because an IFTHEN structure is nested in the loop, there is a double indentation of the program.

5.23 Each record in a student file contains, among other data, the student's NAME and HEIGHT. Write a pseudocode program which finds the name and height of the tallest student. (Compare with Problem 5.10.)

The pseudocode program follows, where again we have two read statements.

```
TALL = 0
TALLMAN = 'JOHN DOE'
read NAME, HEIGHT
DOUNTIL End of File
    IF TALL < HEIGHT
        TALL = HEIGHT
        TALLMAN = NAME
    ENDIF
    read next NAME, HEIGHT
ENDDO
write 'THE TALLEST STUDENT IS!', TALLMAN, 'WITH HEIGHT', TALL
END
```

Here, TALL is a numerical variable, initialized by TALL = 0, but TALLMAN is a nonnumerical variable, initialized by TALLMAN = 'JOHN DOE'. At the end of any iteration, TALL and TALLMAN contain, respectively, the height and name of the tallest student yet processed; hence the final values of TALL and TALLMAN will yield the required height and name of the tallest student.

Supplementary Problems

COMPUTER PROGRAMS; VARIABLES, CONSTANTS

5.24 BASIC, COBOL, FORTRAN are (a) compiler languages, (b) high-level languages, (c) machine-independent languages, (d) all of the above.

5.25 A compiler translates a source program into an object program. (True or false.)

5.26 A data item whose value may change during the course of a program is called a (a) _____, whereas a data item whose value normally does not change during the course of a program is called a (b) _____.

5.27 Which of the following are unacceptable as variable-names in BASIC?

(a) B7   (c) 3D   (e) BONUS   (g) Z   (i) NUMBER
(b) C34   (d) F%   (f) EMPLOYEE   (h) A*B   (j) Y2

5.28 Which of the character strings in Problem 5.27 are unacceptable as variable-names in FORTRAN?

5.29 Which of the following are unacceptable as numeric constants?

(a) 333.444   (c) 79999   (e) 234.5E−15   (g) 50¢
(b) 4,000.00   (d) $95.75   (f) 2 000   (h) 7.7E111

5.30 Write a BASIC formula for each algebraic expression:

(a) $x^2 + y^2$ (c) $\dfrac{2ab}{c+d}$ (e) $\dfrac{x}{yz}$
(b) $(x+y)^2$ (d) t^{n+1} (f) $(x^2 + 2xy - y^2)^5$

5.31 Write a FORTRAN formula for t^{n+1} (a) using parentheses, (b) without parentheses.

5.32   Evaluate each BASIC expression:

(a) $5+3*8-4/2$ (c) $5+3*((8-4)/2)$ (e) $(5+3)*(8-4/2)$

(b) $(5+3)*(8-4)/2$ (d) $(5+3)*((8-4)/2)$ (f) $5+(3*8-4)/2$

5.33   Evaluate each BASIC expression:

(a) $5+2\uparrow3-2*3\uparrow2$ (c) $(5+2)\uparrow3-(2*3)\uparrow2$

(b) $5+2\uparrow(3-2)*3\uparrow2$ (d) $5+(2\uparrow3-2)*3\uparrow2$

FLOWCHARTS, PSEUDOCODE PROGRAMS

5.34   The perimeter P and the AREA of a triangle whose sides have lengths a, b, c are given by

$$P = a + b + c \qquad AREA = \sqrt{s(s-a)(s-b)(s-c)}$$

where $s \equiv (a+b+c)/2$.   Draw a flowchart or write a pseudocode program with input a, b, c and output P and AREA.

5.35   An automobile dealership uses CODE = 1 for new automobiles, CODE = 2 for used automobiles, and CODE = 3 for separate accessories.   A salesman's commissions are as follows: on a new automobile, 3% of the selling price but with a maximum of $300; on a used automobile, 5% of the selling price but with a minimum of $75; on accessories, 6%.   Draw a flowchart or write a pseudocode program with input CODE and PRICE and output COMMISSION.

5.36   The monthly charge for local telephone calls is as follows:

> $8.00 for up to 100 calls
>   plus 6¢ per call for any of the next 100 calls
>   plus 4¢ per call for any calls beyond 200

Draw a flowchart or write a pseudocode program with input the number of LOCAL calls and output CHARGE.

5.37   Three positive numbers a, b, c can be the lengths of the sides of a triangle provided each number is less than the sum of the other two, i.e.

$$a < b + c \qquad b < a + c \qquad c < a + b$$

Draw a flowchart or write a pseudocode program with input a, b, c and output 'YES' or 'NO', according as a triangle can or cannot be formed.

5.38   Find the outputs of the flowchart in Fig. 5-25 for the inputs (a) X = 15, Y = 20; (b) X = 15, Y = 5; (c) X = 2, Y = 18.

5.39   Write the pseudocode program corresponding to the flowchart in Fig. 5-25.

5.40   In which of the process symbols below can the two statements be interchanged without changing the resulting values of the variables?

```
C = A
D = B
```
(a)

```
C = A
C = B
```
(b)

```
C = A
D = A
```
(c)

```
C = A
A = B
```
(d)

```
C = A
D = C
```
(e)

```
C = A
A = C
```
(f)

5.41 Find the outputs of the following pseudocode program for the inputs (a) A = 15, B = 22; (b) A = 18, B = 7; (c) A = 9, B = 7; (d) A = 2, B = 5; (e) A = 6, B = 3.

```
read A, B
IF A < B
    IF A < 10
        X = A + B
    ELSE
        X = B − A
    ENDIF
ELSEIF B < 5
    X = A × B
ELSEIF A < 15
    X = A²
ELSE
    X = B²
ENDIF
write X
END
```

5.42 Draw the flowchart that corresponds to the pseudocode program of Problem 5.41.

LOOP STRUCTURES

5.43 With reference to Problem 5.8, draw a flowchart or write a pseudocode program which lists the names of the A students (those with scores of 90 or above) and finds the number and percentage of such students.

5.44 Write a pseudocode program using the DOWHILE structure that finds all pairs of positive integers m, n such that

$$m^2 + 2n^2 < 100$$

5.45 Draw a flowchart or write a pseudocode program which finds the largest and smallest of 25 distinct numbers.

5.46 Draw a flowchart or write a pseudocode program which finds the second-largest of 25 distinct numbers.

5.47 Suppose that y is the following function of t:

$$y = 2t^3 − t^2 − 37t + 36$$

Write a pseudocode program which outputs y corresponding to each t from −5 to 5 in steps of 0.25, and finds the maximum and the minimum of the calculated values of y.

5.48 Recall that $n!$ (read: n factorial) is defined for a positive integer n by

$$n! = 1 \times 2 \times 3 \times 4 \times \cdots \times (n − 1) \times n$$

Draw a flowchart using a DO loop, with input n and output $n!$.

5.49 Find the number of cycles and the value of the index K for each cycle if a DO loop is headed by:

(a) Do K = 1 to 20 by 3	(c) Do K = 7 to 5 by 2
(b) Do K = 4 to 9 by 2	(d) Do K = 15 to 10 by −2

Answers to Supplementary Problems

5.24   (*d*)

5.25   True.

5.26   (*a*) variable, (*b*) constant

5.30
(*a*) $X\uparrow2+Y\uparrow2$
(*b*) $(X+Y)\uparrow2$
(*c*) $2*A*B/(C+D)$
(*d*) $T\uparrow(N+1)$
(*e*) $X/(Y*Z)$
(*f*) $(X\uparrow2+2*X*Y-Y\uparrow2)\uparrow5$

5.31   (*a*) $T**(N+1)$, (*b*) $T**N*T$

5.32   (*a*) 27, (*b*) 16, (*c*) 11, (*d*) 16, (*e*) 48, (*f*) 15

5.33   (*a*) -5, (*b*) 23, (*c*) 307, (*d*) 59

5.34
```
read A, B, C
P = A + B + C
S = P/2
AREA = √S(S − A)(S − B)(S − C)
write P, AREA
END
```

5.35
```
read CODE, PRICE
IF CODE = 1
    COMMISSION = 0.03 × PRICE
    IF COMMISSION > 300
        COMMISSION = 300
    ENDIF
ELSEIF CODE = 2
    COMMISSION = 0.05 × PRICE
    IF COMMISSION < 75
        COMMISSION = 75
    ENDIF
ELSE
    COMMISSION = 0.06 × PRICE
ENDIF
write COMMISSION
END
```

5.36
```
read LOCAL
IF LOCAL > 200
    CHARGE = 14 + 0.04 × (LOCAL − 200)
ELSEIF LOCAL > 100
    CHARGE = 8 + 0.06 × (LOCAL − 100)
ELSE
    CHARGE = 8
ENDIF
write CHARGE
END
```

5.27   (*b*), (*c*), (*d*), (*e*), (*f*), (*h*), (*i*)

5.28   (*c*), (*d*), (*f*), (*h*)

5.29   (*b*), (*d*), (*f*), (*g*), (*h*)

5.37
```
read A, B, C
IF A < B + C and B < A + C and C < A + B
    write 'YES'
ELSE
    write 'NO'
ENDIF
END
```

5.38   (*a*) 55, (*b*) 60, (*c*) 63

5.39
```
read A, B
IF A < B
    A = A + 2 × B
    IF A < 20
        B = B + 10
    ELSE
        B = B + 5 × A
    ENDIF
ELSE
    A = A² + B
    B = B + 5 × A
ENDIF
write A, B
END
```

5.40   (*a*), (*c*)

5.41   (*a*) 7, (*b*) 49, (*c*) 81, (*d*) 7, (*e*) 18

5.42   See Fig. 5-29.

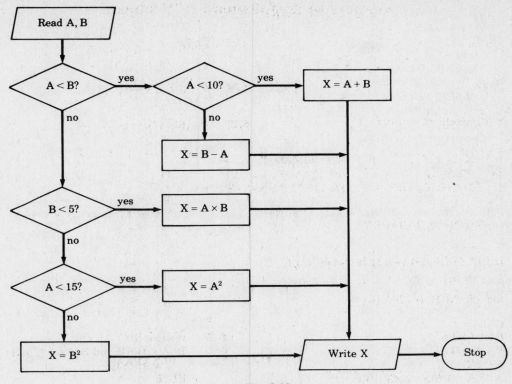

Fig. 5-29

5.43
```
N = 0
A = 0
read NAME, SCORE
DOUNTIL End of File
      IF SCORE ≥ 90
            write NAME
            A = A + 1
      ENDIF
      N = N + 1
      read next NAME, SCORE
ENDDO
PERCENT = 100 × A/N
write 'NUMBER OF A STUDENTS IS', A
write 'PERCENTAGE OF A STUDENTS IS', PERCENT
END
```

5.44
```
M = 1
N = 1
DOWHILE M ≤ 9
      DOWHILE N ≤ 7
            IF M² + 2 × N² < 100
                  write M, N
            ENDIF
            N = N + 1
      ENDDO
      M = M + 1
ENDDO
END
```

5.45   read LARGE
      SMALL = LARGE
      DO K = 1 to 24
          read A
          IF LARGE < A
             LARGE = A
          ELSEIF A < SMALL
             SMALL = A
          ENDIF
          N = N + 1
      ENDDO
      write LARGE, SMALL
      END

(Observe that we initialized both LARGE and SMALL with the first number; hence the loop is cycled only 24 times.)

5.46   read FIRST, SECOND
      IF FIRST < SECOND
          Interchange values in FIRST and SECOND
      ENDIF
      DO K = 1 to 23
          read A
          IF FIRST < A
             SECOND = FIRST
             FIRST = A
          ELSEIF SECOND < A
             SECOND = A
          ENDIF
      ENDDO
      write SECOND
      END

[Observe that FIRST and SECOND were initialized with the first two numbers; hence the loop is cycled only 23 times.   The macroinstruction in the third line could be expanded as in Fig. 5-24(b).]

5.47   T = −5
      MAX = $2 \times T^3 - T^2 - 37 \times T + 36$
      MIN = MAX
      DOWHILE T ≤ 5
          Y = $2 \times T^3 - T^2 - 37 \times T + 36$
          write T, Y
          IF MAX < Y
             MAX = Y
          ELSEIF Y < MIN
             MIN = Y
          ENDIF
          T = T + 0.25
      ENDDO
      write MIN, MAX
      END

(Observe that we initialized MAX and MIN with the first value of y.)

5.48   See Fig. 5-30.   Observe that we initialize PRODUCT at 1.

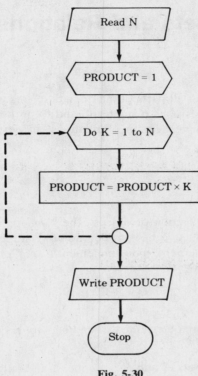

Fig. 5-30

5.49   (*a*) Seven times; for K = 1, 4, 7, 10, 13, 16, 19.   (*b*) Three times; for K = 4, 6, 8.   (*c*) Either once, for K = 7, or not at all, depending on the compiler.   (*d*) Three times; for K = 15, 13, 11.

Chapter 6

Sets and Relations

6.1 INTRODUCTION

The concept of a *set* appears in all parts of mathematics. As we shall see, properties of sets are very similar to properties of propositions (Chapter 4) and to properties of *logic circuits* (Chapter 7). The chapter concludes with an investigation of relations; in particular, *equivalence relations* and *functions*.

6.2 SETS AND ELEMENTS

A *set* may be viewed as a collection of objects, the *elements* or *members* of the set. We will ordinarily use capital letters, A, B, X, Y, ..., to denote sets, and lowercase letters, a, b, x, y, ..., to denote elements of sets. The statement "p is an element of A," or, equivalently, "p belongs to A," is written

$$p \in A$$

The negation of $p \in A$ is written $p \notin A$.

The fact that a set is completely determined when its members are specified is formally stated as the principle of extension.

Principle of Extension: Two sets A and B are equal if and only if they have the same members.

As usual, we write $A = B$ if the sets A and B are equal, and we write $A \neq B$ if the sets are not equal.

There are essentially two ways to specify a particular set. One way, if it is possible, is to list its members. For example,

$$A = \{a, e, i, o, u\}$$

denotes the set A whose elements are the letters a, e, i, o, u. Note that the elements are separated by commas and enclosed in braces { }. The second way is to state the property which characterizes the elements in the set. For example,

$$B = \{x : x \text{ is an integer}, x > 0\}$$

which reads "B is the set of x such that x is an integer and x is greater than 0," denotes the set B whose elements are the positive integers. A letter, usually x, is used to denote a typical member of the set; the colon is read as "such that" and the comma as "and."

EXAMPLE 6.1

(a) The set A above can also be written as

$$A = \{x : x \text{ is a letter in the English alphabet}, x \text{ is a vowel}\}$$

Observe that $b \notin A$, $e \in A$ and $p \notin A$.

(b) We could not list all the elements of the above set B, although frequently we specify the set by writing

$$B = \{1, 2, 3, \ldots\}$$

where we assume that everyone knows what we mean. Observe that $8 \in B$ but $-6 \notin B$.

(c) Let $E = \{x : x^2 - 3x + 2 = 0\}$. In other words, E consists of those numbers which are solutions of the equation $x^2 - 3x + 2 = 0$, sometimes called the *solution set* of the given equation. Since the solutions of the equation are 1 and 2, we could also write $E = \{1, 2\}$.

(d) Let $E = \{x : x^2 - 3x + 2 = 0\}$, $F = \{2, 1\}$ and $G = \{1, 2, 2, 1, 6/3\}$. Then $E = F = G$. Observe that a set does not depend on the way in which its elements are displayed. A set remains the same if its elements are repeated or rearranged.

Even if we can list the elements of a set, it may not be practical to do so. For example, we would not list the members of the set of people born in the world during the year 1976 although theoretically it is possible to compile such a list. That is, we describe a set by listing its elements only if the set contains a few elements; otherwise we describe a set by the property which characterizes its elements.

The fact that we can describe a set in terms of a property is formally stated as the *principle of abstraction*.

Principle of Abstraction: Given any set U and any property P, there is a set A such that the elements of A are exactly those members of U which have the property P.

6.3 UNIVERSAL SET, EMPTY SET

In the application of the theory of sets, the members of all sets under investigation usually belong to some fixed large set called the *universal set* or *universe of discourse*. For example, in plane geometry, the universal set consists of all the points in the plane; and in human population studies the universal set consists of all the people in the world. We will let the symbol

$$U$$

denote the universal set unless otherwise stated or implied.

For a given set U and a property P, there may not be any elements of U which have property P. For example, the set

$$S = \{x : x \text{ is a positive integer, } x^2 = 3\}$$

has no elements since no positive integer has the required property.

The set with no elements is called the *empty set* or *null set* and is denoted by

$$\emptyset$$

From the principle of extension, it follows that there is only one empty set. In other words, if S and T are both empty, then $S = T$ since they have exactly the same elements, namely, none.

6.4 SUBSETS

If every element of a set A is also an element of a set B, then A is called a *subset* of B. We also say that A is *contained in B* or that *B contains A*. This relationship is written

$$A \subset B \qquad \text{or} \qquad B \supset A$$

If A is not a subset of B, i.e. if at least one element of A does not belong to B, we write $A \not\subset B$ or $B \not\supset A$.

EXAMPLE 6.2

(a) Consider the sets

$$A = \{1, 3, 4, 5, 8, 9\} \qquad B = \{1, 2, 3, 5, 7\} \qquad C = \{1, 5\}$$

Then $C \subset A$ and $C \subset B$ since 1 and 5, the elements of C, are also members of A and B. But $B \not\subset A$ since some of its elements, e.g. 2 and 7, do not belong to A. Furthermore, since the elements of A, B and C must also belong to the universal set U, we have that U must at least contain the set $\{1, 2, 3, 4, 5, 7, 8, 9\}$.

(b) Some sets of numbers occur very often and so we use special symbols for them. Unless otherwise specified, we will let:

$\mathbf{N}$ = the set of positive integers: 1, 2, 3, . . .

$\mathbf{Z}$ = the set of integers: . . . , −2, −1, 0, 1, 2, . . .

$\mathbf{Q}$ = the set of rational numbers

$\mathbf{R}$ = the set of real numbers

The above sets are related as follows:

$$\mathbf{N} \subset \mathbf{Z} \subset \mathbf{Q} \subset \mathbf{R}$$

(c) The set $E = \{2, 4, 6\}$ is a subset of the set $F = \{6, 2, 4\}$, since each number 2, 4 and 6 belonging to E also belongs to F. In fact, $E = F$. In a similar manner it can be shown that every set is a subset of itself.

Every set A is a subset of the universal set U since, by definition, all the members of A belong to U. Also the empty set $\emptyset$ is a subset of A.

As noted above, every set A is a subset of itself since, trivially, the elements of A belong to A.

If every element of a set A belongs to a set B, and every element of B belongs to a set C, then clearly every element of A belongs to C. In other words, if $A \subset B$ and $B \subset C$, then $A \subset C$.

If $A \subset B$ and $B \subset A$, then A and B have the same elements, i.e. $A = B$. Conversely, if $A = B$, then $A \subset B$ and $B \subset A$ since every set is a subset of itself.

We state the above results formally:

Theorem 6.1: (i) For any set A, we have $\emptyset \subset A \subset U$.

(ii) For any set A, we have $A \subset A$.

(iii) If $A \subset B$ and $B \subset C$, then $A \subset C$.

(iv) $A = B$ if and only if $A \subset B$ and $B \subset A$.

Remark: If $A \subset B$, it is still possible that $A = B$. Some authors write $A \subseteq B$ to indicate that A is a subset of B, and write $A \subset B$ to indicate that A is a subset of B but is not equal to B.

6.5 VENN DIAGRAMS

A *Venn diagram* is a pictorial representation of sets by sets of points in the plane. The universal set U is represented by the interior of a rectangle, and the other sets are represented by disks lying within the rectangle. If $A \subset B$, then the disk representing A will be entirely within the disk representing B as in Fig. 6-1(a). If A and B are *disjoint*, i.e. have no elements in common, then the disk representing A will be separated from the disk representing B as in Fig. 6-1(b).

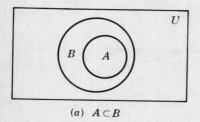

(a) $A \subset B$

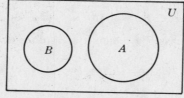

(b) A and B are disjoint.

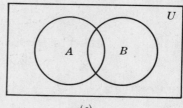

(c)

Fig. 6-1

However, if A and B are two arbitrary sets, it is possible that some objects are in A but not B, some are in B but not A, some are in both A and B, and some are in neither A nor B; hence in general we represent A and B as in Fig. 6-1(c).

Many verbal statements can be translated into equivalent statements about sets which can be described by Venn diagrams. Hence Venn diagrams can very often be used to determine whether an argument is valid (Section 4.9).

EXAMPLE 6.3 Examine the following argument adapted from a book on logic by Lewis Carroll, the author of *Alice in Wonderland*:

S_1:   My saucepans are the only things I have that are made of tin.
S_2:   I find all your presents very useful.
S_3:   None of my saucepans is of the slightest use.

. . . .

S: Your presents to me are not made of tin.

By S_1 the tin objects are contained in the set of saucepans; hence draw the Venn diagram of Fig. 6-2.

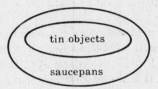

Fig. 6-2

By S_3 the set of saucepans and the set of useful things are disjoint; hence draw Fig. 6-3.

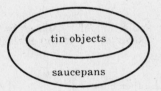 

Fig. 6-3

By S_2 the set of "your presents" is a subset of the set of useful things; hence draw Fig. 6-4.

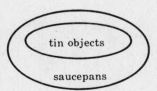 

Fig. 6-4

According to Fig. 6-4, the conclusion S indeed follows from the premises, because the set of "your presents" is disjoint from the set of tin objects. The given argument is therefore valid.

6.6 UNION AND INTERSECTION

The *union* of two sets A and B, denoted by $A \cup B$, is the set of all elements which belong to A or to B:

$$A \cup B = \{x : x \in A \text{ or } x \in B\}$$

Here "or" is used in the sense of and/or.   Figure 6-5(a) is a Venn diagram in which $A \cup B$ is shaded.

The *intersection* of two sets A and B, denoted by $A \cap B$, is the set of elements which belong to both A and B:

$$A \cap B = \{x : x \in A, x \in B\}$$

Figure 6-5(b) is a Venn diagram in which $A \cap B$ is shaded.

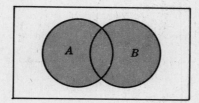
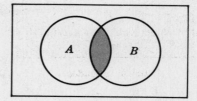

(a) $A \cup B$ is shaded.   (b) $A \cap B$ is shaded.

Fig. 6-5

EXAMPLE 6.4

(a) Let $A = \{1, 2, 3, 4\}$, $B = \{3, 4, 5, 6, 7\}$, $C = \{2, 3, 5, 7\}$.   Then

$$A \cup B = \{1, 2, 3, 4, 5, 6, 7\} \qquad A \cap B = \{3, 4\}$$
$$A \cup C = \{1, 2, 3, 4, 5, 7\} \qquad A \cap C = \{2, 3\}$$
$$B \cup C = \{2, 3, 4, 5, 6, 7\} \qquad B \cap C = \{3, 5, 7\}$$

(b) Let M denote the set of male students in a university C, and let F denote the set of female students in C.   Then

$$M \cup F = C$$

since each student in C belongs to either M or F.   On the other hand,

$$M \cap F = \emptyset$$

since no student belongs to both M and F.

The operation of set inclusion is closely related to the operations of union and intersection, as shown by the following theorem.

Theorem 6.2:   The following are equivalent: $A \subset B$, $A \cap B = A$, $A \cup B = B$.

6.7 COMPLEMENTS

Recall that all sets under consideration at a particular time are subsets of a fixed universal set U. The *absolute complement* or, simply, *complement* of a set A, denoted by A^c, is the set of elements which belong to U but which do not belong to A:

$$A^c = \{x : x \in U, x \notin A\}$$

Some texts denote the complement of A by A' or $\bar{A}$. Figure 6-6(a) is a Venn diagram in which A^c is shaded.

The *relative complement* of a set B with respect to a set A or, simply, the *difference* of A and B, denoted by $A \setminus B$, is the set of elements which belong to A but which do not belong to B:

$$A \setminus B = \{x : x \in A, x \notin B\}$$

The set $A \setminus B$ is read "A minus B". Many texts denote $A \setminus B$ by $A - B$ or $A \sim B$. Figure 6-6(b) is a Venn diagram in which $A \setminus B$ is shaded.

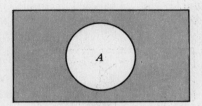

(a) A^c is shaded.

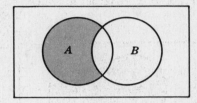

(b) $A \setminus B$ is shaded.

Fig. 6-6

EXAMPLE 6.5

(a) Let $U = \{1, 2, 3, \ldots\}$, the positive integers, be the universal set. Let $A = \{1, 2, 3, 4\}$, $B = \{3, 4, 5, 6, 7\}$, and let $E = \{2, 4, 6, 8, \ldots\}$, the even numbers. Then

$$A^c = \{5, 6, 7, 8, \ldots\} \qquad B^c = \{1, 2, 8, 9, 10, \ldots\}$$

and $E^c = \{1, 3, 5, 7, \ldots\}$, the odd numbers.

(b) We show by means of Venn diagrams that the complement of the union of two sets, A and B, is equal to the intersection of the complements of A and B; that is,

$$(A \cup B)^c = A^c \cap B^c$$

From the Venn diagram for $A \cup B$, in Fig. 6-5(a), we see that $(A \cup B)^c$ is represented by the shaded area in Fig. 6-7(a). To find $A^c \cap B^c$, the area in both A^c and B^c, we shade A^c with strokes in one direction and B^c with strokes in another direction as in Fig. 6-7(b). Then $A^c \cap B^c$ is represented by the cross-hatched area which is shaded in Fig. 6-7(c). Since $(A \cup B)^c$ and $A^c \cap B^c$ are represented by the same area, they are equal. This property of sets is called *DeMorgan's law*.

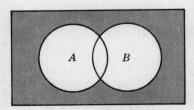

(a) $(A \cup B)^c$ is shaded.

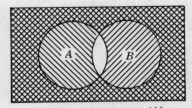

(b) A^c is shaded with ////.
B^c is shaded with \\\\\\.

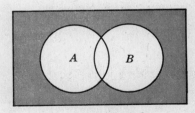

(c) $A^c \cap B^c$ is shaded.

Fig. 6-7

6.8   ALGEBRA OF SETS; DUALITY

Sets, under the above operations of union, intersection, and complement, satisfy the laws listed in Table 6-1.   Observe the similarity between these laws and those of the algebra of propositions listed in Table 4-1—a similarity rooted in the analogy between the set operations $\cup$, $\cap$, and complement, and the logical connectives $\vee$, $\wedge$, and negation.

Table 6-1.   Laws of the Algebra of Sets

Idempotent Laws	
$1a.\quad A \cup A = A$	$1b.\quad A \cap A = A$
Associative Laws	
$2a.\quad (A \cup B) \cup C = A \cup (B \cup C)$	$2b.\quad (A \cap B) \cap C = A \cap (B \cap C)$
Commutative Laws	
$3a.\quad A \cup B = B \cup A$	$3b.\quad A \cap B = B \cap A$
Distributive Laws	
$4a.\quad A \cup (B \cap C) = (A \cup B) \cap (A \cup C)$	$4b.\quad A \cap (B \cup C) = (A \cap B) \cup (A \cap C)$
Identity Laws	
$5a.\quad A \cup \emptyset = A$	$5b.\quad A \cap U = A$
$6a.\quad A \cup U = U$	$6b.\quad A \cap \emptyset = \emptyset$
Involution Law	
$7.\quad (A^c)^c = A$	
Complement Laws	
$8a.\quad A \cup A^c = U$	$8b.\quad A \cap A^c = \emptyset$
$9a.\quad U^c = \emptyset$	$9b.\quad \emptyset^c = U$
DeMorgan's Laws	
$10a.\quad (A \cup B)^c = A^c \cap B^c$	$10b.\quad (A \cap B)^c = A^c \cup B^c$

The reader may have wondered why the identities in Table 6-1 are arranged in pairs, as, for example, 2a and 2b.   We now consider the principle behind this arrangement.   Suppose E is an identity of set algebra.   The *dual* E^* of E is the expression obtained by replacing each occurrence of $\cup$, $\cap$, U and $\emptyset$ in E by $\cap$, $\cup$, $\emptyset$ and U respectively.   For example, the dual of

$$(U \cap A) \cup (B \cap A) = A \quad \text{is} \quad (\emptyset \cup A) \cap (B \cup A) = A$$

Observe that the pairs of laws in Table 6-1 are duals of each other.   It is a fact of set algebra, called the *principle of duality*, that if E is an identity then its dual E^* is also an identity.

6.9   FINITE SETS, COUNTING PRINCIPLE

A set is said to be *finite* if it contains exactly m distinct elements where m denotes some nonnegative integer.   Otherwise, a set is said to be *infinite*.   For example, the empty set $\emptyset$ and the set of letters of the English alphabet are finite sets, whereas the set of even positive integers, $\{2, 4, 6, \ldots\}$, is infinite.

If a set A is finite, we let $n(A)$ denote the number of elements of A.   Some texts use $\#(A)$ instead of $n(A)$.

Lemma: If A and B are disjoint finite sets, then $A \cup B$ is finite and

$$n(A \cup B) = n(A) + n(B)$$

Proof. In counting the elements of $A \cup B$, first count those that are in A. There are $n(A)$ of these. The only other elements of $A \cup B$ are those that are in B but not in A. But since A and B are disjoint, no element of B is in A, so there are $n(B)$ elements that are in B but not in A. Therefore, $n(A \cup B) = n(A) + n(B)$.

We also have a formula for $n(A \cup B)$ even when A and B are not disjoint.

Theorem 6.3: If A and B are finite sets, then $A \cup B$ and $A \cap B$ are finite and

$$n(A \cup B) = n(A) + n(B) - n(A \cap B)$$

We can apply this result to obtain a similar formula for any finite number, k, of finite sets. Thus, for $k = 3$, we have

Corollary 6.4: If A, B and C are finite sets, then so is $A \cup B \cup C$, and

$$n(A \cup B \cup C) = n(A) + n(B) + n(C) - n(A \cap B) - n(A \cap C) - n(B \cap C) + n(A \cap B \cap C)$$

EXAMPLE 6.6 Suppose that 100 of the 120 mathematics students at a college take at least one of the languages French, German, and Russian. Also suppose:

> 65 study French
> 45 study German
> 42 study Russian
> 20 study French and German
> 25 study French and Russian
> 15 study German and Russian

Let F, G and R denote the sets of students studying French, German and Russian respectively. We wish to find the number of students who study all three languages, and to fill in the correct number of students in each of the eight regions of the Venn diagram shown in Fig. 6-8(a).

By Corollary 6.4,

$$n(F \cup G \cup R) = n(F) + n(G) + n(R) - n(F \cap G) - n(F \cap R) - n(G \cap R) + n(F \cap G \cap R)$$

Now, $n(F \cup G \cup R) = 100$, because 100 of the students study at least one of the languages. Substituting,

$$100 = 65 + 45 + 42 - 20 - 25 - 15 + n(F \cap G \cap R)$$

and so, $n(F \cap G \cap R) = 8$, i.e. 8 students study all three languages.

We now use this result to fill in the Venn diagram. We have:

> 8 study all three languages
>
> $20 - 8 = 12$ study French and German but not Russian
> $25 - 8 = 17$ study French and Russian but not German
> $15 - 8 = 7$ study German and Russian but not French
>
> $65 - 12 - 8 - 17 = 28$ study only French
> $45 - 12 - 8 - 7 = 18$ study only German
> $42 - 17 - 8 - 7 = 10$ study only Russian
>
> $120 - 100 = 20$ do not study any of the languages

Thus the completed diagram is Fig. 6-8(b). Observe that

$$28 + 18 + 10 = 56$$

students study exactly one of the three languages.

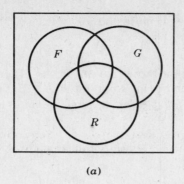
(a)

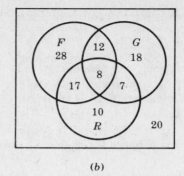
(b)

Fig. 6-8

6.10 CLASSES OF SETS, POWER SETS, PARTITIONS

Given a set S, we might wish to talk about some of its subsets. Thus we would be considering a set of sets. Whenever such a situation occurs, to avoid confusion we will speak of a *class* of sets or *collection* of sets rather than a set of sets. If we wish to consider some of the sets in a given class of sets, then we speak of a *subclass* or *subcollection*.

EXAMPLE 6.7 Suppose $S = \{1, 2, 3, 4\}$. Let $\mathcal{A}$ be the class of subsets of S which contain exactly three elements of S. Then

$$\mathcal{A} = [\{1, 2, 3\}, \{1, 2, 4\}, \{1, 3, 4\}, \{2, 3, 4\}]$$

The elements of $\mathcal{A}$ are the sets $\{1, 2, 3\}$, $\{1, 2, 4\}$, $\{1, 3, 4\}$, and $\{2, 3, 4\}$.

Let $\mathcal{B}$ be the class of subsets of S which contain 2 and two other elements of S. Then

$$\mathcal{B} = [\{1, 2, 3\}, \{1, 2, 4\}, \{2, 3, 4\}]$$

The elements of $\mathcal{B}$ are the sets $\{1, 2, 3\}$, $\{1, 2, 4\}$, and $\{2, 3, 4\}$. Thus $\mathcal{B}$ is a subclass of $\mathcal{A}$, since every element of $\mathcal{B}$ is also an element of $\mathcal{A}$. (To avoid confusion, we will sometimes enclose the sets of a class in brackets instead of braces.)

For a given set S, we may speak of the collection of all subsets of S. This collection is called the *power set* of S, and we will denote it by $\mathscr{P}(S)$. For example, if $S = \{1, 2, 3\}$, then

$$\mathscr{P}(S) = [\varnothing, \{1\}, \{2\}, \{3\}, \{1, 2\}, \{1, 3\}, \{2, 3\}, S]$$

Note that the empty set $\varnothing$ belongs to $\mathscr{P}(S)$, since $\varnothing$ is a subset of S. Similarly, S belongs to $\mathscr{P}(S)$. Note also that $\mathscr{P}(S)$ has $2^3 = 8$ elements. This is true in general; that is, if S is finite then $\mathscr{P}(S)$ contains

$$2^{n(S)}$$

elements. For this reason the power set of S is sometimes denoted by 2^S.

Now let S be a nonempty set. A partition of S is a subdivision of S into nonoverlapping, nonempty subsets. Precisely, a *partition* of S is a collection $\{A_i\}$ of nonempty subsets of S such that:

(i)   Each a in S belongs to one of the A_i.

(ii) The sets of $\{A_i\}$ are mutually disjoint; that is, if

$$A_i \neq A_j \text{ then } A_i \cap A_j = \varnothing.$$

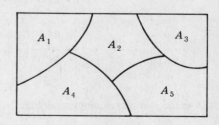

Fig. 6-9

The subsets in a partition are called *cells*. Figure 6-9 is a Venn diagram of a partition of the rectangular set S of points into five cells, A_1, A_2, A_3, A_4, and A_5.

EXAMPLE 6.8 Consider the following collections of subsets of $S = \{1, 2, \ldots, 8, 9\}$:

(i)   [{1, 3, 5}, {2, 6}, {4, 8, 9}]
(ii)   [{1, 3, 5}, {2, 4, 6, 8}, {5, 7, 9}]
(iii) [{1, 3, 5}, {2, 4, 6, 8}, {7, 9}]

Then (i) is not a partition of S since 7 in S does not belong to any of the subsets. Furthermore, (ii) is not a partition of S since {1, 3, 5} and {5, 7, 9} are not disjoint. On the other hand, (iii) is a partition of S.

6.11 ORDERED PAIRS, PRODUCT SETS

The order of the elements of a set with two elements is immaterial, e.g.

$$\{3, 5\} = \{5, 3\}$$

On the other hand, an *ordered pair* consists of two elements, of which one is designated as the first element and the other as the second element. Such an ordered pair is written (a, b), where a is the first element and b is the second element. Two ordered pairs (a, b) and (c, d) are equal if and only if $a = c$ and $b = d$.

Consider two arbitrary sets A and B. The set of all ordered pairs (a, b) where $a \in A$ and $b \in B$ is called the *product*, or *cartesian product*, of A and B. A short designation of this product is $A \times B$, which is read "A cross B". By definition

$$A \times B = \{(a, b) : a \in A, b \in B\}$$

One frequently writes A^2 instead of $A \times A$.

EXAMPLE 6.9

(a) **R** denotes the set of real numbers and so $\mathbf{R}^2 = \mathbf{R} \times \mathbf{R}$ is the set of ordered pairs of real numbers. The reader is familiar with the geometrical representation of $\mathbf{R}^2$ as points in the plane as in Fig. 6-10. Here each point P represents an ordered pair (a, b) of real numbers and vice versa; the vertical line through P meets the x axis at a, and the horizontal line through P meets the y axis at b. $\mathbf{R}^2$ is frequently called the *cartesian plane*.

(b) Let $A = \{1, 2, 3\}$ and $B = \{a, b\}$. Then

$$A \times B = \{(1, a), (1, b), (2, a), (2, b), (3, a), (3, b)\}$$

Since A and B do not contain many elements, it is possible to represent $A \times B$ by a coordinate diagram as shown in Fig. 6-11. Here the vertical lines through the points of A and the horizontal lines through the points of B meet in 6 points which represent $A \times B$ in the obvious way. The point P is the ordered pair $(2, b)$.

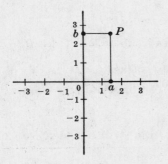

Fig. 6-10

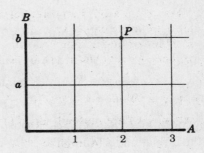

Fig. 6-11

Observe in Example 6.9(b) that

$$n(A \times B) = 6 = 2 \cdot 3 = n(A) \cdot n(B)$$

where $n(A)$ = number of elements in A. In fact, $n(A \times B) = n(A) \cdot n(B)$ for any finite sets A and B. This follows from the observation that, for an ordered pair (a, b) in $A \times B$, there are $n(A)$ possibilities for a, and for each of these there are $n(B)$ possibilities for b. In particular, if A or B is empty, then $A \times B$ is empty.

The idea of a product of sets can be extended to any finite number of sets. For any sets $A_1, A_2, \ldots, A_n$, the set of all ordered n-tuples $(a_1, a_2, \ldots, a_n)$, where $a_1 \in A_1, a_2 \in A_2, \ldots, a_n \in A_n$, is called the *product* of the sets $A_1, \ldots, A_n$ and is denoted by

$$A_1 \times A_2 \times \cdots \times A_n \qquad \text{or} \qquad \prod_{i=1}^{n} A_i$$

Just as we write A^2 instead of $A \times A$, so we write A^n instead of $A \times A \times \cdots \times A$ where there are n factors all equal to A. For example, $\mathbf{R}^3 = \mathbf{R} \times \mathbf{R} \times \mathbf{R}$ denotes the usual three-dimensional space.

6.12   RELATIONS

Let A and B be sets. A *binary relation*, R, *from A to B* assigns to each ordered pair (a, b) in $A \times B$ exactly one of the following statements:

(i)   "a is related to b", written $a\,R\,b$
(ii) "a is not related to b", written $a\,\not\!R\,b$

A relation from a set A to the same set A is called a relation *on A*.

Since we will deal mainly with binary relations, the word "relation" will mean binary relation unless otherwise specified.

EXAMPLE 6.10

(a)   Set inclusion is a relation on any class of sets. For, given any pair of sets A and B, either $A \subset B$ or $A \not\subset B$.

(b)   Marriage is a relation from the set M of men to the set W of women. For, given any man $m \in M$ and any woman $w \in W$, either m is married to w or m is not married to w.

(c)   Order, symbolized by "$<$", or, equivalently, the sentence "x is less than y", is a relation on any set of real numbers. For, given any ordered pair (a, b) of real numbers, either

$$a < b \qquad \text{or} \qquad a \not< b$$

(d)   A familiar relation on the set $\mathbf{Z}$ of integers is "m divides n". A common notation for this relation is to write $m \,|\, n$ when m divides n. Thus $6\,|\,30$ but $7 \nmid 25$.

(e)   Perpendicularity is a relation on the set of lines in the plane. For, given any pair of lines a and b, either a is perpendicular to b or a is not perpendicular to b.

Now any relation R from a set A to a set B uniquely defines a subset R^* of $A \times B$ as follows:

$$R^* = \{(a, b) : a \text{ is related to } b\} = \{(a, b) : a\,R\,b\}$$

Conversely, any subset R^* of $A \times B$ uniquely defines a relation R from A to B as follows:

$$a\,R\,b \text{ whenever } (a, b) \in R^*$$

In view of this one-to-one correspondence between relations R from A to B and subsets of $A \times B$, we redefine a relation as follows:

Definition: A relation R from A to B is a subset of $A \times B$.

The *domain* of a relation R is the set of all first elements of the ordered pairs which belong to R, and the *range* of R is the set of second elements.

EXAMPLE 6.11

(a) Let $A = \{1, 2, 3\}$, and $R = \{(1, 2), (1, 3), (3, 2)\}$. Then R is a relation on A since it is a subset of $A \times A$. With respect to this relation,

$$1\,R\,2, \quad 1\,R\,3, \quad 3\,R\,2, \quad \text{but} \quad 1\,\not R\,1, \quad 2\,\not R\,1, \quad 2\,\not R\,2, \quad 2\,\not R\,3, \quad 3\,\not R\,1, \quad 3\,\not R\,3$$

The domain of R is $\{1, 3\}$ and the range of R is $\{2, 3\}$.

(b) Let $A = \{$eggs, milk, corn$\}$ and $B = \{$cows, goats, hens$\}$. We can define a relation R from A to B by $(a, b) \in R$ if a is produced by b. In other words,

$$R = \{(\text{eggs, hens}), (\text{milk, cows}), (\text{milk, goats})\}$$

With respect to this relation,

$$\text{eggs } R \text{ hens}, \quad \text{milk } R \text{ cows}, \quad \text{etc.}$$

(c) Suppose we say that two countries are *adjacent* if they have some part of their boundaries in common. Then "is adjacent to" is a relation R on the countries of the earth. Thus

$$(\text{Italy, Switzerland}) \in R \quad \text{but} \quad (\text{Canada, Mexico}) \notin R$$

(d) Let A be any set. An important relation on A is that of *equality*,

$$\{(a, a) : a \in A\}$$

which is usually denoted by "$=$". This relation is also called the *identity relation* on A.

(e) Again let A be any set. Then $A \times A$ and the empty set $\varnothing$ are also relations on A, since they are subsets of $A \times A$. They are called the *universal relation* and the *empty relation*, respectively.

Let R be any relation from A to B. The *inverse* of R, denoted by R^{-1}, is the relation from B to A which consists of those ordered pairs which when reversed belong to R:

$$R^{-1} = \{(b, a) : (a, b) \in R\}$$

For example, the inverse of the relation $R = \{(1, 2), (1, 3), (2, 3)\}$ on $A = \{1, 2, 3\}$ is

$$R^{-1} = \{(2, 1), (3, 1), (3, 2)\}$$

(also on A). Clearly, if R is any relation, then $(R^{-1})^{-1} = R$. Also, the domain of R^{-1} equals the range of R, and the range of R^{-1} equals the domain of R.

6.13 PICTORIAL REPRESENTATIONS OF RELATIONS

We first consider a relation S on the set $\mathbf{R}$ of real numbers, i.e. S is a subset of $\mathbf{R}^2 = \mathbf{R} \times \mathbf{R}$. Since $\mathbf{R}^2$ can be represented by the set of points in the plane, we can picture S by emphasizing those points in the plane which belong to S. The pictorial representation of the relation is sometimes called the *graph* of the relation.

Frequently, the relation S consists of all ordered pairs of real numbers which satisfy some given equation

$$E(x, y) = 0$$

We usually identify the relation with the equation, i.e. we speak of "the relation $E(x, y) = 0$." For example, consider the equation $x^2 + y^2 = 25$. The relation consists of all ordered pairs (x_0, y_0) that satisfy the equation. The graph of the relation is the circle with center at the origin and radius 5, which is drawn in Fig. 6-12.

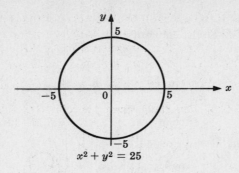

$$x^2 + y^2 = 25$$

Fig. 6-12

Next we consider a relation R from A to B, where A and B are finite sets.   There are a number of ways of picturing such relations:

(1)   Draw the coordinate diagram of $A \times B$ as in Fig. 6-11.

(2)   Form a rectangular array whose rows are labeled by the elements of A and whose columns are labeled by the elements of B.   Put a 1 or 0 in each position of the array according as $a \in A$ is or is not related to $b \in B$.   This array is called the *matrix* of the relation.

(3)   Write down the elements of A and the elements of B in two disjoint disks, and then draw an arrow from $a \in A$ to $b \in B$ whenever a is related to b.   This picture will be called the *arrow diagram* of the relation.

EXAMPLE 6.12

(*a*)   Let R be the following relation from $A = \{1, 2, 3\}$ to $B = \{a, b\}$:

$$R = \{(1, a), (1, b), (3, a)\}$$

Figure 6-13 pictures this relation in the above three ways.

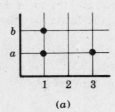   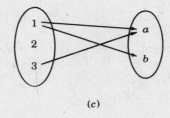

	a	b
1	1	1
2	0	0
3	1	0

(*a*)   (*b*)   (*c*)

Fig. 6-13

(*b*)   Figure 6-14 pictures the relation of Example 6.11(*b*) in the three ways.

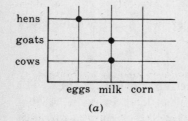   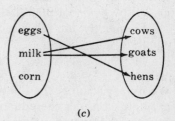

	cows	goats	hens
eggs	0	0	1
milk	1	1	0
corn	0	0	0

(*a*)   (*b*)   (*c*)

Fig. 6-14

There is another way of picturing a relation when it is from a finite set to itself. We write down the elements of the set and then draw an arrow from an element x to an element y whenever x is related to y. This diagram is called the *directed graph* of the relation. Figure 6-15 gives the directed graph of the following relation R on the set $A = \{1, 2, 3, 4\}$:

$$R = \{(1, 2), (2, 2), (2, 4), (3, 2), (3, 4), (4, 1), (4, 3)\}$$

Observe that there is an arrow from 2 to itself, since 2 is related to 2 under R.

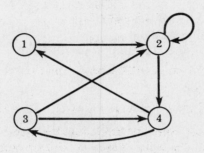

Fig. 6-15

6.14 EQUIVALENCE RELATIONS

A relation $\sim$ on a set S is called an *equivalence relation* if it has the following three properties:

(1)   For each a in S, we have $a \sim a$.

(2)   If $a \sim b$, then $b \sim a$.

(3)   If $a \sim b$ and $b \sim c$, then $a \sim c$.

The first property is called *reflexive*, the second property is called *symmetric*, and the third property is called *transitive*. Thus a relation on a set is an equivalence relation if it is reflexive, symmetric, and transitive.

The general idea behind an equivalence relation is that it is a classification of objects which are in some way "alike." In fact, the relation $=$ of equality on any set S is an equivalence relation; that is, (1) $a = a$ for every a in S; (2) if $a = b$, then $b = a$; and (3) if $a = b$ and $b = c$, then $a = c$. Other equivalence relations follow.

EXAMPLE 6.13

(*a*)   Consider the set L of lines and the set T of triangles in the Euclidean plane. The relation "is parallel to or identical to" is an equivalence relation on L, and congruence and similarity are equivalence relations on T.

(*b*)   The classification of animals by species, that is, the relation "is of the same species as," is an equivalence relation on the set of animals.

(*c*)   The relation $\subset$ of set inclusion is not an equivalence relation. It is reflexive and transitive, but it is not symmetric since $A \subset B$ does not imply $B \subset A$.

(*d*)   Let m be a fixed positive integer. Two integers a and b are said to be *congruent modulo m*, written

$$a \equiv b \ (\text{mod } m)$$

if m divides $a - b$. For example, for $m = 4$ we have $11 \equiv 3 \ (\text{mod } 4)$ since 4 divides $11 - 3$, and $22 \equiv 6 \ (\text{mod } 4)$ since 4 divides $22 - 6$. This relation of congruence modulo m is an equivalence relation.

Suppose $\sim$ is an equivalence relation on a set S. For each a in S, let $[a]$ denote the set of elements to which a is related:

$$[a] = \{x : a \sim x\}$$

We call $[a]$ the *equivalence class* of a in S. The collection of all equivalence classes of elements of S, denoted by $S/\sim$, is called the *quotient* of S by $\sim$. The fundamental property of a quotient set is contained in the following theorem.

Theorem 6.5:   Let $\sim$ be an equivalence relation on a set S. Then the quotient set $S/\sim$ is a partition of S. Specifically:
   (i)   For each a in S, we have $a \in [a]$.
   (ii) $[a] = [b]$ if and only if $a \sim b$.
   (iii) If $[a] \neq [b]$, then $[a]$ and $[b]$ are disjoint.

EXAMPLE 6.14

(a) Let $S = \{1, 2, 3\}$. The relation

$$R = \{(1, 1), (1, 2), (2, 1), (3, 3)\}$$

is an equivalence relation on S. Under the relation R,

$$[1] = \{1, 2\}, \qquad [2] = \{1, 2\} \qquad \text{and} \qquad [3] = \{3\}$$

Observe that $[1] = [2]$ and $\{[1], [3]\}$ is a partition of S.

(b) Let R_5 be the relation on $\mathbf{Z}$, the set of integers, defined by

$$x \equiv y \ (\text{mod } 5)$$

[See Example 6.13(d).] There are exactly five distinct equivalence classes in $\mathbf{Z}/R_5$:

$$A_0 = \{\ldots, -10, -5, 0, 5, 10, \ldots\}$$
$$A_1 = \{\ldots, -9, -4, 1, 6, 11, \ldots\}$$
$$A_2 = \{\ldots, -8, -3, 2, 7, 12, \ldots\}$$
$$A_3 = \{\ldots, -7, -2, 3, 8, 13, \ldots\}$$
$$A_4 = \{\ldots, -6, -1, 4, 9, 14, \ldots\}$$

Observe that each integer x, which is uniquely expressible in the form $x = 5q + r$ where $0 \leq r < 5$, is a member of the equivalence class A_r, where r is the remainder. Note that the equivalence classes are pairwise disjoint and that

$$\mathbf{Z} = A_0 \cup A_1 \cup A_2 \cup A_3 \cup A_4$$

6.15   FUNCTIONS

Suppose that to each element of a set A there is assigned a unique element of a set B; the collection of such assignments is called a *function* from A into B. The set A is called the *domain* of the function, and the set B is called the *codomain*.

One ordinarily uses a symbol to denote a function. For example, let f denote a function from A into B. Then we write

$$f : A \to B$$

which is read: "f is a function from A into B," or "f takes (or maps) A into B." If $a \in A$, then we let $f(a)$ (read "f of a") denote the unique element of B which f assigns to a; it is called the *image* of a under f, or the *value* of f at a. The set of all image values is called the *range* or *image* of f.

Frequently, a function can be expressed by means of a mathematical formula. For example, consider the function which sends each real number into its square. We may describe this function by writing

$$f(x) = x^2$$

Here x is called a *variable*, and the letter f denotes the function.

Remark: Whenever a function is given by such a formula in terms of a variable x, we assume, unless it is otherwise stated, that the domain of the function is **R** (or the largest subset of **R** for which the formula has meaning) and the codomain is **R**.

EXAMPLE 6.15

(a) Consider the function $f(x) = x^3$, i.e. f assigns to each real number its cube. Then the image of 2 is 8, and so we may write $f(2) = 8$.

(b) Let f assign to each country in the world its capital city. Here the domain of f is the set of countries in the world; the codomain is the list of cities of the world. The image of France is Paris; that is, f(France) = Paris.

(c) Figure 6-16 defines a function f from $A = \{a, b, c, d\}$ into $B = \{r, s, t, u\}$ in the obvious way. Here

$$f(a) = s, \qquad f(b) = u, \qquad f(c) = r, \qquad f(d) = s$$

The image of f is the set of image values, $\{r, s, u\}$. Note that t does not belong to the image of f because t is not the image of any element under f.

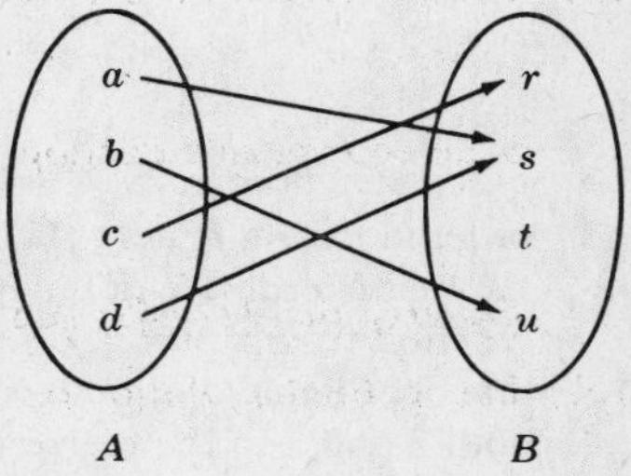

(d) Let A be any set. The function from A into A which assigns to each element that element itself is called the *identity function* on A and is usually denoted by 1_A or simply 1. In other words

$$1_A(a) = a$$

for every element a in A.

Fig. 6-16

There is another point of view from which functions may be considered. First of all, every function $f : A \rightarrow B$ gives rise to a relation from A to B called the *graph of f* and defined by

$$\text{Graph of } f = \{(a, b) : a \in A, b = f(a)\}$$

Two functions $f : A \rightarrow B$ and $g : A \rightarrow B$ are defined to be equal, written $f = g$, if $f(a) = g(a)$ for every $a \in A$; that is, if they have the same graph. Accordingly, we do not distinguish between a function and its graph. Now, such a graph relation has the property that each a in A belongs to a unique ordered pair (a, b) in the relation. On the other hand, any relation f from A to B that has this property gives rise to a function $f : A \rightarrow B$, where $f(a) = b$ for each (a, b) in f. Consequently, one may equivalently define a function as follows:

Definition: A function $f : A \rightarrow B$ is a relation from A to B (i.e. a subset of $A \times B$) such that each $a \in A$ belongs to a unique ordered pair (a, b) in f.

Remark: Since the graph of f, as defined above, is a relation, we can draw its picture by method (1) of Section 6.13; the picture itself is also called "the graph of f." The defining condition of a function translates into the geometrical condition that no two points of this graph shall lie on the same vertical line.

EXAMPLE 6.16

(a) Consider the function $f : A \rightarrow B$ defined by Fig. 6-16. The graph of f is the following set of ordered pairs:

$$\text{Graph of } f = \{(a, s), (b, u), (c, r), (d, s)\}$$

This relation is pictured on the coordinate diagram of $A \times B$ in Fig. 6-17.

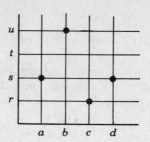

Fig. 6-17

(b)   Consider the following relations on the set $A = \{1, 2, 3\}$:

$$f = \{(1, 3), (2, 3), (3, 1)\}$$
$$g = \{(1, 2), (3, 1)\}$$
$$h = \{(1, 3), (2, 1), (1, 2), (3, 1)\}$$

f is a function from A into A since each member of A appears as the first coordinate in exactly one ordered pair in f; here $f(1) = 3$, $f(2) = 3$ and $f(3) = 1$. g is not a function from A into A since $2 \in A$ is not the first coordinate of any pair in g and so g does not assign any image to 2. (Note, however, that g *is* a function from a subset of A into A.)   Also h is not a function from A into A since $1 \in A$ appears as the first coordinate of two distinct ordered pairs in h, $(1, 3)$ and $(1, 2)$.   If h is to be a function it cannot assign both 3 and 2 to the element $1 \in A$.

(c)   By a real polynomial function, we mean a function $f : \mathbf{R} \to \mathbf{R}$ of the form

$$f(x) = a_n x^n + a_{n-1} x^{n-1} + \cdots + a_1 x + a_0$$

where the a_i are real numbers.   Since $\mathbf{R}$ is an infinite set, it would be impossible to plot each point of the graph.   However, the graph of such a function can be approximated by first plotting some of its points and then drawing a smooth curve through these points.   The points are usually obtained from a table where various values are assigned to x and the corresponding values of $f(x)$ computed.   Figure 6-18 illustrates this technique for the function $f(x) = x^2 - 2x - 3$.

x	$f(x)$
-2	5
-1	0
0	-3
1	-4
2	-3
3	0
4	5

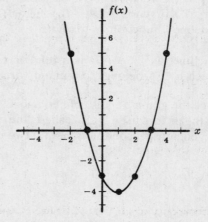

Graph of $f(x) = x^2 - 2x - 3$

Fig. 6-18

Consider the following equation in variables x and y:

$$y = x^2 - 2x - 3$$

By the discussion in Section 6.13, this equation gives rise to a relation on **R**.   Because each value of x gives a unique value of y, the relation is a function; in fact, it is the same function as the function in Fig. 6-18.   Generally speaking, whenever a function is given by means of an equation of the form

$$y = f(x)$$

we say that y is a function of x, and we call x the *independent variable* and y the *dependent variable*.

Solved Problems

SETS, SET OPERATIONS

6.1   Which of these sets are equal: $\{r, t, s\}$, $\{s, t, r, s\}$, $\{t, s, t, r\}$, $\{s, r, s, t\}$?

They are all equal.   Order and/or repetition do not change a set.

6.2   List the elements of the following sets; here $\mathbf{N} = \{1, 2, 3, \ldots\}$.

(a) $A = \{x : x \in \mathbf{N}, 3 < x < 12\}$
(b) $B = \{x : x \in \mathbf{N}, x \text{ is even}, x < 15\}$
(c) $C = \{x : x \in \mathbf{N}, 4 + x = 3\}$

(a) A consists of the positive integers between 3 and 12; hence

$$A = \{4, 5, 6, 7, 8, 9, 10, 11\}$$

(b) B consists of the even positive integers less than 15; hence

$$B = \{2, 4, 6, 8, 10, 12, 14\}$$

(c)   There are no positive integers which satisfy the condition $4 + x = 3$; hence C contains no elements. In other words, $C = \emptyset$, the empty set.

6.3   Consider the following sets:

$$\emptyset, \qquad A = \{1\}, \qquad B = \{1, 3\}, \qquad C = \{1, 5, 9\}, \qquad D = \{1, 2, 3, 4, 5\},$$
$$E = \{1, 3, 5, 7, 9\}, \qquad U = \{1, 2, \ldots, 8, 9\}$$

Insert the correct symbol $\subset$ or $\not\subset$ between each pair of sets:

(a) $\emptyset, A$ (c) B, C (e) C, D (g) D, E
(b) A, B (d) B, E (f) C, E (h) D, U

(a) $\emptyset \subset A$ because $\emptyset$ is a subset of every set.

(b) $A \subset B$ because 1 is the only element of A and it belongs to B.

(c) $B \not\subset C$ because $3 \in B$ but $3 \notin C$.

(d) $B \subset E$ because the elements of B also belong to E.

(e) $C \not\subset D$ because $9 \in C$ but $9 \notin D$.

(f) $C \subset E$ because the elements of C also belong to E.

(g) $D \not\subset E$ because $2 \in D$ but $2 \notin E$.

(h) $D \subset U$ because the elements of D also belong to U.

In Problems 6.4 through 6.6, assume $U = \{1, 2, \ldots, 8, 9\}$ and

$$A = \{1, 2, 3, 4, 5\}$$
$$B = \{4, 5, 6, 7\}$$
$$C = \{5, 6, 7, 8, 9\}$$
$$D = \{1, 3, 5, 7, 9\}$$
$$E = \{2, 4, 6, 8\}$$
$$F = \{1, 5, 9\}$$

6.4   Find: (a) $A \cup B$ and $A \cap B$, (b) $B \cup D$ and $B \cap D$, (c) $A \cup C$ and $A \cap C$.

Recall that the union $X \cup Y$ consists of those elements in either X or Y (or both), and that the intersection $X \cap Y$ consists of those elements in both X and Y.

(a) $A \cup B = \{1, 2, 3, 4, 5, 6, 7\}$, $A \cap B = \{4, 5\}$.

(b) $B \cup D = \{1, 3, 4, 5, 6, 7, 9\}$, $B \cap D = \{5, 7\}$.

(c) $A \cup C = \{1, 2, 3, 4, 5, 6, 7, 8, 9\} = U$, $A \cap C = \{5\}$.

6.5   Find: (a) A^c, B^c, and D^c; (b) $A \setminus B$, $B \setminus A$, and $F \setminus D$.

Recall that the complement X^c consists of those elements in the universal set U which do not belong to X, and the difference $X \setminus Y$ consists of those elements in X which do not belong to Y.

(a) $A^c = \{6, 7, 8, 9\}$, $B^c = \{1, 2, 3, 8, 9\}$, $D^c = \{2, 4, 6, 8\} = E$.

(b) $A \setminus B = \{1, 2, 3\}$, $B \setminus A = \{6, 7\}$, $F \setminus D = \emptyset$.

6.6   Find: (a) $A \cap (B \cup E)$, (b) $(B \cap F) \cup (C \cap E)$, (c) $(A \cap D) \setminus B$.

(a)   First compute $B \cup E = \{2, 4, 5, 6, 7, 8\}$.   Then $A \cap (B \cup E) = \{2, 4, 5\}$.

(b) $B \cap F = \{5\}$ and $C \cap E = \{6, 8\}$.   So $(B \cap F) \cup (C \cap E) = \{5, 6, 8\}$.

(c) $A \cap D = \{1, 3, 5\}$.   Now $(A \cap D) \setminus B = \{1, 3\}$.

6.7   Show that we can have $A \cap B = A \cap C$ without $B = C$.

Let $A = \{1, 2\}$, $B = \{2, 3\}$ and $C = \{2, 4\}$.   Then $A \cap B = \{2\}$ and $A \cap C = \{2\}$.   Thus $A \cap B = A \cap C$ but $B \neq C$.

ALGEBRA OF SETS, VENN DIAGRAMS

6.8 Write the dual of each set equation:

$$(a) \quad (A \cup B) \cap (A \cup B^c) = A \cup \emptyset \qquad (b) \quad (A \cap U) \cup (B \cap A) = A$$

Replace each occurrence of $\cup$, $\cap$, U and $\emptyset$ by $\cap$, $\cup$, $\emptyset$ and U respectively:

$$(a) \quad (A \cap B) \cup (A \cap B^c) = A \cap U \qquad (b) \quad (A \cup \emptyset) \cap (B \cup A) = A$$

6.9 Prove the identity $(U \cap A) \cup (B \cap A) = A$ by use of Table 6-1.

Statement	Reason
$(U \cap A) \cup (B \cap A) = (A \cap U) \cup (A \cap B)$	Commutative law 3a.
$= A \cap (U \cup B)$	Distributive law 4b.
$= A \cap (B \cup U)$	Commutative law 3a.
$= A \cap U$	Identity law 6a.
$= A$	Identity law 5b.

6.10 Prove the identity $(\emptyset \cup A) \cap (B \cup A) = A$.

This is the dual of the identity proved in Problem 6.9 and hence is true by the principle of duality. In other words, replacing each step in the proof in Problem 6.9 by the dual statement gives a proof of this identity.

6.11 Shade the set $A \cap B^c$ in the Venn diagram of Fig. 6-19(a).

Shade A with strokes in one direction (////), and shade B^c, the area outside B, with strokes in another direction (\\\\), as in Fig. 6-19(b). The cross-hatched area in Fig. 6-19(b) is the intersection, $A \cap B^c$, which is shaded in Fig. 6-19(c). Observe that $A \cap B^c = A \setminus B$.

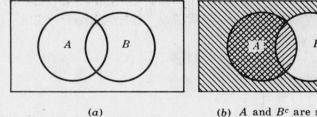
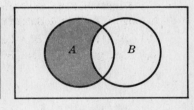

(a)   (b) A and B^c are shaded.   (c) $A \cap B^c$ is shaded.

Fig. 6-19

6.12 Illustrate the distributive law $A \cap (B \cup C) = (A \cap B) \cup (A \cap C)$ with Venn diagrams.

Draw three intersecting circles labeled A, B and C, as in Fig. 6-20(a). Now, as in Fig. 6-20(b), shade A with strokes in one direction and shade $B \cup C$ with strokes in another direction; the cross-hatched area is $A \cap (B \cup C)$, as in Fig. 6-20(c). Next shade $A \cap B$ and then $A \cap C$, as in Fig. 6-20(d); the total area shaded is $(A \cap B) \cup (A \cap C)$, as in Fig. 6-20(e).

As expected from the distributive law, $A \cap (B \cup C)$ and $(A \cap B) \cup (A \cap C)$ are both represented by the same set of points.

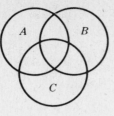

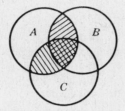

(a)

(b) *A* and *B* ∪ *C* are shaded.

(c) *A* ∩ (*B* ∪ *C*) is shaded.

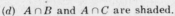

(d) *A* ∩ *B* and *A* ∩ *C* are shaded.

(e) (*A* ∩ *B*) ∪ (*A* ∩ *C*) is shaded.

Fig. 6-20

6.13   In a survey of 60 people, it was found that 25 read *Newsweek* magazine, 26 read *Time* and 26 read *Fortune*.   Also 9 read both *Newsweek* and *Fortune*, 11 read both *Newsweek* and *Time*, 8 read both *Time* and *Fortune*, and 8 read none of the three magazines.

(a)   Find the number of people who read all three magazines.

(b)   Fill in the correct number of readers in each of the eight regions of the Venn diagram of Fig. 6-21.   Here *N*, *T* and *F* denote the sets of people who read *Newsweek*, *Time* and *Fortune* respectively.

(c)   Determine the number of people who read exactly one magazine.

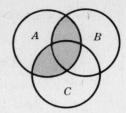

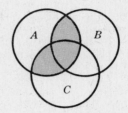

Fig. 6-21

(a)   We have $n(N \cup T \cup F) = 60 - 8 = 52$, because 8 people read none of the magazines.   By Corollary 6.4,

$$n(N \cup T \cup F) = n(N) + n(T) + n(F) - n(N \cap T) - n(N \cap F) - n(T \cap F) + n(N \cap T \cap F)$$

or $52 = 25 + 26 + 26 - 11 - 9 - 8 + n(N \cap T \cap F)$

whence $n(N \cap T \cap F) = 3$.

(b)   The required Venn diagram, Fig. 6-22, is obtained as follows:

3 read all three magazines,
$11 - 3 = 8$ read *Newsweek* and *Time* but not all three
          magazines,
$9 - 3 = 6$ read *Newsweek* and *Fortune* but not all three
          magazines,
$8 - 3 = 5$ read *Time* and *Fortune* but not all three magazines,
$25 - 8 - 6 - 3 = 8$ read only *Newsweek*,
$26 - 8 - 5 - 3 = 10$ read only *Time*,
$26 - 6 - 5 - 3 = 12$ read only *Fortune*.

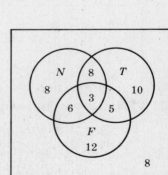

(c)   We have that $8 + 10 + 12 = 30$ read only one magazine.

Fig. 6-22

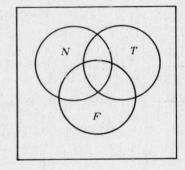

6.14   Consider the following assumptions:

S_1:   Poets are happy people.
S_2:   Every doctor is wealthy.
S_3:   No one who is happy is also wealthy.

Determine the validity of each of the following conclusions: (*a*) No poet is wealthy.   (*b*) Doctors are happy people.   (*c*) No one can be both a poet and a doctor.

By S_1 the set of poets is contained in the set of happy people, and by S_3 the set of happy people is disjoint from the set of wealthy people.   Hence draw the Venn diagram of Fig. 6-23.

   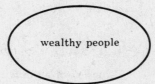

Fig. 6-23

By S_2 the set of doctors is contained in the set of wealthy people.   So draw the Venn diagram of Fig. 6-24.   From this diagram it is obvious that (*a*) and (*c*) are valid conclusions whereas (*b*) is not valid.

   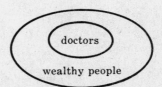

Fig. 6-24

CLASSES OF SETS, PARTITIONS

6.15   Determine the power set $\mathscr{P}(S)$ of $S = \{a, b, c, d\}$.

The elements of $\mathscr{P}(S)$ are the subsets of S.   Hence:

$$\mathscr{P}(S) = [S, \{a, b, c\}, \{a, b, d\}, \{a, c, d\}, \{b, c, d\}, \{a, b\}, \{a, c\}, \{a, d\},$$
$$\{b, c\}, \{b, d\}, \{c, d\}, \{a\}, \{b\}, \{c\}, \{d\}, \varnothing]$$

Observe that $\mathscr{P}(S)$ has $2^4 = 16$ elements.

6.16   Given $X = \{1, 2, 3, 4, 5, 6, 7, 8, 9\}$.   Determine which of the following are partitions of X:

(*a*)   [{1, 3, 6}, {2, 8}, {5, 7, 9}]   (*c*)   [{2, 4, 5, 8}, {1, 9}, {3, 6, 7}]
(*b*)   [{1, 5, 7}, {2, 4, 8, 9}, {3, 5, 6}]   (*d*)   [{1, 2, 7}, {3, 5}, {4, 6, 8, 9}, {3, 5}]

(*a*)   No; because $4 \in X$ does not belong to any cell.   In other words, X is not the union of the cells.
(*b*)   No; because $5 \in X$ belongs to two distinct cells, $\{1, 5, 7\}$ and $\{3, 5, 6\}$.   In other words, the two distinct cells are not disjoint.
(*c*)   Yes; because each element of X belongs to exactly one cell.   In other words, the cells are disjoint and their union is X.
(*d*)   Yes.   Although 3 and 5 appear in two places, the cells are not distinct.

6.17   Find all the partitions of $X = \{a, b, c, d\}$.

Note first that each partition of X contains either 1, 2, 3, or 4 distinct sets.   The partitions are as follows:

(1) $[\{a, b, c, d\}]$
(2) $[\{a\}, \{b, c, d\}], [\{b\}, \{a, c, d\}], [\{c\}, \{a, b, d\}], [\{d\}, \{a, b, c\}],$
      $[\{a, b\}, \{c, d\}], [\{a, c\}, \{b, d\}], [\{a, d\}, \{b, c\}]$
(3) $[\{a\}, \{b\}, \{c, d\}], [\{a\}, \{c\}, \{b, d\}], [\{a\}, \{d\}, \{b, c\}],$
      $[\{b\}, \{c\}, \{a, d\}], [\{b\}, \{d\}, \{a, c\}], [\{c\}, \{d\}, \{a, b\}]$
(4) $[\{a\}, \{b\}, \{c\}, \{d\}]$

There are fifteen different partitions of X.   For a generalization of this result, see Problem 11.18.

RELATIONS

6.18   Given $A = \{1, 2\}$, $B = \{x, y, z\}$ and $C = \{3, 4\}$.   Find $A \times B \times C$.

$A \times B \times C$ consists of all ordered triplets (a, b, c) where $a \in A$, $b \in B$, $c \in C$.   These elements of $A \times B \times C$ can be systematically obtained from a *tree diagram* (Fig. 6-25).   The elements of $A \times B \times C$ are precisely the 12 ordered triplets to the right of the tree diagram.

Observe that $n(A) = 2$, $n(B) = 3$, and $n(C) = 2$ and, as expected,

$$n(A \times B \times C) = 12 = n(A) \cdot n(B) \cdot n(C)$$

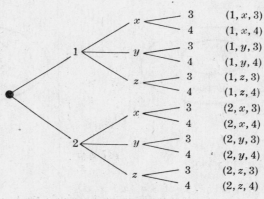

Fig. 6-25

6.19   Given $(2x, x + y) = (6, 2)$.   Find x and y.

Two ordered pairs are equal if and only if the corresponding components are equal.   Hence we obtain the equations

$$2x = 6 \qquad \text{and} \qquad x + y = 2$$

The two equations yield $x = 3$ and $y = -1$.

6.20   Let $M = \{a, b, c, d\}$ and let R be the relation on M consisting of those points which are displayed in the coordinate diagram of $M \times M$, Fig. 6-26.

(a) Find all the elements in M which are related to b; that is, $\{x : (x, b) \in R\}$.

(b) Find all those elements in M to which d is related; that is, $\{x : (d, x) \in R\}$.

(c) Find the inverse relation R^{-1}.

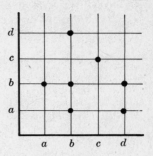

(a) The horizontal line through b contains all points of R in which b appears as the second element: (a, b), (b, b) and (d, b). Hence the desired set is $\{a, b, d\}$.

(b) The vertical line through d contains all the points of R in which d appears as the first element: (d, a) and (d, b). Hence $\{a, b\}$ is the desired set.

Fig. 6-26

(c) First write R as a set of ordered pairs, and then write the pairs in reverse order.

$$R^{-1} = \{(a, b), (a, d), (b, a), (b, b), (b, d), (c, c), (d, b)\}$$

6.21 Given $A = \{1, 2, 3, 4\}$ and $B = \{x, y, z\}$.   Consider the following relation from A to B:

$$R = \{(1, y), (1, z), (3, y), (4, x), (4, z)\}$$

(a) Plot R on a coordinate diagram of $A \times B$.   (b) Determine the matrix of the relation.   (c) Draw the arrow diagram of R.   (d) Find the inverse relation R^{-1} of R.   (e) Determine the domain and range of R.

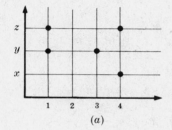

(a)

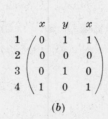

(b)

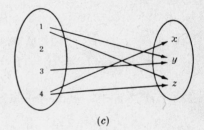

(c)

Fig. 6-27

(a) See Fig. 6-27(a).

(b) See Fig. 6-27(b).   Observe that the rows of the matrix are labeled by the elements of A and the columns by the elements of B.   Also observe that the entry in the matrix corresponding to $a \in A$ and $b \in B$ is 1 if a is related to b and 0 otherwise.

(c) See Fig. 6-27(c).   Observe that there is an arrow from a in A to b in B when and only when a is related to b, i.e. (a, b) belongs to R.

(d) Reverse the ordered pairs of R to obtain the ordered pairs of R^{-1}:

$$R^{-1} = \{(y, 1), (z, 1), (y, 3), (x, 4), (z, 4)\}$$

Observe that by reversing the arrows in Fig. 6-27(c) we obtain the arrow diagram of R^{-1}, and that by interchanging rows and columns of Fig. 6-27(b) we obtain the matrix of R^{-1}.

(e) The domain of R consists of the first elements of the ordered pairs of R, and the range of R consists of the second elements.   Thus,

domain of $R = \{1, 3, 4\}$ and   range of $R = \{x, y, z\}$

6.22 Let $A = \{1, 2, 3, 4, 6\}$, and let R be the relation on A defined by "x divides y", indicated as $x \mid y$.   ($x \mid y$ if and only if there exists an integer z such that $xz = y$.)

(a) Write R as a set of ordered pairs.

(b) Plot R on a coordinate diagram of $A \times A$, and draw its directed graph.

(c) Find the inverse relation R^{-1} of R, and describe R^{-1} in words.

(a) Find sequentially those numbers in A divisible by 1, 2, 3, 4, 6.   These are:

$$1|1, \; 1|2, \; 1|3, \; 1|4, \; 1|6, \; 2|2, \; 2|4, \; 2|6, \; 3|3, \; 3|6, \; 4|4, \; 6|6$$

Hence $R = \{(1, 1), (1, 2), (1, 3), (1, 4), (1, 6), (2, 2), (2, 4), (2, 6), (3, 3), (3, 6), (4, 4), (6, 6)\}$.

(b) See Fig. 6-28.

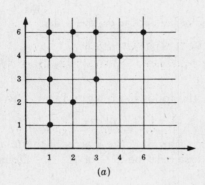

(a)   (b)

Fig. 6-28

(c) Reverse the ordered pairs of R to obtain the ordered pairs of R^{-1}:

$$R^{-1} = \{(1, 1), (2, 1), (3, 1), (4, 1), (6, 1), (2, 2), (4, 2), (6, 2), (3, 3), (6, 3), (4, 4), (6, 6)\}$$

R^{-1} can be described by the statement "x is a multiple of y".

6.23 Let R and S be the following relations on $A = \{1, 2, 3\}$:

$$R = \{(1, 1), (1, 2), (2, 3), (3, 1), (3, 3)\} \qquad S = \{(1, 2), (1, 3), (2, 1), (3, 3)\}$$

Find $R \cap S$, $R \cup S$ and R^c.

Treat R and S simply as sets, and take the usual intersection and union.   For R^c, use the fact that $A \times A$ is the universal relation on A.

$$R \cap S = \{(1, 2), (3, 3)\}$$
$$R \cup S = \{(1, 1), (1, 2), (1, 3), (2, 1), (2, 3), (3, 1), (3, 3)\}$$
$$R^c = \{(1, 3), (2, 1), (2, 2), (3, 2)\}$$

6.24 Consider the following five relations on the set $A = \{1, 2, 3\}$.

$$R = \{(1, 1), (1, 2), (1, 3), (3, 3)\}$$
$$S = \{(1, 1), (1, 2), (2, 1), (2, 2), (3, 3)\}$$
$$T = \{(1, 1), (1, 2), (2, 2), (2, 3)\}$$
$$\emptyset = \text{empty relation}$$
$$A \times A = \text{universal relation}$$

Determine whether or not each of the above relations on A is (a) reflexive, (b) symmetric, (c) transitive, (d) an equivalence relation.

(a) R is not reflexive since $2 \in A$ but $(2, 2) \notin R$. T is not reflexive since $(3, 3) \notin T$ and, similarly, $\emptyset$ is not reflexive. S and $A \times A$ are reflexive.

(b) R is not symmetric since $(1, 2) \in R$ but $(2, 1) \notin R$, and similarly T is not symmetric. S, $\emptyset$ and $A \times A$ are symmetric.

(c) T is not transitive since $(1, 2)$ and $(2, 3)$ belong to T, but $(1, 3)$ does not belong to T. The other four relations are transitive.

(d) Only S and $A \times A$ are equivalence relations, since only they are reflexive, symmetric, and transitive.

6.25 Let R be the following equivalence relation on the set $A = \{1, 2, 3, 4, 5, 6\}$:

$$R = \{(1, 1), (1, 5), (2, 2), (2, 3), (2, 6), (3, 2), (3, 3),$$
$$(3, 6), (4, 4), (5, 1), (5, 5), (6, 2), (6, 3), (6, 6)\}$$

Find the partition of A *induced by* R, i.e. find the equivalence classes composing A/R.

Those elements related to 1 are 1 and 5; hence

$$[1] = \{1, 5\}$$

We pick an element which does not belong to [1], say 2. Those elements related to 2 are 2, 3 and 6; hence

$$[2] = \{2, 3, 6\}$$

The only element which does not belong to [1] or [2] is 4. The only element related to 4 is 4. Thus

$$[4] = \{4\}$$

Accordingly $[\{1, 5\}, \{2, 3, 6\}, \{4\}]$

is the partition of A induced by R.

FUNCTIONS

6.26 For each diagram in Fig. 6-29, state whether or not it defines a function from $A = \{a, b, c\}$ into $B = \{x, y, z\}$.

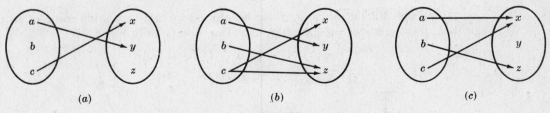

(a)   (b)   (c)

Fig. 6-29

(a) No. There is nothing assigned to the element $b \in A$.
(b) No. Two elements, x and z, are assigned to $c \in A$.
(c) Yes.

6.27 Let $A = \{1, 2, 3, 4, 5\}$ and let $f : A \to A$ be the function defined in Fig. 6-30. (a) Find the image of f. (b) Find the graph of f, i.e. write f as a set of ordered pairs.

(a) The image of f consists of all the image values. Now only 2, 3 and 5 appear as the images of elements of A; hence $f(A) = \{2, 3, 5\}$.

(b) The ordered pairs $(a, f(a))$, where $a \in A$, form the graph of f. Now $f(1) = 3$, $f(2) = 5$, $f(3) = 5$, $f(4) = 2$ and $f(5) = 3$; hence

$$f = \{(1, 3), (2, 5), (3, 5), (4, 2), (5, 3)\}$$

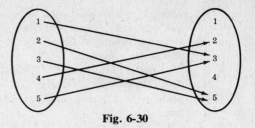

Fig. 6-30

6.28 Determine if the relation (a) $f = \{(2, 3), (1, 4), (2, 1), (3, 2), (4, 4)\}$, (b) $g = \{(3, 1), (4, 2), (1, 1)\}$, (c) $h = \{(2, 1), (3, 4), (1, 4), (2, 1), (4, 4)\}$, is a function from $X = \{1, 2, 3, 4\}$ into X.

Recall that a subset f of $X \times X$ is a function $f : X \to X$ if and only if each $a \in X$ appears as the first coordinate in exactly one ordered pair in f.

(a) No. Two different ordered pairs $(2, 3)$ and $(2, 1)$ in f have the same number 2 as their first coordinate.

(b) No. The element $2 \in X$ does not appear as the first coordinate in any ordered pair in g.

(c) Yes. Although $2 \in X$ appears as the first coordinate in two ordered pairs in h, these pairs are equal.

6.29 Let A be the set of students in a school. Determine which of the following assignments define functions on A. (a) To each student, the student's age. (b) To each student, the student's teacher. (c) To each student, the student's sex. (d) To each student, the student's spouse.

A collection of assignments is a function on A if and only if each element a in A is assigned exactly one element (of another set B). Thus:

(a) Yes, because each student has one and only one age.

(b) Yes, if each student has only one teacher; no, if any student has more than one teacher.

(c) Yes.

(d) No, if any student is not married.

6.30 Let $f : \mathbf{R} \to \mathbf{R}$ be defined by $f(x) = x^3$. (a) Find: (i) $f(3)$, (ii) $f(-2)$, (iii) $f(y)$, (iv) $f(y + 1)$. (b) Plot f on a coordinate diagram of $\mathbf{R} \times \mathbf{R}$.

(a) (i) $f(3) = 3^3 = 27$, (ii) $f(-2) = (-2)^3 = -8$, (iii) $f(y) = (y)^3 = y^3$
    (iv) $f(y + 1) = (y + 1)^3 = y^3 + 3y^2 + 3y + 1$

(b) Since f is a polynomial function, it can be plotted by first tabulating some of its points and then drawing a smooth curve through these points, as in Fig. 6-31.

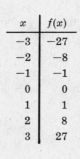

x	$f(x)$
-3	-27
-2	-8
-1	-1
0	0
1	1
2	8
3	27

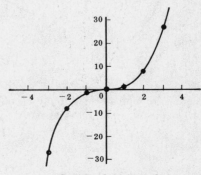

Graph of $f(x) = x^3$

Fig. 6-31

6.31   Sketch the graphs of

(a) $f(x) = 3x - 2$ (b) $g(x) = x^2 + x - 6$ (c) $h(x) = x^3 - 3x^2 - x + 3$

In (a) the function is linear; only two points (three as a check) are needed to sketch its graph.   Set up a table with three values of x, say, $x = -2, 0, 2$ and find the corresponding values of $f(x)$:

$$f(-2) = 3(-2) - 2 = -8 \qquad f(0) = 3(0) - 2 = -2 \qquad f(2) = 3(2) - 2 = 4$$

Draw the line through these points as in Fig. 6-32.

x	$f(x)$
-2	-8
0	-2
2	4

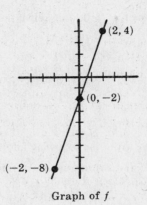

Graph of f

Fig. 6-32

In (b) and (c), set up a table of values for x and then find the corresponding values of the function. Plot the points in a coordinate diagram, and then draw a smooth continuous curve through the points as in Fig. 6-33.

x	$g(x)$
-4	6
-3	0
-2	-4
-1	-6
0	-6
1	-4
2	0
3	6

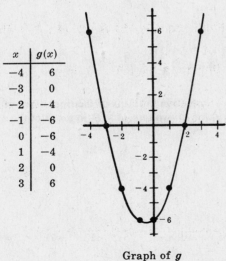

Graph of g

x	$h(x)$
-2	-15
-1	0
0	3
1	0
2	-3
3	0
4	15

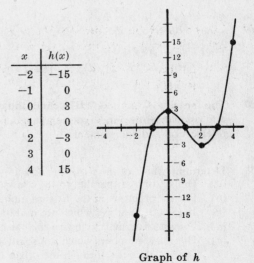

Graph of h

Fig. 6-33

Supplementary Problems

SETS, SET OPERATIONS

6.32   Which of the following sets are equal?

$$\{1, 2\}, \{1, 3\}, \{2, 1\}, \{3, 1, 3\}, \{1, 2, 1\}$$
$$A = \{x : x^2 - 4x + 3 = 0\} \qquad C = \{x : x \in \mathbf{N}, x < 3\}$$
$$B = \{x : x^2 - 3x + 2 = 0\} \qquad D = \{x : x \in \mathbf{N}, x \text{ is odd}, x < 5\}$$

6.33   List the elements of the following sets if the universal set is $U = \{a, b, c, \ldots, y, z\}$.   Which of the sets, if any, are equal?

$$A = \{x : x \text{ is a vowel}\}$$
$$B = \{x : x \text{ is a letter in the word "little"}\}$$
$$C = \{x : x \text{ precedes } f \text{ in the alphabet}\}$$
$$D = \{x : x \text{ is a letter in the word "title"}\}$$

6.34   Let $A = \{1, 2, \ldots, 8, 9\}$, $B = \{2, 4, 6, 8\}$, $C = \{1, 3, 5, 7, 9\}$, $D = \{3, 4, 5\}$ and $E = \{3, 5\}$.   Which sets can equal X if we are given the following information?

(a) X and B are disjoint.   (b) $X \subset D$ but $X \not\subset B$.   (c) $X \subset A$ but $X \not\subset C$.
(d) $X \subset C$ but $X \not\subset A$.

6.35   Consider the following sets:

$$\varnothing, \qquad A = \{a\}, \qquad B = \{c, d\}, \qquad C = \{a, b, c, d\}, \qquad D = \{a, b\}, \qquad E = \{a, b, c, d, e\}$$

Insert the correct symbol, $\subset$ or $\not\subset$, between each pair of sets:

(a) $\varnothing, A$ (c) A, B (e) B, C (g) C, D
(b) D, E (d) D, A (f) D, C (h) B, D

6.36   Given: $A = \{a, b, c, d, e\}$ $C = \{b, c, e, g, h\}$
             $B = \{a, b, d, f, g\}$ $D = \{d, e, f, g, h\}$

Find: (a) $A \cup B$ (c) $C \setminus D$ (e) $(A \cap D) \cup B$
        (b) $B \cap C$ (d) $A \cap (B \cup D)$ (f) $B \cap C \cap D$

6.37   The formula $A \setminus B = A \cap B^c$ defines the difference operation in terms of the operations of intersection and complement.   Find a formula that defines the union of two sets, $A \cup B$, in terms of the operations of intersection and complement.

6.38   Determine which of the following sets are finite:
(a)   The set of lines parallel to the x axis.
(b)   The set of letters in the English alphabet.
(c)   The set of numbers which are multiples of 5.
(d)   The set of animals living on the earth.
(e)   The set of numbers which are solutions of the equation $x^{27} + 26x^{18} - 17x^{11} + 7x^3 - 10 = 0$.
(f)   The set of circles through the origin $(0, 0)$.

ALGEBRA OF SETS, VENN DIAGRAMS

6.39   Write the dual of each set equation:
(a) $A \cup (A \cap B) = A$
(b) $(A \cap B) \cup (A^c \cap B) \cup (A \cap B^c) \cup (A^c \cap B^c) = U$

6.40 Use the laws of Table 6-1 to prove each set identity:

  (a) $(A \cap B) \cup (A \cap B^c) = A$

  (b) $A \cup (A \cap B) = A$

  (c) $A \cup B = (A \cap B^c) \cup (A^c \cap B) \cup (A \cap B)$

6.41 The Venn diagram of Fig. 6-34 shows sets A, B, and C.   Shade the following sets:

  (a) $A \cap B \cap C$ (c) $A \cup (B \cap C)$ (e) $A^c \cap (B \cup C)$

  (b) $A \cap B^c \cap C$ (d) $C \cap (A \cup B^c)$ (f) $(A^c \cap B) \setminus C$

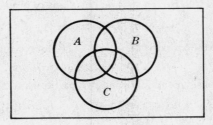

Fig. 6-34

6.42 A survey of 100 students produced the following statistics:

  32 study mathematics,
  20 study physics,
  45 study biology,
  15 study mathematics and biology,
   7 study mathematics and physics,
  10 study physics and biology,
  30 do not study any of the three subjects.

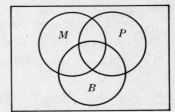

Fig. 6-35

  (a) Find the number of students studying all three subjects.
  (b) Fill in the number of students in each of the eight regions of the Venn
       diagram (Fig. 6-35), where M, P and B denote the sets of students
       studying mathematics, physics and biology respectively.
  (c) Find the number of students taking exactly one of the three subjects.

6.43 Employ Venn diagrams to find a conclusion yielded by the following set of premises (adapted from a book on logic by Lewis Carroll, the author of *Alice in Wonderland*).

  S_1: Babies are illogical.
  S_2: Nobody is despised who can manage a crocodile.
  S_3: Illogical people are despised.

6.44 Consider the following premises:

  S_1: I planted all my expensive trees last year.
  S_2: All my fruit trees are in my orchard.
  S_3: No tree in my orchard was planted last year.

Determine whether or not each of the following is the conclusion in a valid argument:

  (a)   The fruit trees were planted last year.
  (b)   No expensive tree is in the orchard.
  (c)   No fruit tree is expensive.

CLASSES OF SETS, PARTITIONS

6.45   Find the power set, $\mathscr{P}(S)$, of $S = \{1, 2, 3, 4, 5\}$.

6.46   Let $W = \{1, 2, 3, 4, 5, 6\}$.   Determine which of the following are partitions of W:

        (a)   [{1, 3, 5}, {2, 4}, {3, 6}]   (c)   [{1, 5}, {2}, {4}, {1, 5}, {3, 6}]

        (b)   [{1, 5}, {2}, {3, 6}]   (d)   [{1, 2, 3, 4, 5, 6}]

6.47   Find all partitions of $V = \{1, 2, 3\}$.

6.48   Let $[A_1, A_2, \ldots, A_m]$ and $[B_1, B_2, \ldots, B_n]$ be partitions of a set X.   Show that the collection of sets

$$[A_i \cap B_j : i = 1, \ldots, m, \ j = 1, \ldots, n] \setminus \emptyset$$

is also a partition (called the *cross partition*) of X.   (Observe that we have excluded the empty set $\emptyset$.)

6.49   Let $S = \{1, 2, 3, 4, 5\}$.   Consider the following two partitions of S:

$$A = [\{1, 3, 4\}, \{2, 5\}] \qquad B = [\{1, 3\}, \{2, 4\}, \{5\}]$$

Find the cross partition of S.

PRODUCT SETS

6.50   Let $A = \{a, b, c, d\}$.   Find the ordered pairs corresponding to the points P_1, P_2, P_3 and P_4 which appear in the adjacent coordinate diagram of $A \times A$ (Fig. 6-36).

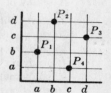

6.51   Find x and y if: (a) $(x + 2, 4) = (5, 2x + y)$, (b) $(y - 2, 2x + 1) = (x - 1, y + 2)$.

6.52   Let $W = \{$Mark, Eric, Paul$\}$ and let $V = \{$Eric, David$\}$.   Find: (a) $W \times V$, (b) $V \times W$, (c) $V \times V$.

Fig. 6-36

6.53   Let $S = \{a, b, c\}$, $T = \{b, c, d\}$ and $W = \{a, d\}$.   Construct the tree diagram of $S \times T \times W$ and then exhibit $S \times T \times W$.

RELATIONS

6.54   Let R be the following relation on $A = \{1, 2, 3, 4\}$: $R = \{(1, 3), (1, 4), (3, 2), (3, 3), (3, 4)\}$.

        (a)   Find the domain and range of R.   (c)   Find the matrix, M_R, of R.

        (b)   Draw the directed graph of R.   (d)   Find R^{-1}.

6.55   Let R be the relation on $C = \{1, 2, 3, 4, 5\}$ given by the set of points displayed in the coordinate diagram of $C \times C$ in Fig. 6-37.

    (1)   State whether each is true or false: (a) $1 R 4$, (b) $2 R 5$, (c) $3 \not R 1$, (d) $5 \not R 3$.

    (2)   Find the elements in each of the following subsets of C:

        (a) $\{x : 3 R x\}$ (c) $\{x : (x, 2) \notin R\}$

        (b) $\{x : (4, x) \in R\}$ (d) $\{x : x R 5\}$

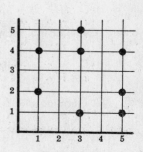

    (3)   Find (a) the domain of R, (b) the range of R, (c) R^{-1}.

    (4)   Draw the directed graph of R.

Fig. 6-37

6.56   Let R and S be the following relations on $B = \{a, b, c, d\}$:

$$R = \{(a, a), (a, c), (c, b), (c, d), (d, b)\}$$
$$S = \{(b, a), (c, c), (c, d), (d, a)\}$$

Find $R \cup S$, $R \cap S$, and $(R \cup S)^c$.

6.57   Let R be the relation on the positive integers $\mathbf{N}$ defined by the equation $x + 3y = 12$; that is,

$$R = \{(x, y) : x + 3y = 12\}$$

(a)   Write R as a set of ordered pairs.
(b)   Find (i) domain of R, (ii) range of R, and (iii) R^{-1}.

EQUIVALENCE RELATIONS

6.58   Let $W = \{1, 2, 3, 4\}$.   Consider the following relations on W:

$$R_1 = \{(1, 1), (2, 1)\} \qquad R_4 = \{(1, 1), (2, 2), (3, 3)\}$$
$$R_2 = \{(1, 3), (2, 3), (4, 1)\} \quad R_5 = \{(1, 3), (2, 4)\}$$
$$R_3 = \{(3, 4)\}$$

Determine which relations are (a) reflexive, (b) symmetric, (c) transitive, (d) equivalence relations.

6.59   Let $S = \{1, 2, 3, 4, 5\}$.   The following is an equivalence relation on S:

$$R = \{(1, 1), (1, 2), (2, 1), (2, 2), (3, 3), (4, 4), (4, 5), (5, 4), (5, 5)\}$$

Find the partition of S induced by R.

6.60   Let $S = \{1, 2, 3, \ldots, 19, 20\}$.   Let R be the equivalence relation on S defined by $x \equiv y \pmod 5$, that is, $x - y$ is divisible by 5.   Find the partition of S induced by R, i.e. the quotient set S/R.

6.61   Let $A = \{1, 2, 3, \ldots, 9\}$ and let $\sim$ be the relation on $A \times A$ defined by

$$(a, b) \sim (c, d) \qquad \text{if} \qquad a + d = b + c$$

(a)   Prove that $\sim$ is an equivalence relation.
(b)   Find $[(2, 5)]$, the equivalence class of $(2, 5)$.

6.62   Prove that if R is an equivalence relation on a set A, then R^{-1} is also an equivalence relation on A.

6.63   Let R and S be nonempty relations on a set A.   Assuming A has at least three elements, state whether each of the following statements is true or false.   If it is false, give a counterexample on the set $A = \{1, 2, 3\}$:
(a)   If R and S are symmetric then $R \cap S$ is symmetric.
(b)   If R and S are symmetric then $R \cup S$ is symmetric.
(c)   If R and S are reflexive then $R \cap S$ is reflexive.
(d)   If R and S are reflexive then $R \cup S$ is reflexive.
(e)   If R and S are transitive then $R \cup S$ is transitive.
(f)   If R is reflexive then $R \cap R^{-1}$ is not empty.
(g)   If R is symmetric then $R \cap R^{-1}$ is not empty.

FUNCTIONS

6.64   State whether each diagram of Fig. 6-38 defines a function from $\{1, 2, 3\}$ into $\{4, 5, 6\}$.

6.65   Define each function from $\mathbf{R}$ into $\mathbf{R}$ by a formula:

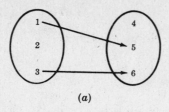
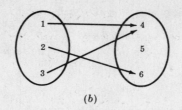
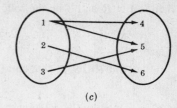

Fig. 6-38

(a) To each number let f assign its square plus 3.

(b) To each number let g assign its cube plus twice the number.

6.66 Let $W = \{a, b, c, d\}$. Determine whether each set of ordered pairs is a function from W into W.

(a) $\{(b, a), (c, d), (d, a), (c, d), (a, d)\}$ (c) $\{(a, b), (b, b), (c, b), (d, b)\}$

(b) $\{(d, d), (c, a), (a, b), (d, b)\}$ (d) $\{(a, a), (b, a), (a, b), c, d)\}$

6.67 Let the function g assign to each name in the set {Betty, Martin, David, Alan, Rebecca} the number of different letters needed to spell the name. Find the graph of g, i.e. write g as a set of ordered pairs.

6.68 Let $V = \{1, 2, 3, 4\}$. Determine whether the set of points in each coordinate diagram of $V \times V$ (Fig. 6-39) is a function from V into V.

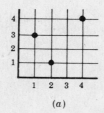
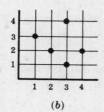
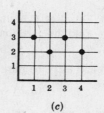
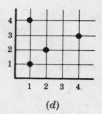

Fig. 6-39

6.69 Let $W = \{1, 2, 3, 4\}$ and let $g : W \to W$ be defined by Fig. 6-40. (a) Write g as a set of ordered pairs. (b) Find the image of g.

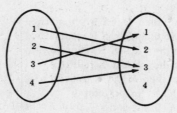

Fig. 6-40

6.70 Consider functions $f : A \to B$ and $g : B \to C$. The *composition* of f and g, written $g \circ f$, is the function from A into C defined by

$$(g \circ f)(a) = g(f(a))$$

Find the composition functions $f \circ g$, $h \circ f$, and $g \circ g$ (written g^2), for the functions f, g, and h pictured in Fig. 6-41.

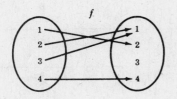 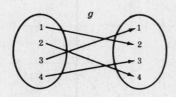 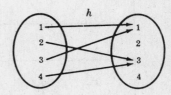

Fig. 6-41

6.71   Consider the functions $f(x) = x^2 + 3x + 1$ and $g(x) = 2x - 3$.   Find formulas defining the composition functions (a) $f \circ g$, (b) $g \circ f$.

6.72   Determine the number of different functions from $\{a, b\}$ into $\{1, 2, 3\}$.

6.73   Sketch the graph of each function:

$$(a) \quad f(x) = \tfrac{1}{2}x - 1 \qquad (b) \quad g(x) = x^3 - 3x + 2 \qquad (c) \quad h(x) = \frac{1}{x}$$

Answers to Supplementary Problems

6.32 $\{1, 2\} = \{2, 1\} = \{1, 2, 1\} = B = C$, $\{1, 3\} = \{3, 1, 3\} = A = D$

6.33 $A = \{a, e, i, o, u\}$, $B = D = \{l, i, t, e\}$, $C = \{a, b, c, d, e\}$

6.34 (a) C and E, (b) D and E, (c) A, B and D, (d) none

6.35 (a) $\emptyset \subset A$, (b) $D \subseteq E$, (c) $A \not\subset B$, (d) $D \not\subset A$, (e) $B \subset C$, (f) $D \subset C$, (g) $C \not\subset D$, (h) $B \not\subset D$

6.36 (a) $A \cup B = \{a, b, c, d, e, f, g\}$ (c) $C \setminus D = \{b, c\}$ (e) $(A \cap D) \cup B = \{a, b, d, e, f, g\}$
(b) $B \cap C = \{b, g\}$ (d) $A \cap (B \cup D) = \{a, b, d, e\}$ (f) $B \cap C \cap D = \{g\}$

6.37 $A \cup B = (A^c \cap B^c)^c$

6.38 (a) infinite, (b) finite, (c) infinite, (d) finite, (e) finite, (f) infinite

6.39 (a) $A \cap (A \cup B) = A$, (b) $(A \cup B) \cap (A^c \cup B) \cap (A \cup B^c) \cap (A^c \cup B^c) = \emptyset$

6.41   See Fig. 6-42.

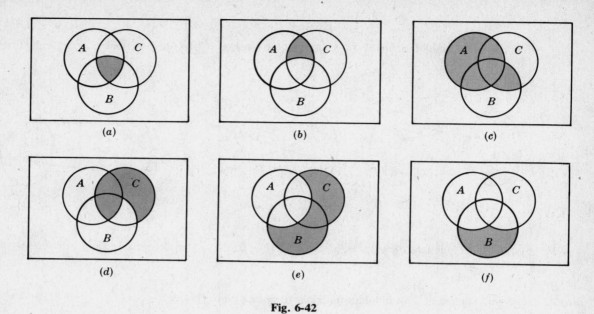

Fig. 6-42

6.42   (a) 5.   (b) See Fig. 6-43.   (c) 48.

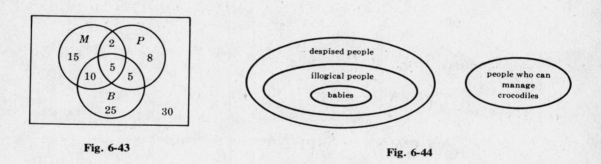

Fig. 6-43

Fig. 6-44

6.43   The premises indicate the Venn diagram of Fig. 6-44.   The evident conclusion is: "Babies cannot manage crocodiles."   (One may also conclude that people who can manage crocodiles are logical.)

6.44   The premises indicate the Venn diagram of Fig. 6-45.   (a) no, (b) yes, (c) yes.

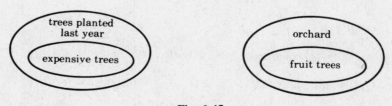

Fig. 6-45

6.45 $\mathscr{P}(S) = [\varnothing, \{1\}, \{2\}, \{3\}, \{4\}, \{5\}, \{1, 2\}, \{1, 3\}, \{1, 4\}, \{1, 5\},$
$\{2, 3\}, \{2, 4\}, \{2, 5\}, \{3, 4\}, \{3, 5\}, \{4, 5\}, \{1, 2, 3\},$
$\{1, 2, 4\}, \{1, 2, 5\}, \{2, 3, 4\}, \{2, 3, 5\}, \{3, 4, 5\}, \{1, 3, 4\},$
$\{1, 3, 5\}, \{1, 4, 5\}, \{2, 4, 5\}, \{1, 2, 3, 4\}, \{1, 2, 3, 5\},$
$\{1, 2, 4, 5\}, \{1, 3, 4, 5\}, \{2, 3, 4, 5\}, S]$

There are $2^5 = 32$ sets in $\mathscr{P}(S)$.

6.46   (c) and (d)

6.47   There are five: $[\{1, 2, 3\}], [\{1\}, \{2, 3\}], [\{2\}, \{1, 3\}], [\{3\}, \{1, 2\}],$ and $[\{1\}, \{2\}, \{3\}]$.

6.49 $[\{1, 3\}, \{2\}, \{4\}, \{5\}]$

6.50 $P_1 = (a, b), P_2 = (b, d), P_3 = (d, c), P_4 = (c, a)$

6.51   (a) $x = 3$, $y = -2$, (b) $x = 2$, $y = 3$

6.52   (a) $W \times V = \{(\text{Mark, Eric}), (\text{Mark, David}), (\text{Eric, Eric}), (\text{Eric, David}), (\text{Paul, Eric}), (\text{Paul, David})\}$
(b) $V \times W = \{(\text{Eric, Mark}), (\text{David, Mark}), (\text{Eric, Eric}), (\text{David, Eric}), (\text{Eric, Paul}), (\text{David, Paul})\}$
(c) $V \times V = \{(\text{Eric, Eric}), (\text{Eric, David}), (\text{David, Eric}), (\text{David, David})\}$

6.53   See Fig. 6-46. $S \times T \times W = \{(a, b, a), (a, b, d), (a, c, a), (a, c, d), (a, d, a), (a, d, d),$
$(b, b, a), (b, b, d), (b, c, a), (b, c, d), (b, d, a), (b, d, d),$
$(c, b, a), (c, b, d), (c, c, a), (c, c, d), (c, d, a), (c, d, d)\}$

6.54   (a)   domain = $\{1, 3\}$, range = $\{2, 3, 4\}$
(b)   See Fig. 6-47.
(c) $M_R = \begin{pmatrix} 0 & 0 & 1 & 1 \\ 0 & 0 & 0 & 0 \\ 0 & 1 & 1 & 1 \\ 0 & 0 & 0 & 0 \end{pmatrix}$
(d) $R^{-1} = \{(3, 1), (4, 1), (2, 3), (3, 3), (4, 3)\}$

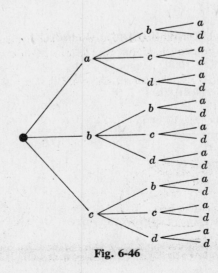

Fig. 6-46

6.55   (1)   (a) True, (b) False, (c) False, (d) True
       (2)   (a) {1, 4, 5}, (b) Ø, (c) {2, 3, 4}, (d) {3}
       (3)   (a) {1, 3, 5}, (b) {1, 2, 4, 5}, (c) R^{-1} = {(1, 3), (1, 5), (2, 1), (2, 5), (4, 1), (4, 3), (4, 5), (5, 3)}
       (4)   See Fig. 6-48.

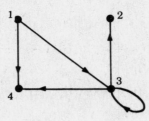

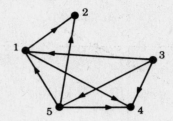

Fig. 6-47   Fig. 6-48

6.56 $R \cup S$ = {(a, a), (a, c), (b, a), (c, b), (c, c), (c, d), (d, a), (d, b)}
       $R \cap S$ = {(c, d)}
       $(R \cup S)^c$ = {(a, b), (a, d), (b, b), (b, c), (b, d), (c, a), (d, c), (d, d)}

6.57   (a) {(9, 1), (6, 2), (3, 3)}; (b) (i) {9, 6, 3}, (ii) {1, 2, 3}; (iii) {(1, 9), (2, 6), (3, 3)}, or the relation defined by the
       equation $3x + y = 12$.

6.58   (a) none, (b) R_4, (c) all except R_2, (d) none

6.59 S/R = [{1, 2}, {3}, {4, 5}]

6.60   [{1, 6, 11, 16}, {2, 7, 12, 17}, {3, 8, 13, 18}, {4, 9, 14, 19}, {5, 10, 15, 20}]

6.61   (b) {(1, 4), (2, 5), (3, 6), (4, 7), (5, 8), (6, 9)}

6.63   All are true except (e); choose R = {(1, 2)}, S = {(2, 3)}.

6.64   (a) No, (b) Yes, (c) No

6.65   (a) $f(x) = x^2 + 3$, (b) $g(x) = x^3 + 2x$

6.66   (a) Yes, (b) No, (c) Yes, (d) No

6.67 g = {(Betty, 4), (Martin, 6), (David, 4), (Alan, 3), (Rebecca, 5)}

6.68   (a) No, (b) No, (c) Yes, (d) No

6.69   (a) g = {(1, 2), (2, 3), (3, 1), (4, 3)}, (b) {1, 2, 3}

6.70

x	$(f \circ g)(x)$	$(h \circ f)(x)$	$g^2(x)$
1	1	3	4
2	4	1	3
3	2	1	2
4	1	3	1

6.71   (a) $(f \circ g)(x) = 4x^2 - 6x + 1$, (b) $(g \circ f)(x) = 2x^2 + 6x - 1$

6.72   nine

6.73   See Fig. 6-49.

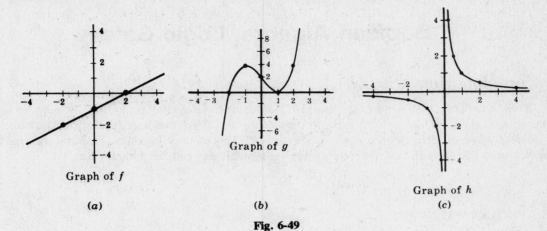

Graph of f

(a)

Graph of g

(b)

Graph of h

(c)

Fig. 6-49

Chapter 7

Boolean Algebra, Logic Gates

7.1 INTRODUCTION

Both sets (Chapter 6) and propositions (Chapter 4) have similar properties, as may be seen by comparing Tables 6-1 and 4-1. These properties are used to define a mathematical structure called a *Boolean algebra*, after George Boole (1813–1864). This chapter examines Boolean algebra first in the abstract, and then in two concrete examples, switching circuits and logic gates.

7.2 BOOLEAN ALGEBRA

Let B be a set on which are defined two binary operations, $+$ and $*$, and a unary operation, denoted $'$; let 0 and 1 denote two distinct elements of B. Then the sextuplet

$$\langle B, +, *, ', 0, 1 \rangle$$

is called a *Boolean algebra* if the following axioms hold for any elements a, b, c of the set B:

[$\mathbf{B_1}$] Commutative Laws:
(1a) $a + b = b + a$ (1b) $a * b = b * a$

[$\mathbf{B_2}$] Distributive Laws:
(2a) $a + (b * c) = (a + b) * (a + c)$ (2b) $a * (b + c) = (a * b) + (a * c)$

[$\mathbf{B_3}$] Identity Laws:
(3a) $a + 0 = a$ (3b) $a * 1 = a$

[$\mathbf{B_4}$] Complement Laws:
(4a) $a + a' = 1$ (4b) $a * a' = 0$

The above Boolean algebra is usually denoted by B alone when the operations are understood.

The element 0 is called the *zero* element, the element 1 is called the *unit* element, and a' is called the *complement* of a. The results of the operations $+$ and $*$ are called the *sum* and the *product*, respectively. We will frequently drop the symbol $*$ and use juxtaposition instead. Then (2b) and (2a) are written:

(2b) $a(b + c) = ab + ac$ (2a) $a + bc = (a + b)(a + c)$

The first is a familiar identity, but the second is not an identity in ordinary algebra.

We adopt the usual convention that, unless we are guided by parentheses, $'$ has precedence over $*$, and $*$ has precedence over $+$. For example,

$$a + b * c \text{ means } a + (b * c) \text{ and not } (a + b) * c$$

$$a * b' \text{ means } a * (b') \text{ and not } (a * b)'$$

Of course when $a + b * c$ is written $a + bc$ then the meaning is clear.

EXAMPLE 7.1

(a) Let B be the set of two elements, $\{0, 1\}$, with operations $+$ and $*$ as defined in Fig. 7-1. Suppose that complements are defined by $1' = 0$ and $0' = 1$. Then B is a Boolean algebra.

$$
\begin{array}{c|cc}
+ & 1 & 0 \\
\hline
1 & 1 & 1 \\
0 & 1 & 0
\end{array}
\qquad
\begin{array}{c|cc}
* & 1 & 0 \\
\hline
1 & 1 & 0 \\
0 & 0 & 0
\end{array}
$$

(a)   (b)

Fig. 7-1

(b) This is a generalization of (a). Let B_n denote the set of n-bit sequences. Define the sum, product, and complement of these sequences bit by bit as in (a). For example, given the elements

$$a = 1101010 \qquad b = 1011011$$

of B_7, we have

$$a + b = 1111011 \qquad a * b = 1001010 \qquad a' = 0010101$$

That is, in a given position, $a + b$ contains 1 if a or b contains 1; $a * b$ contains 1 if a and b contain 1; and a' contains 1 if a does not contain 1, i.e. if a contains 0. Then B_n is a Boolean algebra.

(c) Let $\mathscr{C}$ be a collection of sets closed under union, intersection, and complement. Then $\mathscr{C}$ is a Boolean algebra, with the empty set $\emptyset$ as the zero element and the universal set U as the unit element.

(d) Let Π be the set of propositions. Then Π is a Boolean algebra under the operations $\vee$ and $\wedge$, with negation $\sim$ being the complement. (Propositions in Π that are logically equivalent, i.e. have the same truth table, are taken to be identical.) As seen from Table 4-1, a contradiction f is the zero element, and a tautology t is the unit element.

(e) Let $D_{70} = \{1, 2, 5, 7, 10, 14, 35, 70\}$, the divisors of 70. Define $+$, $*$ and $'$ on D_{70} by

$$a + b \equiv \operatorname{lcm}(a, b) = \text{least common multiple of } a \text{ and } b$$

$$a * b \equiv \gcd(a, b) = \text{greatest common divisor of } a \text{ and } b$$

$$a' \equiv \frac{70}{a}$$

For example,

$$10 + 14 = \operatorname{lcm}(10, 14) = 70 \qquad 10 * 14 = \gcd(10, 14) = 2 \qquad 10' = \frac{70}{10} = 7$$

Then D_{70} is a Boolean algebra, with 1 as the zero element and 70 as the unit element.

7.3 DUALITY

The *dual* of any statement in a Boolean algebra B is the statement obtained by interchanging the operations $+$ and $*$, and interchanging the corresponding identity elements 0 and 1, in the original statement. For example, the dual of

$$(1 + a) * (b + 0) = b \qquad \text{is} \qquad (0 * a) + (b * 1) = b$$

Observe the symmetry in the axioms of a Boolean algebra B. That is, the dual of the set of axioms of B is the same as the original set of axioms. Accordingly, we have

Theorem 7.1 (Principle of Duality): The dual of any theorem in a Boolean algebra is also a theorem.

In other words, if any statement is a consequence of the axioms of a Boolean algebra, then the dual is also a consequence of those axioms since the dual statement can be proven by using the dual of each step of the proof of the original statement.

7.4 BASIC THEOREMS

Using the axioms $[\mathbf{B_1}]$ through $[\mathbf{B_4}]$, we prove (Problem 7.15) the following theorem.

Theorem 7.2: Let a, b, c be any elements in a Boolean algebra B.

    (i)   Idempotent Laws:

        (5a) $a + a = a$ (5b) $a * a = a$

    (ii)   Boundedness Laws:

        (6a) $a + 1 = 1$ (6b) $a * 0 = 0$

    (iii) Absorption Laws:

        (7a) $a + (a * b) = a$ (7b) $a * (a + b) = a$

    (iv) Associative Laws:

        (8a) $(a + b) + c = a + (b + c)$ (8b) $(a * b) * c = a * (b * c)$

Theorem 7.2 and our axioms still do not reflect all the properties of sets listed in Table 6-1. The next two theorems give us the remaining properties.

Theorem 7.3: Let a be any element in a Boolean algebra B.

    (i)   (Uniqueness of Complement)

        If $a + x = 1$ and $a * x = 0$, then $x = a'$.

    (ii)   (Involution Law) $(a')' = a$

    (iii) (9a) $0' = 1$ (9b) $1' = 0$

Theorem 7.4 (DeMorgan's laws): (10a) $(a + b)' = a' * b'$ (10b) $(a * b)' = a' + b'$

We prove these theorems in Problems 7.16 and 7.17.

7.5 ORDER AND BOOLEAN ALGEBRAS

A relation $\precsim$ on a set S is called a *partial ordering* on S if it has the following three properties:

(1) $a \precsim a$ for every a in S.

(2) If $a \precsim b$ and $b \precsim a$, then $a = b$.

(3) If $a \precsim b$ and $b \precsim c$, then $a \precsim c$.

A set S together with a partial ordering is called a *partially ordered set*, or *poset*. In such a case, $a \precsim b$ is read "a precedes b." We also write

$$a < b \text{ (read "}a\text{ strictly precedes }b\text{") if } a \precsim b \text{ but } a \neq b$$
$$a \succsim b \text{ (read "}a\text{ succeeds }b\text{") if } b \precsim a$$
$$a > b \text{ (read "}a\text{ strictly succeeds }b\text{") if } b < a$$

The term "partial" is used in defining a poset S because there may be elements a and b of S which are *noncomparable*, i.e. such that neither $a \precsim b$ nor $b \precsim a$. If, on the other hand, every pair of elements of S are comparable, then S is said to be *totally ordered*, or *linearly ordered*, and S is called a *chain*.

EXAMPLE 7.2

(a) Let $\mathscr{S}$ be any class of sets. The relation of set inclusion $\subset$ is a partial order on $\mathscr{S}$. For $A \subset A$ for any set in $\mathscr{S}$, if $A \subset B$ and $B \subset A$ then $A = B$, and if $A \subset B$ and $B \subset C$ then $A \subset C$.

(b) Consider the positive integers $\mathbf{N}$. We say that "a divides b," written $a \mid b$, if there is an integer c such that $ac = b$. For example, $2 \mid 4$, $3 \mid 12$, $7 \mid 21$, and so on. This relation of divisibility is a partial order on $\mathbf{N}$. Observe that 3 and 5 are noncomparable, since neither divides the other.

(c)   The relation $\leq$ is also a partial order on the positive integers **N**.   (In fact $\leq$ is a partial order on any subset of the real numbers.)   This relation is sometimes called the *usual order* on **N**.   Note that **N** is totally ordered by $\leq$; i.e. for any two integers a and b, either $a \leq b$ or $b \leq a$.

The notion of a partially ordered set comes up in the context of Boolean algebras because of the following theorem:

Theorem 7.5:   Let B be a Boolean algebra.   Then B is a partially ordered set, where $a \precsim b$ if and only if $a + b = b$.

EXAMPLE 7.3

(a)   Let B be any Boolean algebra.   Then, for any element a of B,

$$0 \precsim a \precsim 1$$

since $0 + a = a$ and $a + 1 = 1$.

(b)   Consider the Boolean algebra of sets [Example 7.1(c)].   Then set A precedes set B if A is a subset of B.

(c)   Consider the Boolean algebra of the propositional calculus [Example 7.1(d)].   Then proposition P precedes proposition Q if P logically implies Q (see Section 4.10).

Finite partially ordered sets S and, in particular, finite Boolean algebras S, can be pictured by diagrams as follows.   An element b of S is said to be an *immediate successor* of an element a, written $a \lessdot\mathrel{\mkern-5mu}\lessdot b$, if $a < b$ but there is no element x of S such that $a < x < b$.   In the diagram of S the elements are represented by points and there is an arrow, or a line slanting upward, from an element a to an element b whenever $a \lessdot\mathrel{\mkern-5mu}\lessdot b$.   In case S is a Boolean algebra, the zero element will be at the bottom of the diagram and the unit element will be at the top of the diagram.

EXAMPLE 7.4   Let $A = \{a, b, c\}$ and let $\mathscr{P}(A)$ be the collection of all subsets of A:

$$\mathscr{P}(A) = [A, \{a, b\}, \{a, c\}, \{b, c\}, \{a\}, \{b\}, \{c\}, \varnothing]$$

Then $\mathscr{P}(A)$ is a Boolean algebra of sets whose diagram is given in Fig. 7-2.   Observe that $\varnothing$ is at the bottom of the diagram and A is at the top of the diagram.

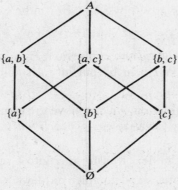

Fig. 7-2

Let B be a Boolean algebra.   An element a of B is called an *atom* of B if it is an immediate successor of the zero element.   Thus, in Example 7.4, the atoms are the three single-element sets $\{a\}$, $\{b\}$, and $\{c\}$.   Observe that $\mathscr{P}(A)$ contains $2^3 = 8$ elements, and that every nonempty set in $\mathscr{P}(A)$ is the union of a unique collection of atoms.   This result holds true in general; that is,

Theorem 7.6:   Let B be a finite Boolean algebra having n atoms.   Then B has 2^n elements, and every nonzero element of B is the sum of a unique set of atoms.

7.6 BOOLEAN EXPRESSIONS; SUM-OF-PRODUCTS FORM

Consider a set of variables (or letters or symbols), say $x_1, x_2, \ldots, x_n$. By a *Boolean expression E* in these variables, sometimes written $E(x_1, \ldots, x_n)$, we mean any variable or any expression built up from the variables using the Boolean operations $+$, $*$ and $'$. For example,

$$E = (x + y'z)' + (xyz' + x'y)' \qquad \text{and} \qquad F = ((xy'z' + y)' + x'z)'$$

are Boolean expressions in x, y and z.

A *literal* is a variable or complemented variable, e.g. x, x', y, y'. By a *fundamental product* we mean a literal or a product of two or more literals in which no two literals involve the same variable. For example, xz', $xy'z$, x, y', yz', $x'yz$ are fundamental products. However, $xyx'z$ and $xyzy$ are not fundamental products; the first contains x and x', and the second contains y in two places. Observe that

$$xyx'z = xx'yz = 0yz = 0$$

(since $x * x' = 0$, by the complement law) and

$$xyzy = xyyz = xyz$$

(since $y * y = y$, by the idempotent law). In other words, any Boolean product can be reduced to either 0 or a fundamental product.

One fundamental product, P_1, is said to be *included in* or *contained in* another fundamental product, P_2, if the literals of P_1 are also literals of P_2. For example, $x'z$ is included in $x'yz$, since x' and z are literals in $x'yz$. However, $x'z$ is not contained in $xy'z$, since x' is not a literal in $xy'z$. In case P_1 is included in P_2, then by the absorption law

$$P_1 + P_2 = P_1$$

For example, $x'z + x'yz = x'z$.

A Boolean expression E is said to be in a *sum-of-products* form or a *minterm* form if E is a fundamental product or the sum of two or more fundamental products none of which is included in another. For example, consider the expressions

$$E_1 = xz' + y'z + xyz' \qquad \text{and} \qquad E_2 = xz' + x'yz' + xy'z$$

Although the first expression, E_1, is a sum of products, it is not in a sum-of-products form, since xz' is contained in xyz'. However, by the absorption law, E_1 can be expressed as

$$E_1 = xz' + y'z + xyz' = xz' + xyz' + y'z = xz' + y'z$$

which is a sum-of-products form. The second expression, E_2, is already in a sum-of-products form.

Any nonzero Boolean expression E can be put into a sum-of-products form by the following procedure:

(1) Using DeMorgan's laws and involution, we can move the complement operation into any parentheses until finally it applies only to variables. Then E will consist only of sums and products of literals.

(2) Using the distributive law, we can next transform E into a sum of products.

(3) Using the commutative, idempotent, and complement laws, we can transform each product in E into 0 or a fundamental product. Finally, using the absorption law, we can put E into a sum-of-products form.

EXAMPLE 7.5 Consider the Boolean expression $E = ((ab)'c)'((a' + c)(b' + c'))'$. Applying the above algorithm,

(1) $$E = ((ab)'' + c')((a' + c)' + (b' + c')') = (ab + c')(ac' + bc)$$

(2) $$E = abac' + abbc + ac'c' + bcc'$$

(3) $$E = abc' + abc + ac' + 0 = ac' + abc$$

A (nonzero) Boolean expression $E(x_1, x_2, \ldots, x_n)$ is said to be in *complete sum-of-products form* if E is in a sum-of-products form and each product involves all the variables (we note that there are a maximum of 2^n such products). Any sum-of-products Boolean expression E can be put into complete sum-of-products form. In fact, if a fundamental product P of E does not involve x_i, then we may multiply P by $x_i + x_i'$; this is permissible, since $x_i + x_i' = 1$. We continue until all the products involve all the variables. A further consideration shows that the complete sum-of-products form for E is unique. In summary:

Theorem 7.7:   Every nonzero Boolean expression $E(x_1, x_2, \ldots, x_n)$ can be put into complete sum-of-products form, and such a representation is unique.

7.7 LOGIC GATES

This section begins the study of certain types of circuits called *logic circuits*. These circuits may be viewed as machines which contain one or more input devices and exactly one output device. At any instant of time each input device holds exactly one bit of information, i.e. 0 or 1; these data are processed by the circuit to yield as an output one bit, i.e. 0 or 1, at the output device. Thus the input devices may be assigned sequences of bits (where all sequences have the same number of bits) which are processed by the circuit one bit at a time to produce an output sequence with the same number of bits.

One may interpret a bit as a step-voltage across an input/output device; indeed, a sequence of bits, say 100010, is often pictured as follows:

$$\overline{1} \, \lfloor 000 \, \overline{\lfloor 11} \, \rfloor 0$$

One may assume that the circuit always processes the sequence from left to right or from right to left. Unless otherwise stated, we shall make the former assumption.

Logic circuits are built up from certain elementary circuits called *logic gates*, three of which will now be investigated. Section 7.8 will examine logic circuits in general.

OR Gate

Figure 7-3(a) indicates an OR gate with inputs A and B and output Y. We denote the output of an OR gate as

$$Y = A + B$$

where "addition" is defined by Fig. 7-1(a). That is, $Y = 1$ if $A = 1$ or $B = 1$, and $Y = 0$ only if both $A = 0$ and $B = 0$. For example, suppose A and B are assigned the following sequences of bits:

$$A = 11000110$$
$$B = 10010101$$

Then the OR gate will produce the sequence

$$A + B = 11010111$$

(This result can be obtained by reading A and B from right to left or from left to right.)

Usually we want to know the output of a logic circuit for all the different possible combinations of input bits. For two inputs, A and B, the *special sequences* which give these different combinations contain four bits, as follows:

$$A = 0011 \qquad\qquad A = 0011$$
$$\qquad\qquad \text{or} \qquad\qquad\qquad\qquad \text{or} \qquad \ldots$$
$$B = 0101 \qquad\qquad B = 1001$$

(In general, special sequences for n inputs will contain 2^n bits.) The value of the output for these special sequences is called the *truth table* for the circuit. Figure 7-3(b) is the truth table for the OR gate of Fig. 7-3(a); the sequences in this truth table are written vertically rather than horizontally. When the number of inputs becomes large, we shall usually write such sequences horizontally, as above.

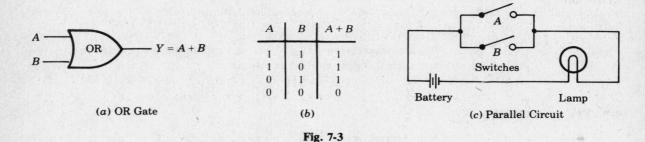

A	B	$A+B$
1	1	1
1	0	1
0	1	1
0	0	0

(a) OR Gate   (b)   (c) Parallel Circuit

Fig. 7-3

Figure 7-3(c) illustrates, for the OR gate of Fig. 7-3(a), the close relationship between logic circuits and electrical switching circuits. An electrical switching circuit normally contains some source of energy (say, a battery), an output device (say, a lamp), and one or more switches—all connected by wires. A switch is a two-state device that is either closed (on) or open (off), and current can pass through the switch only when the switch is closed. In Fig. 7-3(c) two switches, A and B, are connected in parallel. Observe that the lamp will light if switch A is closed, or if switch B is closed, or if both switches are closed. But this is precisely the property described by the truth table for the OR gate, where 1 denotes that the switch (A, B) or lamp ($A+B$) is on and 0 denotes that it is off.

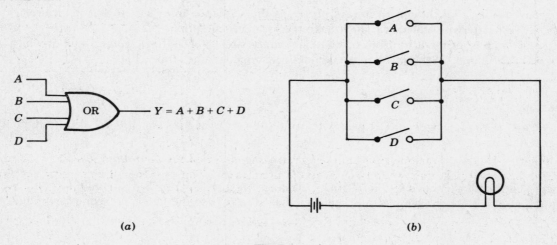

(a)   (b)

Fig. 7-4

OR gates may have more than two inputs. Figure 7-4(a) shows an OR gate with four inputs, A, B, C, and D, and output

$$Y = A + B + C + D$$

The output is 0 if and only if all inputs are 0. Thus, the four input sequences

$$A = 10000101$$
$$B = 10100001$$
$$C = 00100100$$
$$D = 10010101$$

give

$$Y = 10110101$$

as the output sequence.   The analogous switching circuit appears in Fig. 7-4(*b*); clearly, the lamp will be on if and only if one (or more) of the four switches is on.

AND Gate

Figure 7-5(*a*) pictures an AND gate, with inputs A and B and output Y.   We designate the output of an AND gate as the product of the inputs,

$$Y = A \cdot B$$

or, simply, $Y = AB$.   The value of Y is determined by the "multiplication" table of Fig. 7-1(*b*). That is, $Y = 1$ when both $A = 1$ and $B = 1$; otherwise, $Y = 0$.   For example, suppose A and B are assigned the following sequences of bits:

$$A = 11000110$$
$$B = 01101101$$

Then the AND gate will produce the sequence

$$A \cdot B = 01000100$$

The truth table for this AND gate appears in Fig. 7-5(*b*), where again the sequences are written vertically.

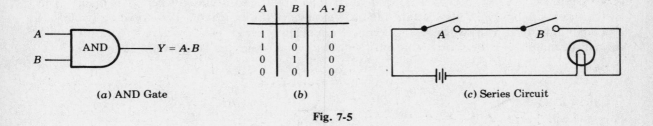

A	B	$A \cdot B$
1	1	1
1	0	0
0	1	0
0	0	0

(a) AND Gate   (b)   (c) Series Circuit

Fig. 7-5

Figure 7-5(*c*) is a switching circuit showing two switches, A and B, connected in series.   Observe that the lamp will light only when both A and B are closed.   This is exactly the property described by the truth table for the AND gate, if again we let 1 denote that the circuit element is on and 0 denote that it is off.

An AND gate may also have more than two inputs.   Figure 7-6(*a*) shows an AND gate with four inputs, A, B, C, and D, and output

$$Y = A \cdot B \cdot C \cdot D \qquad \text{or} \qquad Y = ABCD$$

The output is 1 if and only if all inputs are 1.   Thus, the four input sequences

$$A = 11100111$$
$$B = 01111011$$
$$C = 01110011$$
$$D = 11101110$$

give

$$Y = 01100010$$

as the output sequence.   The analogous switching circuit appears in Fig. 7-6(*b*); clearly, the lamp will be on only when all the switches are on.

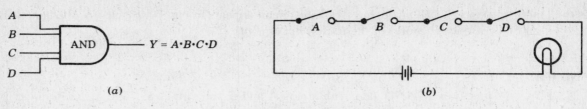

Fig. 7-6

NOT Gate

Figure 7-7(a) shows a NOT gate, also called an *inverter*, with input A and output Y. The NOT gate can have only one input, and its output is denoted by putting a bar over the input:

$$Y = \bar{A}$$

The value of the output Y is the opposite (ones complement) of the value of the input A; i.e. $Y = 1$ when $A = 0$, and $Y = 0$ when $A = 1$. Thus, if A is assigned the sequence of bits

$$A = 11000110$$

the NOT gate will produce the sequence

$$\bar{A} = 00111001$$

The truth table for the NOT gate appears in Fig. 7-7(b).

Switching circuits also contain the analog of the NOT gate. Specifically, along with any switch A we can include a switch $\bar{A}$ that is open when A is closed and is closed when A is open. This switch $\bar{A}$, pictured in Fig. 7-7(c), is called the *complement* of the switch A. (We could also realize $\bar{A}$ as a lamp in parallel with switch A, the combination being in series with a battery. With the switch closed, the lamp would be shorted out (off); with the switch open, the lamp would be on.)

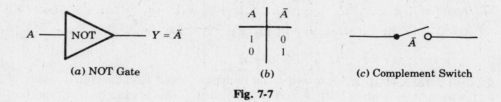

(a) NOT Gate   (b)   (c) Complement Switch

Fig. 7-7

7.8 LOGIC CIRCUITS

The truth tables for the OR, AND, and NOT gates, Figs. 7-3(b), 7-5(b), and 7-7(b), are respectively identical to those for the propositions $p \vee q$ (disjunction, "p or q"), $p \wedge q$ (conjunction, "p and q"), and $\sim p$ (negation, "not p"), which appear in Sections 4.3, 4.2, and 4.4. The only difference is that 1 and 0 are used here instead of T and F. Thus logic circuits, of which these gates are the circuit elements, satisfy the same laws as do propositions and hence they form a Boolean algebra. We state this result formally.

Theorem 7.8:   Logic circuits form a Boolean algebra.

Logic circuits come in various patterns. We will mainly be concerned with a pattern that corresponds to a Boolean sum-of-products expression. Specifically, an *AND-OR circuit* has several inputs, with some of the inputs or their complements fed into each AND gate. The outputs of all the AND gates are fed into a single OR gate which gives the output for the circuit. (In the limiting cases, there may be a single AND gate and no OR gate, or a single OR gate and no AND gate.)   Figure 7-8(a) is a typical AND-OR circuit, with three inputs, A, B, and C.   (Frequently, for economy of space, we omit the word from the interior of the gate symbol.)

Given any logic circuit L, we want to know the effect of L on an arbitrary input; this is usually specified by means of a truth table.   The truth table of L is obtained by first writing L as a Boolean expression $L(A, B, C, \ldots)$, with inputs $A, B, C, \ldots$, and then calculating the truth table step by step, as we did for the truth tables of propositions.   The Boolean expression itself is obtained from the circuit by tracing the inputs through all the gates, labeling when necessary each gate with its inputs and output.   In an AND-OR circuit, we need only label each AND gate with its inputs and output, and then label the output of the OR gate, which is the output of the circuit, with the sum of the outputs of the AND gates.

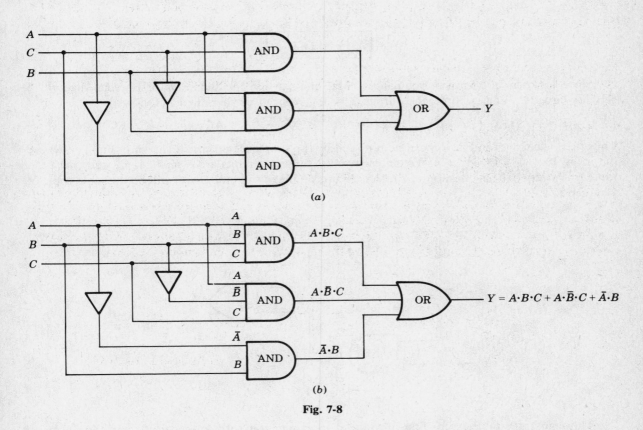

Fig. 7-8

EXAMPLE 7.6   Consider the logic circuit in Fig. 7-8(a).   We label the first AND gate with inputs A, B, and C, and output $A \cdot B \cdot C$; the second AND gate with inputs A, $\bar{B}$, and C, and output $A \cdot \bar{B} \cdot C$; and the third AND gate with inputs $\bar{A}$ and B, and output $\bar{A} \cdot B$.   See Fig. 7-8(b).   Then the output of the OR gate, which is the output of the circuit, is the Boolean expression

$$Y = A \cdot B \cdot C + A \cdot \bar{B} \cdot C + \bar{A} \cdot B$$

Observe that this is a Boolean sum-of-products.

Now the truth table of the circuit may be found by substituting in the Boolean expression the three special sequences

$$A = 00001111$$
$$B = 00110011$$
$$C = 01010101$$

A given bit in $A \cdot B \cdot C$ will be 1 if and only if A, B, and C have a 1 in that position.   Thus,

$$A \cdot B \cdot C = 00000001$$

Similarly,
$$A \cdot \bar{B} \cdot C = 00000100$$
$$\bar{A} \cdot B = 00110000$$

Hence,
$$Y = 00110101$$

is the output.   The truth table consists of the input sequences together with the output sequence:

A	00001111
B	00110011
C	01010101
Y	00110101

Since logic circuits form a Boolean algebra, one can use the theorems of Boolean algebra to simplify circuits.   For example, the output Y of Fig. 7-8 can be simplified as follows:

$$Y = A \cdot B \cdot C + A \cdot \bar{B} \cdot C + \bar{A} \cdot B = AC(B + \bar{B}) + \bar{A}B = AC \cdot 1 + \bar{A}B = AC + \bar{A}B$$

Thus the logic circuit in Fig. 7-8 can be replaced by the simpler logic circuit shown in Fig. 7-9, whose output is $Y = A \cdot C + \bar{A} \cdot B$.   We emphasize that the two circuits are *equivalent*, i.e. they have the same truth table.   This question of simpler circuits and how to obtain them will be the main topic of Chapter 8.

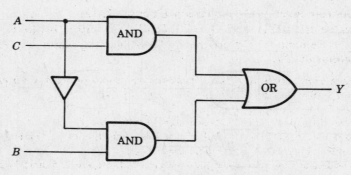

Fig. 7-9

Although truth tables were first introduced in connection with propositions, and then in connection with logic circuits, they are actually a property of Boolean expressions in general.   In fact, the (unique) truth table of a Boolean expression is tantamount to the unique complete sum-of-products form given by Theorem 7.7.   This correspondence arises from the fact (see Problem 7.11) that when any combination of 1s and 0s is assigned to the variables, one and only one of the fundamental products involving all the variables takes on the value 1; all the rest take on the value 0.   Hence, from the truth table we can write down the complete sum-of-products form by inspection; and conversely.

EXAMPLE 7.7   The complete sum-of-products form for the Boolean expression of Example 7.6 is

$$Y = A \cdot B \cdot C + A \cdot \bar{B} \cdot C + \bar{A} \cdot B \cdot (C + \bar{C})$$
$$= A \cdot B \cdot C + A \cdot \bar{B} \cdot C + \bar{A} \cdot B \cdot C + \bar{A} \cdot B \cdot \bar{C}$$

When $A = 1$, $B = 1$, $C = 1$, the first fundamental product, $A \cdot B \cdot C$, and along with it Y, equals 1; all other complete fundamental products equal 0.   Similarly, $Y = 1$ when $A = 1$, $\bar{B} = 1$ or $B = 0$, $C = 1$; when $A = 0$,

$B = 1$, $C = 1$; and when $A = 0$, $B = 1$, $C = 0$. For all other combinations of 1s and 0s, $Y = 0$. Thus we have the truth table

A	$1100\cdots\cdots$
B	$1011\cdots\cdots$
C	$1110\cdots\cdots$
Y	11110000

which, except for the order of the columns, coincides with the truth table found in Example 7.6.

Conversely, starting with the truth table, one reads off the fundamental products corresponding to the 1s in the Y-row and thereby obtains the complete sum-of-products form for Y.

Solved Problems

BOOLEAN ALGEBRA

7.1   Consider the Boolean algebra D_{70} defined in Example 7.1(e).
   (a)   Find the value of: (1) $A = 35 * (2 + 7')$, (2) $B = (35 * 10) + 14'$, (3) $C = (2 + 7) * (14 * 10)'$.
   (b)   How are the elements of D_{70} ordered?   Draw the diagram of D_{70}.
   (c)   Find the atoms of D_{70}.

   (a)   Calculate each expression step by step, using the definitions of $a + b$, $a * b$ and a'.
      (1) $7' = 10$, $2 + 10 = 10$; hence, $A = 35 * 10 = 5$.
      (2) $35 * 10 = 5$, $14' = 5$; hence, $B = 5 + 5 = 5$.
      (3) $2 + 7 = 14$, $14 * 10 = 2$, $2' = 35$; hence, $C = 14 * 35 = 7$.
   (b)   Note that $a + b = \text{lcm}(a, b) = b$ if and only if b is a multiple of a. Thus $2 \lesssim 14$, but 2 and 5 are noncomparable.   Figure 7-10 is the diagram of D_{70}.
   (c)   The atoms of D_{70} are the immediate successors of 1; these are 2, 5, and 7 (the primes in D_{70}).

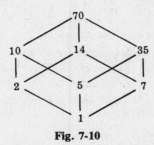

Fig. 7-10

7.2   Write the dual of each Boolean equation: (a) $(a * 1) * (0 + a') = 0$, (b) $a + a'b = a + b$.

   (a)   To obtain the dual equation, interchange $+$ and $*$, and interchange 0 and 1.   Thus

$$(a + 0) + (1 * a') = 1$$

(b) First write the equation using $*$: $a + (a' * b) = a + b$.   Then the dual is: $a * (a' + b) = a * b$, which can be written as

$$a(a' + b) = ab$$

7.3 Put the following Boolean expressions $E(x, y, z)$ into sum-of-products form, and then into complete sum-of-products form: (a) $E_1 = x(y'z)'$, (b) $E_2 = z(x' + y) + y'$.

Use the algorithm given in Section 7.6.
(a) First we have

$$E_1 = x(y'z)' = x(y + z') = xy + xz'$$

and E_1 is in a sum-of-products form. Next we have

$$E_1 = xy + xz' = xy(z + z') + x(y + y')z' = xyz + xyz' + xyz' + xy'z'$$
$$= xyz + xyz' + xy'z'$$

which is in the complete sum-of-products form.
(b) First we have

$$E_2 = z(x' + y) + y' = x'z + yz + y'$$

Then,

$$E_2 = x'z + yz + y' = x'z(y + y') + yz(x + x') + y'(x + x')(z + z')$$
$$= x'yz + x'y'z + xyz + x'yz + xy'z + xy'z' + x'y'z + x'y'z'$$
$$= xyz + xy'z + xy'z' + x'yz + x'y'z + x'y'z'$$

7.4 Express $E(x, y, z) = (x' + y)' + x'y$ in complete sum-of-products form.

We have $E = (x' + y)' + x'y = xy' + x'y$, which would be the complete sum-of-products form of E if E were a Boolean expression in x and y. However, it is specified that E is a Boolean expression in the three variables x, y, and z. Hence,

$$E = xy' + x'y = xy'(z + z') + x'y(z + z') = xy'z + xy'z' + x'yz + x'yz'$$

is the complete sum-of-products form for E.

7.5 Let $E = xy' + xyz' + x'yz'$.   Prove that (a) $xz' + E = E$, (b) $x + E \neq E$, (c) $z' + E \neq E$.

Since the complete sum-of-products form is unique (Theorem 7.7), $A + E = E$, where $A \neq 0$, if and only if the summands in the complete sum-of-products form for A are among the summands in the complete sum-of-products form for E. Hence, first find the complete sum-of-products form for E:

$$E = xy'(z + z') + xyz' + x'yz' = xy'z + xy'z' + xyz' + x'yz'$$

(a) Express xz' in complete sum-of-products form:

$$xz' = xz'(y + y') = xyz' + xy'z'$$

Since the summands of xz' are among those of E, we have $xz' + E = E$.
(b) Express x in complete sum-of-products form:

$$x = x(y + y')(z + z') = xyz + xyz' + xy'z + xy'z'$$

The summand xyz of x is not a summand of E; hence $x + E \neq E$.
(c) Express z' in complete sum-of-products form:

$$z' = z'(x + x')(y + y') = xyz' + xy'z' + x'yz' + x'y'z'$$

The summand $x'y'z'$ of z' is not a summand of E; hence $z' + E \neq E$.

7.6   Consider the Venn diagram of sets A, B, and C given in Fig. 7-11.   Observe that the universal set U (the rectangle) is partitioned into $2^3 = 8$ sets, which are labeled (1) to (8). (a)
Express each of the sets in terms of A, B, and C. (b) In the Boolean algebra $\mathscr{P}(U)$, let
$E(A, B, C)$ be any Boolean expression involving the sets A, B, and C.   Give a geometrical
interpretation of the complete sum-of-products form for E.

(a) Using the Boolean notations AB for $A \cap B$ and A' for A^c, each of the eight sets is a fundamental
product involving A, B, and C:

$$(1) = AB'C' \qquad (3) = A'BC' \qquad (5) = ABC \qquad (7) = A'B'C$$
$$(2) = ABC' \qquad (4) = AB'C \qquad (6) = A'BC \qquad (8) = A'B'C'$$

(b) The Boolean expression E is represented by the union of one or more of the areas (1) through (8)
in Fig. 7-11.   This union (sum) is unique, and yields the complete sum-of-products form for E.

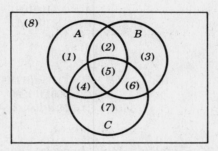

Fig. 7-11

LOGIC CIRCUITS

7.7   Consider the following three pairs of sequences of bits:

   (i)   110001   (ii)   10001111   (iii)   101100111000
        101101   00111100   000111001101

How would each pair of sequences be processed by (a) an OR gate? (b) an AND gate?

(a) Recall that a 0 occurs as an output of an OR gate only where both inputs are 0.   In (i), this occurs
only in the 5th position; in (ii), only in the 2nd position; in (iii), only in the 2nd and 11th positions.
Hence the outputs are

   (i)   111101   (ii)   10111111   (iii)   101111111101

(b) Recall that a 1 occurs as an output of an AND gate only where both inputs are 1.   In (i), this
occurs only in the first and last positions; in (ii), only in the 5th and 6th positions; in (iii), only in the
4th and 9th positions.   Hence the outputs are

   (i)   100001   (ii)   00001100   (iii)   000100001000

7.8   How would a NOT gate process each sequence?

   (i)   110001   (ii)   10001111   (iii)   101100111000

   A NOT gate changes 0 to 1 and 1 to 0.   Hence the outputs are

   (i)   001110   (ii)   01110000   (iii)   010011000111

7.9   Given $A = 1100110110$
$$B = 1110000111$$
$$C = 1010010110$$

find (a) $A + B + C$, (b) $A \cdot B \cdot C$, (c) $C(\bar{A} + B)$, (d) $A(\overline{B + C})$.

(a)   A 0 occurs in a sum, representing an OR gate, only where all the inputs are 0s.   Note that there are three 0s only in the 4th and 7th positions.   Hence

$$A + B + C = 1110110111$$

(b)   A 1 occurs in a product, representing an AND gate, only where all the inputs are 1s.   This happens only in the first, 8th, and 9th positions.   Hence

$$A \cdot B \cdot C = 1000000110$$

(c)   Calculate step by step:

$$\bar{A} = 0011001001$$
$$\bar{A} + B = 1111001111$$
$$C(\bar{A} + B) = 1010000110$$

(d)   Calculate step by step:

$$B + C = 1110010111$$
$$\overline{B + C} = 0001101000$$
$$A(\overline{B + C}) = 0000100000$$

7.10   Given five inputs, A, B, C, D, and E, find special sequences which give all the different possible combinations of input bits.

Each sequence will contain $2^5 = 32$ bits.   One assignment scheme is as follows:

(i)   Let A be assigned $2^4 = 16$ bits which are 0s, followed by $2^4 = 16$ bits which are 1s.
(ii)   Let B be assigned $2^3 = 8$ bits which are 0s, followed by $2^3 = 8$ bits which are 1s; and then repeat once.
(iii) Let C be assigned $2^2 = 4$ bits which are 0s, followed by $2^2 = 4$ bits which are 1s; and then repeat three times.
(iv)   Let D be assigned $2^1 = 2$ bits which are 0s, followed by $2^1 = 2$ bits which are 1s; and then repeat seven times.
(v)   Let E be assigned $2^0 = 1$ bit which is 0, followed by $2^0 = 1$ bit which is 1; and then repeat fifteen times.

The resulting special sequences are

$$A = 00000000000000001111111111111111$$
$$B = 00000000111111110000000011111111$$
$$C = 00001111000011110000111100001111$$
$$D = 00110011001100110011001100110011$$
$$E = 01010101010101010101010101010101$$

(Noting that the columns of the above array are, from left to right, the first 32 binary integers in increasing order, we have another, and perhaps simpler, assignment scheme.)

7.11   Given three inputs, A, B, and C.   Find the truth tables of the eight fundamental products:

$$A \cdot B \cdot C \qquad A \cdot B \cdot \bar{C} \qquad A \cdot \bar{B} \cdot C \qquad A \cdot \bar{B} \cdot \bar{C}$$
$$\bar{A} \cdot B \cdot C \qquad \bar{A} \cdot B \cdot \bar{C} \qquad \bar{A} \cdot \bar{B} \cdot C \qquad \bar{A} \cdot \bar{B} \cdot \bar{C}$$

Note first that the special sequences for A, B, and C each contain $2^3 = 8$ bits.   We have:

$$A = 0\ 0\ 0\ 0\ 1\ 1\ 1\ 1$$
$$B = 0\ 0\ 1\ 1\ 0\ 0\ 1\ 1$$
$$C = 0\ 1\ 0\ 1\ 0\ 1\ 0\ 1$$

Then

$$\bar{A} = 1\ 1\ 1\ 1\ 0\ 0\ 0\ 0$$
$$\bar{B} = 1\ 1\ 0\ 0\ 1\ 1\ 0\ 0$$
$$\bar{C} = 1\ 0\ 1\ 0\ 1\ 0\ 1\ 0$$

and

$$A \cdot B \cdot C = 0\ 0\ 0\ 0\ 0\ 0\ 0\ 1$$
$$A \cdot B \cdot \bar{C} = 0\ 0\ 0\ 0\ 0\ 0\ 1\ 0$$
$$A \cdot \bar{B} \cdot C = 0\ 0\ 0\ 0\ 0\ 1\ 0\ 0$$
$$A \cdot \bar{B} \cdot \bar{C} = 0\ 0\ 0\ 0\ 1\ 0\ 0\ 0$$
$$\bar{A} \cdot B \cdot C = 0\ 0\ 0\ 1\ 0\ 0\ 0\ 0$$
$$\bar{A} \cdot B \cdot \bar{C} = 0\ 0\ 1\ 0\ 0\ 0\ 0\ 0$$
$$\bar{A} \cdot \bar{B} \cdot C = 0\ 1\ 0\ 0\ 0\ 0\ 0\ 0$$
$$\bar{A} \cdot \bar{B} \cdot \bar{C} = 1\ 0\ 0\ 0\ 0\ 0\ 0\ 0$$

It is seen that for each combination of inputs, exactly one of the eight fundamental products assumes the value 1. Thus, for $A = 0$, $B = 1$, $C = 1$ (i.e. $\bar{A} = B = C = 1$), only the product $\bar{A} \cdot B \cdot C$ equals 1. Furthermore, a given product takes on the value 1 for just one combination of inputs; that is, its truth table has 1 in exactly one position and 0s elsewhere.

7.12   Find a Boolean expression and the truth table for the logic circuit in Fig. 7-12(a).

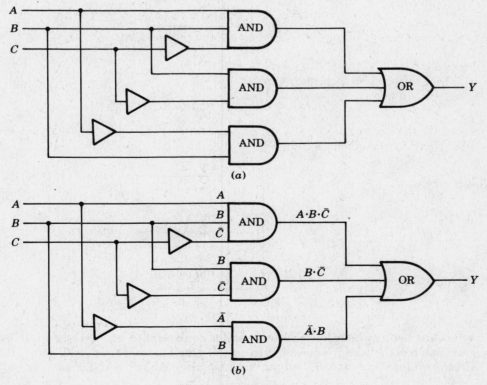

Fig. 7-12

Observe first that the circuit has an AND-OR pattern.   Label the inputs and output of each AND gate as in Fig. 7-12(b).   Then the output Y of the circuit is the sum of the outputs of the AND gates,

$$Y = AB\bar{C} + B\bar{C} + \bar{A}B$$

Method 1.
Since there are three inputs, the truth table of the circuit will consist of 8-bit sequences. We calculate as follows:

$$A = 00001111$$
$$B = 00110011$$
$$C = 01010101$$
$$\bar{A} = 11110000$$
$$\bar{C} = 10101010$$
$$AB\bar{C} = 00000010$$
$$B\bar{C} = 00100010$$
$$\bar{A}B = 00110000$$
$$Y = 00110010$$

That is, the truth table of the circuit is

A	00001111
B	00110011
C	01010101
Y	00110010

Method 2.
The complete sum-of-products form for Y is

$$Y = AB\bar{C} + B\bar{C}(A + \bar{A}) + \bar{A}B(C + \bar{C})$$
$$= AB\bar{C} + \bar{A}B\bar{C} + \bar{A}BC$$

from which, by inspection, the truth table is

A	100······
B	111······
C	001······
Y	11100000

7.13 We define two new gates.   A *NAND gate* is equivalent to an AND gate followed by a NOT gate, and a *NOR gate* is equivalent to an OR gate followed by a NOT gate.   (See Fig. 7-13.) Given two inputs, A and B, find the truth table for the (a) NAND gate, (b) NOR gate.

(a) The output of the NAND gate is $Y = \overline{A \cdot B}$. Calculate the truth table as follows:

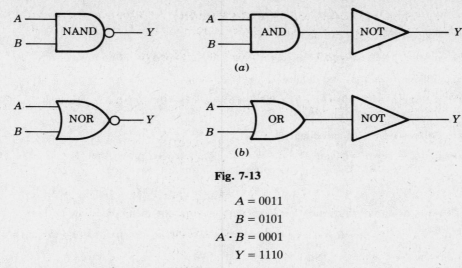

Fig. 7-13

$$A = 0011$$
$$B = 0101$$
$$A \cdot B = 0001$$
$$Y = 1110$$

(b) The output of the NOR gate is $Y = \overline{A + B}$. Calculate its truth table as follows:

$$A = 0011$$
$$B = 0101$$
$$A + B = 0111$$
$$Y = 1000$$

7.14 Determine a Boolean expression for each switching circuit in Fig. 7-14.

Recall that we use a sum for a parallel circuit, and a product for a series circuit. Thus:

$(a)\ A \cdot (B + \bar{A}) \cdot C$ $(b)\ A \cdot (C + \bar{B}) + B \cdot \bar{C}$

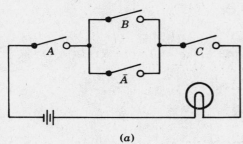
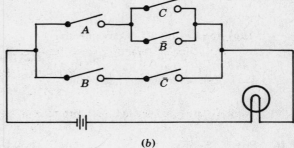

(a)   (b)

Fig. 7-14

PROOF OF THEOREMS

7.15 Prove Theorem 7.2:

(i)   Idempotent Laws:
      $(5a)\ a + a = a$ $(5b)\ a * a = a$
(ii) Boundedness Laws:
      $(6a)\ a + 1 = 1$ $(6b)\ a * 0 = 0$
(iii) Absorption Laws:
      $(7a)\ a + (a * b) = a$ $(7b)\ a * (a + b) = a$
(iv) Associative Laws:
      $(8a)\ (a + b) + c = a + (b + c)$ $(8b)\ (a * b) * c = a * (b * c)$

(5b) $a = a * 1 = a * (a + a') = (a * a) + (a * a') = (a * a) + 0 = a * a$

(5a)   Follows from (5b) and duality.

(6b) $a * 0 = (a * 0) + 0 = (a * 0) + (a * a') = a * (0 + a') = a * (a' + 0) = a * a' = 0$

(6a)   Follows from (6b) and duality.

(7b) $a * (a + b) = (a + 0) * (a + b) = a + (0 * b) = a + (b * 0) = a + 0 = a$, where the boundedness law was used in the next-to-last step.

(7a)   Follows from (7b) and duality.

(8b)   Let $L = (a * b) * c$ and $R = a * (b * c)$. We need to prove that $L = R$. We first prove that $a + L = a + R$. Using the absorption laws in the last two steps,

$$a + L = a + ((a * b) * c) = (a + (a * b)) * (a + c) = a * (a + c) = a$$

Also, using the absorption law in the last step and the idempotent law in the next-to-last step,

$$a + R = a + (a * (b * c)) = (a + a) * (a + (b * c)) = a * (a + (b * c)) = a$$

Thus $a + L = a + R$. Next we show that $a' + L = a' + R$. We have,

$$
\begin{aligned}
a' + L &= a' + ((a * b) * c) \\
&= (a' + (a * b)) * (a' + c) \\
&= ((a' + a) * (a' + b)) * (a' + c) \\
&= (1 * (a' + b)) * (a' + c) \\
&= (a' + b) * (a' + c) \\
&= a' + (b * c)
\end{aligned}
$$

Also,

$$
\begin{aligned}
a' + R &= a' + (a * (b * c)) \\
&= (a' + a) * (a' + (b * c)) \\
&= 1 * (a' + (b * c)) \\
&= a' + (b * c)
\end{aligned}
$$

Thus $a' + L = a' + R$. Consequently,

$$L = L + 0 = L + (a * a') = (L + a) * (L + a') = (a + L) * (a' + L) = (a + R) * (a' + R) = R$$

(8a)   Follows from (8b) and duality.

7.16   Prove Theorem 7.3:

(i)   (Uniqueness of Complement) If $a + x = 1$ and $a * x = 0$, then $x = a'$.

(ii)   (Involution Law) $(a')' = a$

(iii)   (9a) $0' = 1$ (9b) $1' = 0$

(i)   We have:

$$
\begin{aligned}
a' = a' + 0 &= a' + (a * x) = (a' + a) * (a' + x) \\
&= 1 * (a' + x) = a' + x
\end{aligned}
$$

Also,

$$x = x + 0 = x + (a * a') = (x + a) * (x + a') = 1 * (x + a') = x + a'$$

Hence

$$x = x + a' = a' + x = a'$$

(ii)   By definition of complement, $a + a' = 1$ and $a * a' = 0$. By commutativity, $a' + a = 1$ and $a' * a = 0$. By uniqueness of complement, a is the complement of a', that is, $a = (a')'$.

(iii) By boundedness law ($6a$), $0 + 1 = 1$, and by identity axiom ($3b$), $0 * 1 = 0$. By uniqueness of complement, 1 is the complement of 0, that is, $1 = 0'$. By duality, $0 = 1'$.

7.17   Prove Theorem 7.4 (DeMorgan's laws):

$(10a)$ $(a + b)' = a' * b'$ $(10b)$ $(a * b)' = a' + b'$

$(10a)$ We need to show that $(a + b) + (a' * b') = 1$ and $(a + b) * (a' * b') = 0$; then by uniqueness of complement, $a' * b' = (a + b)'$. We have:

$$(a + b) + (a' * b') = b + a + (a' * b') = b + (a + a') * (a + b')$$
$$= b + 1 * (a + b') = b + a + b' = b + b' + a = 1 + a = 1$$

Also,

$$(a + b) * (a' * b') = ((a + b) * a') * b'$$
$$= ((a * a') + (b * a')) * b' = (0 + (b * a')) * b'$$
$$= (b * a') * b' = (b * b') * a' = 0 * a' = 0$$

Thus $a' * b' = (a + b)'$.

$(10b)$ Principle of duality (Theorem 7.1).

Supplementary Problems

BOOLEAN ALGEBRA

7.18   Consider $D_{110} = \{1, 2, 5, 10, 11, 22, 55, 110\}$, the Boolean algebra of the divisors of 110.   The operations are as defined in Example 7.1(e).

(a)   Evaluate

$$X_1 = 2 + 11' \qquad X_3 = 22 * (5 + 10) \qquad X_5 = (55 * 10)' + 2$$
$$X_2 = 5 * 10 + 2 \qquad X_4 = 5' * 55 \qquad X_6 = (2 + 11') * 22$$

(b)   Draw the diagram of D_{110}.
(c)   Find the atoms of D_{110}.

7.19   For the Boolean algebra of Example 7.1(a), evaluate

$$X = 1 * (0 + 1') \qquad Y = (1 + 1) * (0' + 0) \qquad Z = (1' + 0)' + (1 * 0')'$$

7.20   An element M in a Boolean algebra B is called a *maxterm* if the unit element 1 is its only strict successor.

(a)   Find the maxterms of the Boolean algebra in Problem 7.1.
(b)   Find the maxterms of the Boolean algebra in Problem 7.18.
(c)   Show that the complements of atoms are maxterms.   (*Hint*: $x = y$ if and only if $x' = y'$.)

7.21   Write the dual of each Boolean equation:

(a) $a(a' + b) = ab$ (b) $(a + 1)(a + 0) = a$ (c) $(a + b)(b + c) = ac + b$

7.22 Write each Boolean expression $E(x, y, z)$ as a sum of products, and then in complete sum-of-products form:

         (a) $x(xy' + x'y + y'z)$ (c) $(x + y'z)(y + z')$ (e) $(x' + y)' + y'z$

         (b) $(x'y)'(x' + xyz')$ (d) $(x + y)'(xy')'$ (f) $y(x + yz)'$

7.23 Write the following set expressions involving sets A, B, and C as unions of intersections:

         (a) $(A \cup B)^c \cap (C^c \cup B)$ (b) $(B \cap C)^c \cap (A^c \cup C)^c$

LOGIC GATES

7.24 Determine the output of each gate in Fig. 7-15.

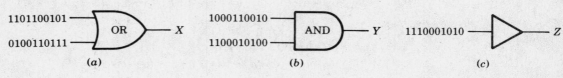

            (a)   (b)   (c)

Fig. 7-15

7.25 If $A = 1100110111$, $B = 0001110110$, $C = 1010110011$, evaluate

     (a) $A + B$ (c) $\bar{A}$ (e) $A \cdot C$ (g) $\bar{A} + B \cdot \bar{C}$

     (b) $A + C$ (d) $\bar{C}$ (f) $B \cdot C$ (h) $B(\bar{A} + C)$

7.26 Given four inputs, A, B, C, and D, find special sequences which give all the different possible combinations of inputs.

7.27 Consider the logic circuit in Fig. 7-16. (a) Give the output Y as a Boolean expression in the inputs A, B, and C. (b) Find the truth table of the circuit.

7.28 Consider the logic circuit in Fig. 7-17. (a) Give the output Y as a Boolean expression in the inputs A, B, and C. (b) Find the truth table of the circuit.

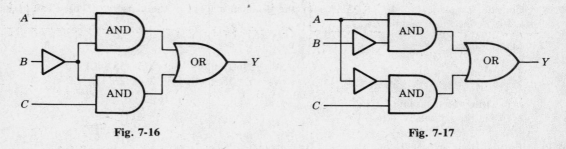

               Fig. 7-16   **Fig. 7-17**

7.29 Consider the logic circuit in Fig. 7-18. (a) Give the output Y as a Boolean expression in the inputs A, B, and C. (b) Find the truth table of the circuit.

7.30 The logic circuit in Fig. 7-19 contains a NAND gate and a NOR gate (Problem 7.13), among others. (a) Write the output Y as a Boolean expression in the inputs A, B, and C. (b) Find the truth table of the circuit.

7.31 Draw the logic circuit corresponding to each Boolean expression:

      (a) $E_1 = A\bar{B} + AB\bar{C}$ (b) $E_2 = \overline{A + BC} + B$ (c) $E_3 = \bar{A}B + \overline{A + C}$

7.32 Find the Boolean expression corresponding to each switching circuit in Fig. 7-20.

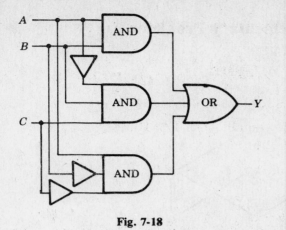

Fig. 7-18

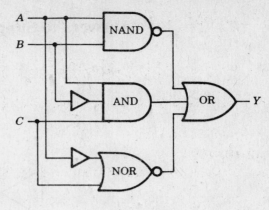

Fig. 7-19

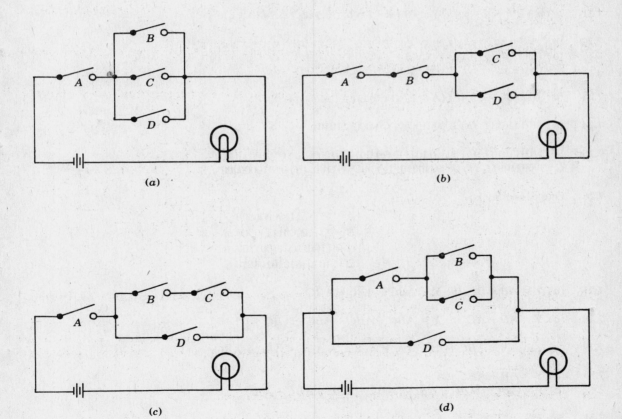

(a) (b)

(c) (d)

Fig. 7-20

Answers to Supplementary Problems

7.18   (a) $X_1 = 10$, $X_2 = 10$, $X_3 = 2$, $X_4 = 11$, $X_5 = 22$, $X_6 = 2$
       (b)   See Fig. 7-21.
       (c)   2, 5, 11

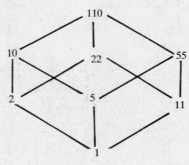

Fig. 7-21

7.19 $X = 0$, $Y = 1$, $Z = 1$

7.20   (a) 10, 14, 35; (b) 10, 22, 55

7.21   (a) $a + a'b = a + b$,   (b) $a * 0 + a * 1 = a$,   (c) $ab + bc = (a + c)b$

7.22   (a) $xy' + xy'z = xy'z' + xy'z$ (d) $x'y' = x'y'z + x'y'z'$
       (b) $xyz' + x'y' = xyz' + x'y'z + x'y'z'$ (e) $xy' + y'z = xy'z + xy'z' + x'y'z$
       (c) $xy + xz' = xyz + xyz' + xy'z'$ (f) $x'yz'$

7.23   (a) $A^c \cap B^c \cap C^c$, (b) $(A \cap B^c \cap C^c) \cup (A \cap C^c)$

7.24   (a) 1101110111, (b) 1000010000, (c) 0001110101

7.25   (a) 1101110111, (b) 1110110111, (c) 0011001000, (d) 0101001100,
       (e) 1000110011, (f) 0000110010, (g) 0011001100, (h) 00010000000

7.26   One possibility is:

$$A = 1111111100000000$$
$$B = 1111000011110000$$
$$C = 1100110011001100$$
$$D = 1010101010101010$$

7.27   (a) $Y = A\bar{B} + \bar{B}C$, (b) $Y = A\bar{B}C + A\bar{B}\bar{C} + \bar{A}\bar{B}C$

7.28   (a) $Y = A\bar{B} + \bar{A}C$, (b) $Y = A\bar{B}C + A\bar{B}\bar{C} + \bar{A}BC + \bar{A}\bar{B}C$

7.29   (a) $Y = AB + \bar{A}BC + A\bar{B}\bar{C}$, (b) $Y = ABC + AB\bar{C} + \bar{A}BC + A\bar{B}\bar{C}$

7.30   (a) $Y = \overline{AB} + A\bar{B}C + \overline{\bar{A} + C}$
       (b) $Y = \bar{A}BC + \bar{A}B\bar{C} + \bar{A}\bar{B}C + \bar{A}\bar{B}\bar{C} + A\bar{B}C + A\bar{B}\bar{C} + AB\bar{C}$

7.31   See Fig. 7-22.

7.32   (a) $A(B + C + D)$, (b) $AB(C + D)$, (c) $A(D + BC)$, (d) $A(B + C) + D$

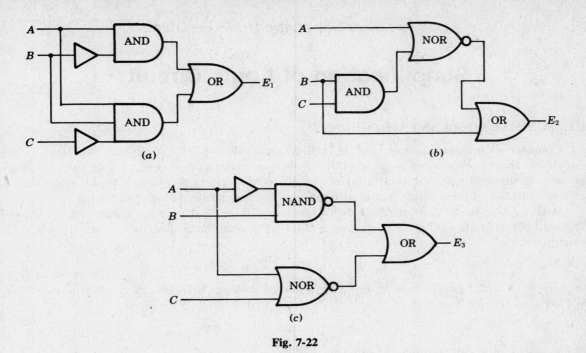

(a)

(b)

(c)

Fig. 7-22

Chapter 8

Simplification of Logic Circuits

8.1 MINIMAL BOOLEAN EXPRESSIONS

Consider a Boolean expression E in a Boolean algebra B. Since E may represent a logic circuit, we may want a representation of E which is in some sense minimal. Here we define and investigate minimal sum-of-products forms for E. (Other types of minimal forms exist, such as minimal product-of-sums forms, but their treatment lies beyond the scope of this book.)

If E is a sum-of-products Boolean expression, we will let E_L denote the number of literals in E (counted according to multiplicity), and we will let E_S denote the number of summands in E. For example, if

$$E = abc' + a'b'd + ab'c'd + a'bcd$$

then $E_L = 14$ and $E_S = 4$. Now let F denote a sum-of-products Boolean expression equivalent to E. We say E is *simpler* than F if

$$E_L \leq F_L \qquad \text{and} \qquad E_S \leq F_S$$

and at least one of the relations is a strict inequality.

> **Definition:** A Boolean expression E is in a *minimal sum-of-products form* (or, simply, is a *minimal sum*) if it is in sum-of-products form and there is no other equivalent expression in sum-of-products form that is simpler than E.

Before we discuss the structure of minimal sums, we need to introduce the notion of prime implicants. A fundamental product P is called a *prime implicant* of a Boolean expression E if

$$P + E = E$$

but no other fundamental product included in P has this property. (Note that, in the Boolean algebra of propositions, the condition $P + E = E$ translates as "P logically implies E"; hence the term "implicant.") For example, Problem 7.5 shows that $P = xz'$ is a prime implicant of

$$E = xy' + xyz' + x'yz'$$

The importance of prime implicants is contained in the following

Theorem 8.1: If a Boolean expression E is in a minimal sum-of-products form, then each summand in E is a prime implicant of E.

The so-called *consensus method*, discussed in Problems 8.3 and 8.4, can be used to represent any Boolean expression as the sum of all its prime implicants. One way of finding a minimal sum for E is to express each prime implicant in complete sum-of-products form, and to delete one by one those prime implicants whose summands appear among the summands of the remaining prime implicants. For example, we show in Problem 8.4 that

$$E = x'z' + xy + x'y' + yz'$$

is expressed as the sum of all its prime implicants. We have

$$x'z' = x'z'(y + y') = x'yz' + x'y'z'$$
$$xy = xy(z + z') = xyz + xyz'$$
$$x'y' = x'y'(z + z') = x'y'z + x'y'z'$$
$$yz' = yz'(x + x') = xyz' + x'yz'$$

194

Now $x'z'$ can be deleted, since its summands, $x'yz'$ and $x'y'z'$, appear among the others.   Thus

$$E = xy + x'y' + yz'$$

and this is a minimal-sum form for E since none of the prime implicants is *superfluous*, i.e. can be deleted without changing E.   Observe that, instead of $x'z'$, we might have eliminated yz'—which shows that the minimal sum for a Boolean expression is not necessarily unique.

The above method of finding minimal-sum forms for Boolean expressions E is direct but inefficient.   In the next section we give a geometrical method of finding minimal-sum forms when the number of variables is not too large.

8.2   KARNAUGH MAPS

Karnaugh maps are pictorial devices for finding prime implicants and minimal-sum forms for Boolean expressions involving at most six variables.   We will only treat the case of two, three or four variables.

In our Karnaugh maps, fundamental products in the same variables will be represented by squares.   We say that two such fundamental products P_1 and P_2 are *adjacent* if P_1 and P_2 differ in exactly one literal, which must be a complemented variable in one product and uncomplemented in the other.   Thus the sum of two adjacent products will be a fundamental product with one less literal.   For example,

$$xyz' + xy'z' = xz'(y + y') = xz'(1) = xz'$$
$$x'yzt + x'yz't = x'yt(z + z') = x'yt(1) = x'yt$$

Note that $x'yzt$ and $xyz't$ are not adjacent.   Also note that xyz' and $xyzt$ will not appear in the same Karnaugh map since they involve different variables.   In the context of Karnaugh maps, we will sometimes use the terms "squares" and "fundamental products" interchangeably.

Case of Two Variables

The Karnaugh map corresponding to Boolean expressions $E(x, y)$ is pictured in Fig. 8-1(a).   We may view the Karnaugh map as a Venn diagram where x is represented by the points in the upper half of the map, shaded in Fig. 8-1(b), and y is represented by points in the left half of the map, shaded in Fig. 8-1(c).   Hence x' is represented by the points in the lower half of the map, and y' is represented by the points in the right half of the map.   Accordingly, the four possible fundamental products with two literals,

$$xy \qquad xy' \qquad x'y \qquad x'y'$$

are represented by the four squares in the map, as labeled in Fig. 8-1(d).   Observe that two such squares are adjacent in the sense defined above if and only if they are geometrically adjacent (have a side in common).

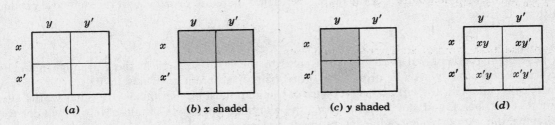

(a)   (b) x shaded   (c) y shaded   (d)

Fig. 8-1

Any complete sum-of-products Boolean expression $E(x, y)$ is represented in a Karnaugh map by placing checks in the appropriate squares. For example,

$$E_1 = xy + xy' \qquad E_2 = xy + x'y + x'y' \qquad E_3 = xy + x'y'$$

are represented respectively in Fig. 8-2(a), (b), and (c). (The loops will be explained later.)

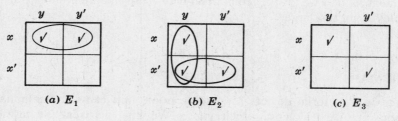

(a) E_1 (b) E_2 (c) E_3

Fig. 8-2

A prime implicant of $E(x, y)$ will be either a pair of adjacent squares or an *isolated square*, i.e. a square which is not adjacent to any other square of $E(x, y)$. For example, E_1 consists of two adjacent squares designated by the loop in Fig. 8-2(a). This pair of adjacent squares represents the variable x, so x is a (the only) prime implicant of E_1 and

the $\qquad\qquad\qquad\qquad\qquad E_1 = x$

is its minimal sum. Observe that E_2 contains two pairs of adjacent squares (designated by the two loops) which include all the squares of E_2. The vertical pair represents y and the horizontal pair x'; so y and x' are prime implicants of E_2 and

$$E_2 = x' + y$$

is its minimal sum. On the other hand, E_3 consists of two isolated squares which represent xy and $x'y'$; hence xy and $x'y'$ are the prime implicants of E_3 and

$$E_3 = xy + x'y'$$

is its minimal sum.

Case of Three Variables

The Karnaugh map corresponding to Boolean expressions $E(x, y, z)$ is pictured in Fig. 8-3(a). Again we may view the Karnaugh map as a Venn diagram, with the variable x still represented by the points in the upper half of the map, as in Fig. 8-3(b), and the variable y still represented by the points in the left half of the map, as in Fig. 8-3(c). The new variable z is represented by the points in the left and right quarters of the map, shaded in Fig. 8-3(d). Hence, x' is represented by the points in the lower half of the map, y' by the points in the right half of the map, and z' by the points in the middle two quarters of the map. Note that there are exactly eight fundamental products with three literals,

$$xyz \qquad xyz' \qquad xy'z \qquad xy'z' \qquad x'yz \qquad x'yz' \qquad x'y'z \qquad x'y'z'$$

and these eight fundamental products correspond to the eight squares in the Karnaugh map, Fig. 8-3(a), in the obvious way. In order that *every* pair of adjacent products in Fig. 8-3(a) may be geometrically adjacent, the left and right edges of the map must be identified. In other words, if we were to cut out, bend, and glue the map along the identified edges, we should obtain a cylinder (Fig. 8-4) having the property that adjacent products are represented by "squares" with one edge in common.

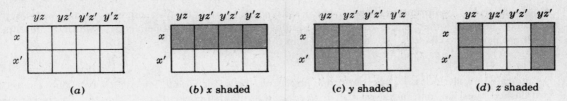

Fig. 8-3

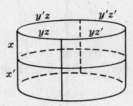

Fig. 8-4

By a *basic rectangle* in the Karnaugh map with three variables, Fig. 8-3(*a*) or Fig. 8-4, we mean a square, two adjacent squares, or four squares which form a one-by-four or a two-by-two rectangle. These basic rectangles correspond to fundamental products of three, two, and one literal, respectively. Moreover, the fundamental product represented by a basic rectangle is the product of just those literals that appear in every square of the rectangle.

Any complete sum-of-products Boolean expression $E(x, y, z)$ will be represented on the Karnaugh map by checking the appropriate squares. A prime implicant of E will be a *maximal basic rectangle* of E, i.e. a basic rectangle which is not contained in any larger basic rectangle. A minimal sum for E will consist of a *minimal cover* of E, i.e. a minimal number of maximal basic rectangles which together include all the squares of E.

EXAMPLE 8.1   Consider the following three complete sum-of-products Boolean expressions in variables x, y and z:

$$E_1 = xyz + xyz' + x'yz' + x'y'z$$
$$E_2 = xyz + xyz' + xy'z + x'yz + x'y'z$$
$$E_3 = xyz + xyz' + x'yz' + x'y'z' + x'y'z$$

E_1, E_2, and E_3 are represented in Fig. 8-5 by checking the appropriate squares in the Karnaugh maps. We show how to use these maps to find minimal sums for the expressions.

(*a*)   Observe that E_1 has three prime implicants (maximal basic rectangles), which are circled; these are xy, yz', and $x'y'z$. All three are needed to cover E_1; hence the minimal sum for E_1 is

$$E_1 = xy + yz' + x'y'z$$

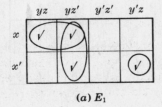

(a) E_1

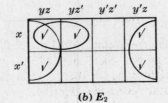

(b) E_2

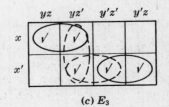

(c) E_3

Fig. 8-5

(*b*) Note that E_2 has two prime implicants, which are circled. One is the two adjacent squares which represents *xy*, and the other is the two-by-two square (spanning the identified edges) which represents *z*. Both are needed to cover E_2, so the minimal sum for E_2 is

$$E_2 = xy + z$$

(*c*) As indicated by the loops, E_3 has four prime implicants, *xy*, *yz'*, *x'z'*, and *x'y'*. However, only one of the two dashed ones, i.e. one of *yz'* or *x'z'*, is needed in a minimal cover of E_3. Thus E_3 has two minimal sums:

$$E_3 = xy + yz' + x'y' = xy + x'z' + x'y'$$

(as previously determined in Section 8.1).

Case of Four Variables

The Karnaugh map corresponding to Boolean expressions $E(x, y, z, t)$ is pictured in Fig. 8-6. Each of the sixteen squares in the map corresponds to one of the sixteen fundamental products

$$xyzt \qquad xyzt' \qquad xyz't' \qquad xyz't \qquad xy'zt \qquad \cdots \qquad x'yz't$$

as indicated by the labels of the row and column of the square. Observe that the top line and the left side are labeled so that adjacent products differ in precisely one literal. Again we must identify the left edge with the right edge (as we did with three variables) but we must also identify the top edge with the bottom edge. (These identifications give rise to a donut-shaped surface called a *torus*, and we may view our map as really being a torus.)

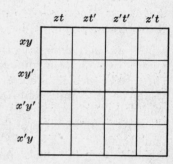

A basic rectangle is a square, two adjacent squares, four squares which form a one-by-four or two-by-two rectangle, or eight squares which form a two-by-four rectangle. These rectangles correspond to fundamental products with four, three, two and one literal respectively. Again, maximal basic rectangles are the prime implicants. The minimization technique for a Boolean expression $E(x, y, z, t)$ is the same as before.

Fig. 8-6

EXAMPLE 8.2 Consider the three Boolean expressions E_1, E_2, E_3 in variables *x*, *y*, *z*, *t* which are given by the Karnaugh maps in Fig. 8-7, e.g.

$$E_1 = xyz't' + xyz't + xy'zt + xy'zt' + x'y'zt + x'y'zt' + x'yz't'$$

We use these maps to find minimal-sum forms.

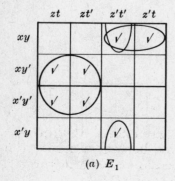
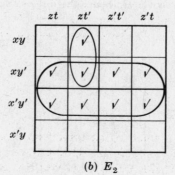
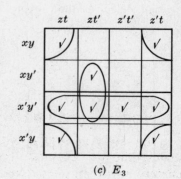

(*a*) E_1 (*b*) E_2 (*c*) E_3

Fig. 8-7

(a) The two-by-two maximal basic rectangle represents $y'z$ since only y' and z appear in all four squares. The horizontal pair of adjacent squares represents xyz', and the adjacent squares overlapping the top and bottom edges represent $yz't'$. As all three rectangles are needed for a minimal cover,

$$E_1 = y'z + xyz' + yz't'$$

is the minimal sum for E_1.

(b) Only y' appears in all eight squares of the two-by-four maximal basic rectangle, and the designated pair of adjacent squares represents xzt'. As both rectangles are needed for a minimal cover,

$$E_2 = y' + xzt'$$

is the minimal sum for E_2.

(c) The four corner squares form a two-by-two maximal basic rectangle which represents yt, since only y and t appear in all the four squares. The four-by-one maximal basic rectangle represents $x'y'$, and the two adjacent squares represent $y'zt'$. As all three rectangles are needed for a minimal cover,

$$E_3 = yt + x'y' + y'zt'$$

is the minimal sum for E_3.

Remark: Suppose a Boolean expression E is a sum of fundamental products. We emphasize that it is not necessary to put E into complete sum-of-products form in order to represent it by a Karnaugh map. Take, for example,

$$E = xy' + xyz + x'y'z' + x'yzt'$$

We simply check all the squares representing each fundamental product. That is, we check all four squares representing xy', the two squares representing xyz, the two squares representing $x'y'z'$, and the one square representing $x'yzt'$, as in Fig. 8-8. (In this particular case, multiple checks do not arise. In general, one omits to check squares that have previously been checked by virtue of belonging to another fundamental product.) A minimal cover of the map consists of the three designated maximal basic rectangles. Hence

$$E = xz + y'z' + yzt'$$

is a minimal sum for E.

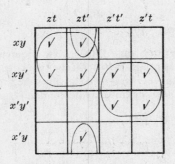

Fig. 8-8

8.3 MINIMAL AND-OR CIRCUITS

The foregoing theory may be applied to an important circuit-design problem, which has two slightly different versions: (1) To construct an AND-OR circuit whose Boolean expression is in minimal-sum form (a *minimal AND-OR circuit*) and which is equivalent to a given logic circuit L. (2) To construct a minimal AND-OR circuit that will have a prescribed truth table.

Example 8.3 shows how version (2) is handled. As for version (1), we might reduce it to version (2) (by putting the Boolean expression for L into complete sum-of-products form or by using special sequences to generate the truth table) or we might simply start with a sum-of-products form for L (obtained, e.g., by the algorithm of Section 7.6). See Problem 8.11.

EXAMPLE 8.3 Design a three-input minimal AND-OR circuit L that will have the truth table

A	00001111
B	00110011
C	01010101
L	11001101

From the truth table we can (see Section 7.8) read off the complete sum-of-products form for L:

$$L = \bar{A} \cdot \bar{B} \cdot \bar{C} + \bar{A} \cdot \bar{B} \cdot C + A \cdot \bar{B} \cdot \bar{C} + A \cdot \bar{B} \cdot C + A \cdot B \cdot C$$

The associated Karnaugh map appears in Fig. 8-9(a). Observe that L has two prime implicants, $\bar{B}$ and AC, in its minimal cover; hence

$$L = \bar{B} + AC$$

is a minimal sum for L. Figure 8-9(b) gives the corresponding minimal AND-OR circuit for L.

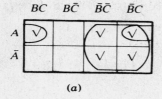

(a)

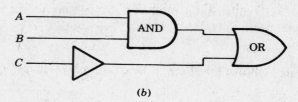

(b)

Fig. 8-9

EXAMPLE 8.4 Consider the NOR gate [Problem 7.13(b)]. Two equivalent (by DeMorgan's law) Boolean expressions for the NOR gate are realized as in Fig. 8-10. Of the two circuits, Fig. 8-10(b) is a minimal AND-OR circuit, because $L = \bar{A} \cdot \bar{B}$ is a minimal sum, yet Fig. 8-10(a) involves the smaller number of circuit elements! This example shows that the "simplest" AND-OR circuit may not remain "simplest" when compared with an equivalent, but not an AND-OR, circuit.

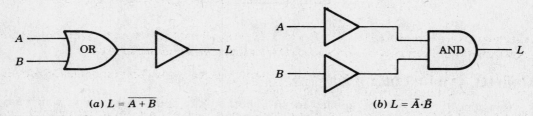

(a) $L = \overline{A + B}$ (b) $L = \bar{A} \cdot \bar{B}$

Fig. 8-10

Solved Problems

MINIMAL SUMS

8.1   Find E_L, the number of literals, and E_S, the number of summands, for each Boolean expression E:

   (a) $E = xy'z + x'z' + yz' + x$ (c) $E = xyt' + x'y'zt + xz't$

   (b) $E = x'y'z + xyz + y + yz' + x'z$ (d) $E = (xy' + z)' + xy'$

   Simply add up the number of literals, distinguishing between complemented and uncomplemented forms, and the number of summands in each expression:

   (a) $E_L = 3 + 2 + 2 + 1 = 8$ $E_S = 4$

   (b) $E_L = 3 + 3 + 1 + 2 + 2 = 11$ $E_S = 5$

   (c) $E_L = 3 + 4 + 3 = 10$ $E_S = 3$

   (d)   Because E is not written as a sum of products, E_L and E_S are not defined.

8.2   Given that E and F are each in a sum-of-products form and are equivalent Boolean expressions, define: (a) E is simpler than F, (b) E is minimal.

   (a) E is simpler than F if $E_L < F_L$ and $E_S \le F_S$, or if $E_L \le F_L$ and $E_S < F_S$.
   (b) E is minimal if there is no equivalent sum-of-products expression which is simpler than E.

8.3   Let P_1 and P_2 be fundamental products such that exactly one variable, say x_k, appears complemented in one of P_1 and P_2 and uncomplemented in the other. Then the *consensus* of P_1 and P_2 is the product (without repetition) of the literals of P_1 and the literals of P_2 after x_k and x'_k are deleted. (We do not define a consensus of $P_1 = x$ and $P_2 = x'$.) (a) Find the consensus of:

   (1) $xyz's$ and $xy't$ (3) $x'yz$ and $x'yt$
   (2) xy' and y (4) $x'yz$ and xyz'

   (b)   Prove that if Q is the consensus of P_1 and P_2, $P_1 + P_2 + Q = P_1 + P_2$.

   (a)   (1) $xz'st$
         (2) x
         (3)   They have no consensus, since no variable appears uncomplemented in one of the products and complemented in the other.
         (4)   They have no consensus, since *both* x and z appear complemented in one of the products and uncomplemented in the other.
   (b)   Since the literals commute, we can assume without loss of generality that

   $$P_1 = a_1 a_2 \cdots a_r t \qquad P_2 = b_1 b_2 \cdots b_s t' \qquad Q = a_1 a_2 \cdots a_r b_1 b_2 \cdots b_s$$

   Now, $Q = Q(t + t') = Qt + Qt'$. Because Qt contains P_1, $P_1 + Qt = P_1$; and because Qt' contains P_2, $P_2 + Qt' = P_2$. Hence

   $$P_1 + P_2 + Q = P_1 + P_2 + Qt + Qt' = (P_1 + Qt) + (P_2 + Qt') = P_1 + P_2$$

8.4   Consider a Boolean expression $E = P_1 + P_2 + \cdots + P_n$ where the P's are fundamental products. Applying the following two steps to E will be called the *consensus method*:

   Step (1):   Delete any fundamental product P_i which includes any other fundamental product P_j. (Permissible by the absorption law.)

   Step (2):   Add the consensus Q of any P_i and P_j providing Q does not include any of the P's. [Permissible by Problem 8.3(b).]

A fundamental theorem in Boolean algebra states that the consensus method, applied to any Boolean sum of products E, will eventually stop, and then E will be the sum of all its prime implicants. Apply the consensus method to

$$E = xyz + x'z' + xyz' + x'y'z + x'yz'$$

We have:

$$E = xyz + x'z' + xyz' + x'y'z \qquad (x'yz' \text{ includes } x'z')$$
$$= xyz + x'z' + xyz' + x'y'z + xy \qquad (\text{Consensus of } xyz \text{ and } xyz')$$
$$= x'z' + x'y'z + xy \qquad (xyz \text{ and } xyz' \text{ include } xy)$$
$$= x'z' + x'y'z + xy + x'y' \qquad (\text{Consensus of } x'z' \text{ and } x'y'z)$$
$$= x'z' + xy + x'y' \qquad (x'y'z \text{ includes } x'y')$$
$$= x'z' + xy + x'y' + yz' \qquad (\text{Consensus of } x'z' \text{ and } xy)$$

Observe that neither step in the consensus method can now be applied. (The consensus of the first two products includes—in fact, equals—the last product; the consensus of the last two products equals the first product.) Hence E is now expressed as the sum of its prime implicants, $x'z'$, xy, $x'y'$, and yz'.

8.5 Use the consensus method to find the prime implicants and a minimal sum for

$$E = xy' + xyz' + x'yz'$$

We have:

$$E = xy' + xyz' + x'yz' + xz' \qquad (\text{Consensus of } xy' \text{ and } xyz')$$
$$= xy' + x'yz' + xz' \qquad (xyz' \text{ includes } xz')$$
$$= xy' + x'yz' + xz' + yz' \qquad (\text{Consensus of } x'yz' \text{ and } xz')$$
$$= xy' + xz' + yz' \qquad (x'yz' \text{ includes } yz')$$

Neither step in the consensus method can now be applied. Hence, xy', xz', and yz' are the prime implicants of E. Writing these prime implicants in complete sum-of-products form, we obtain:

$$xy' = xy'(z + z') = xy'z + xy'z'$$
$$xz' = xz'(y + y') = xyz' + xy'z'$$
$$yz' = yz'(x + x') = xyz' + x'yz'$$

Only the summands xyz' and $xy'z'$ of xz' appear among the other summands and hence xz' can be eliminated as superfluous. Thus

$$E = xy' + yz'$$

is a minimal sum for E.

8.6 Use the consensus method to find the prime implicants and a minimal sum for

$$E = xy + y't + x'yz' + xy'zt'$$

$$E = xy + y't + x'yz' + xy'zt' + xzt' \qquad (\text{Consensus of } xy \text{ and } xy'zt')$$
$$= xy + y't + x'yz' + xzt' \qquad (xy'zt' \text{ includes } xzt')$$
$$= xy + y't + x'yz' + xzt' + yz' \qquad (\text{Consensus of } xy \text{ and } x'yz')$$
$$= xy + y't + xzt' + yz' \qquad (x'yz' \text{ includes } yz')$$
$$= xy + y't + xzt' + yz' + xt \qquad (\text{Consensus of } xy \text{ and } y't)$$
$$= xy + y't + xzt' + yz' + xt + xz \qquad (\text{Consensus of } xzt' \text{ and } xt)$$
$$= xy + y't + yz' + xt + xz \qquad (xzt' \text{ includes } xz)$$
$$= xy + y't + yz' + xt + xz + z't \qquad (\text{Consensus of } y't \text{ and } yz')$$

Neither step in the consensus method can now be applied.   Hence the prime implicants of E are xy, $y't$, yz', xt, xz, and $z't$.   Writing these prime implicants in complete sum-of-products form and deleting one by one those which are superfluous, we finally obtain

$$E = y't + xz + yz'$$

as a minimal sum for E.

KARNAUGH MAPS

8.7   Find the fundamental product P represented by each basic rectangle in the Karnaugh maps shown in Fig. 8-11.

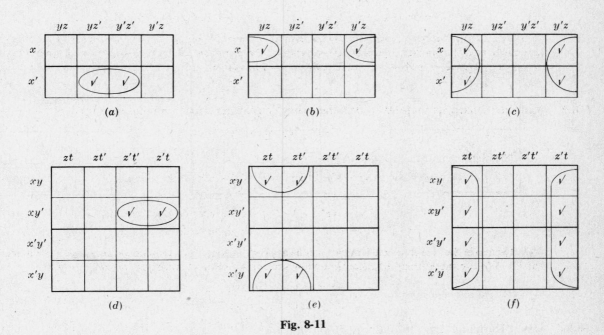

Fig. 8-11

In each case, find those literals which appear in all the squares of the basic rectangle; P is the product of such literals.

(a) x' and z' appear in both squares; hence $P = x'z'$.

(b) x and z appear in both squares; hence $P = xz$.

(c)   Only z appears in all four squares; hence $P = z$.

(d) x, y' and z' appear in both squares; hence $P = xy'z'$.

(e)   Only y and z appear in all four squares; hence $P = yz$.

(f)   Only t appears in all eight squares; hence $P = t$.

8.8   Find a minimal sum for each expression E whose Karnaugh map is shown in Fig. 8-12.

(a)   There are five prime implicants, designated by the four loops and the dashed circle in Fig. 8-13(a). However, the dashed circle is not needed to cover all the squares, whereas the four loops are required. Thus the four loops give the minimal sum for E:

$$E = xzt' + xy'z' + x'y'z + x'z't'$$

(b)   A minimal cover of E is given by the three loops in Fig. 8-13(b).   Hence $E = zt' + xy't' + x'yt$ is a minimal sum for E.

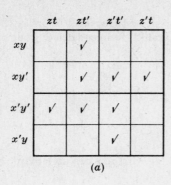 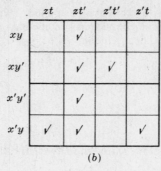 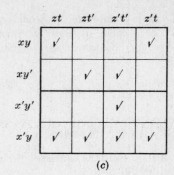

(a) (b) (c)

Fig. 8-12

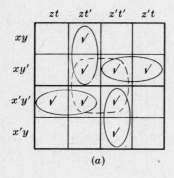 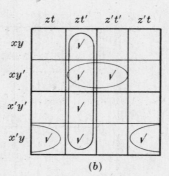 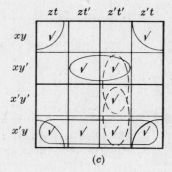

(a) (b) (c)

Fig. 8-13

(c) There are two ways to cover the $x'y'z't'$ square as indicated in Fig. 8-13(c). Hence

$$E = x'y + yt + xy't' + y'z't' = x'y + yt + xy't' + x'z't'$$

are two minimal sums for E.

8.9   Use a Karnaugh map to find a minimal sum for

$$E = y't' + y'z' + yzt' + x'y'zt$$

Check the four squares corresponding to the fundamental product $y't'$, the four squares corresponding to $y'z'$, the two squares corresponding to yzt', and the one square corresponding to $x'y'zt$. This gives the Karnaugh map in Fig. 8-14. A minimal cover consists of the three designated maximal basic rectangles.

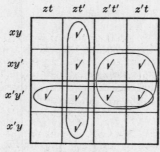

Fig. 8-14

Hence

$$E = zt' + y'z' + x'y'$$

is a minimal sum for E.

MINIMAL AND-OR CIRCUITS

8.10 Draw a minimal AND-OR circuit which yields the following truth table:

A	00001111
B	00110011
C	01010101
L	10101001

From the truth table we have the complete sum-of-products representation

$$L = \bar{A} \cdot \bar{B} \cdot \bar{C} + \bar{A} \cdot B \cdot \bar{C} + A \cdot \bar{B} \cdot \bar{C} + A \cdot B \cdot C$$

The Karnaugh map of L appears in Fig. 8-15(a).   There are three prime implicants, as indicated by the three loops.   Hence

$$L = ABC + \bar{A}\bar{C} + \bar{B}\bar{C}$$

is a minimal sum for L; the corresponding minimal AND-OR circuit appears in Fig. 8-15(b).

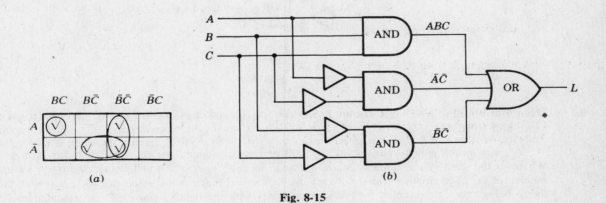

(a)   (b)

Fig. 8-15

8.11 Redesign the circuit L in Fig. 8-16(a) so that it becomes a minimal AND-OR circuit.

First, find the output of the circuit by successively labeling the input(s) and output of each gate until reaching the output of the circuit, as in Fig. 8-16(b).   Next, reduce L to a sum of products by application of the laws of Boolean algebra:

$$L = AB + \overline{(A + B)} + \overline{A}B = AB + \bar{A}\bar{B} + A + \bar{B} = A + \bar{B}$$

where the absorption law has been invoked twice in the last step.   The final expression is obviously a minimal sum for L.   Figure 8-16(c) shows the corresponding minimal AND-OR circuit.

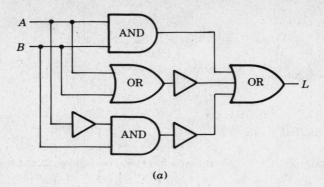

(a)

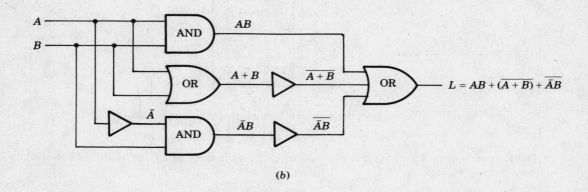

$$L = AB + (\overline{A + B}) + \overline{\overline{A}B}$$

(b)

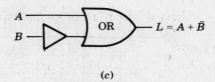

$$L = A + \bar{B}$$

(c)

Fig. 8-16

8.12 Design a minimal AND-OR circuit L whereby three switches, A, B, and C, can control the same hall light.

A given switch may be either "up" (closed) or "down" (open), respectively denoted by 1 and 0. Whatever the state of the three switches, a change in any single switch will change the parity of the number of 1s. The circuit will therefore achieve the desired function if it associates odd parity with the light's being "on" (represented by 1) and even parity with the light's being "off" (represented by 0), according to the truth table.

A	0 0 0 0 1 1 1 1
B	0 0 1 1 0 0 1 1
C	0 1 0 1 0 1 0 1
L	0 1 1 0 1 0 0 1

From the truth table,

$$L = \bar{A}BC + \bar{A}B\bar{C} + A\bar{B}\bar{C} + ABC$$

and this is a minimal sum for L, as may be verified from the Karnaugh map, Fig. 8-17(a). The corresponding minimal AND-OR circuit appears in Fig. 8-17(b).

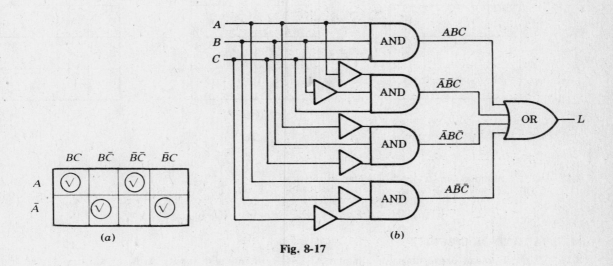

Fig. 8-17

Supplementary Problems

MINIMAL SUMS, KARNAUGH MAPS

8.13 Find the consensus (see Problem 8.3) of each pair of fundamental products:

  (a) xz and xz' (c) xy and $x'zs't$ (e) $xy'zt$ and $xzs't$
  (b) $x'yt$ and $y'z$ (d) $xs't$ and $xy'st$ (f) $xy'zt$ and $x'yst$

8.14 Use the consensus method (see Problem 8.4) to find the prime implicants of:

  (a) $E_1 = xy'z' + x'y + x'y'z' + x'yz$
  (b) $E_2 = xy' + x'z't + xyzt' + x'y'zt'$
  (c) $E_3 = xyzt + xyz't' + xz't' + x'y'z' + x'yz't$

8.15 Find all possible minimal sums for each Boolean expression E given by a Karnaugh map in Fig. 8-18.

	yz	yz'	$y'z'$	$y'z$
x	√		√	√
x'	√	√		

(a)

	yz	yz'	$y'z'$	$y'z$
x	√		√	√
x'	√	√		√

(b)

	yz	yz'	$y'z'$	$y'z$
x	√			√
x'	√	√	√	√

(c)

Fig. 8-18

8.16   Find all possible minimal sums for each Boolean expression E given by a Karnaugh map in Fig. 8-19.

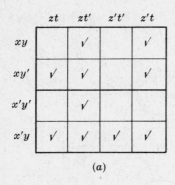

	zt	zt'	z't'	z't
xy		✓		✓
xy'	✓	✓		✓
$x'y'$		✓		
$x'y$	✓	✓	✓	✓

(a)

	zt	zt'	z't'	z't
xy	✓	✓	✓	
xy'		✓	✓	✓
$x'y'$		✓		
$x'y$	✓	✓	✓	

(b)

	zt	zt'	z't'	z't
xy	✓			✓
xy'		✓	✓	
$x'y'$			✓	
$x'y$	✓	✓	✓	✓

(c)

Fig. 8-19

8.17   Find a minimal sum for each Boolean expression:

     (a) $E_1 = xy + x'y + x'y'$ (c) $E_3 = y'z + y'z't' + z't$

     (b) $E_2 = x + x'yz + xy'z'$ (d) $E_4 = y'zt + xzt' + xy'z'$

MINIMAL AND-OR CIRCUITS

8.18   Write Boolean expressions for minimal AND-OR circuits with inputs A, B, C which yield the truth tables Y_1, Y_2, and Y_3 given in Fig. 8-20.

A	1 1 1 1 0 0 0 0
B	1 1 0 0 1 1 0 0
C	1 0 1 0 1 0 1 0
Y_1	1 1 0 0 1 0 0 0
Y_2	1 1 1 0 0 0 1 1
Y_3	0 0 1 1 1 1 0 1

Fig. 8-20

8.19   Redesign the circuit L in Fig. 8-21 so that it becomes a minimal AND-OR circuit.

8.20   Redesign the circuit L in Fig. 8-22 so that it becomes a minimal AND-OR circuit.

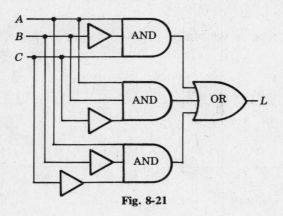

Fig. 8-21

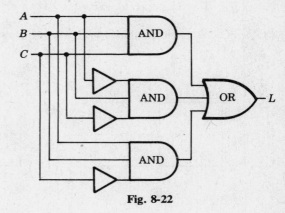

Fig. 8-22

Answers to Supplementary Problems

8.13 (*a*) x, (*b*) $x'zt$, (*c*) $yzs't$, (*d*) $xy't$, (*e*) no consensus, (*f*) no consensus

8.14 (*a*) $x'y$, $x'z$, $y'z'$; (*b*) xy', xzt', $y'zt'$, $x'z't$, $y'z't$; (*c*) $xyzt$, $xz't'$, $y'z't'$, $x'y'z'$, $x'z't$

8.15 (*a*) $E = xy' + x'y + yz = xy' + x'y + xz$, (*b*) $E = xy' + x'y + z$, (*c*) $E = x' + z$

8.16 (*a*) $E = x'y + zt' + xz't + xy'z = x'y + zt' + xz't + xy't$
     (*b*) $E = yz + yt' + zt' + xy'z'$
     (*c*) $E = x'y = yt + xy't' + x'zt = x'y + yt + xy't' + y'zt$

8.17 (*a*) $E_1 = x' + y$ (*c*) $E_3 = y' + z't$
     (*b*) $E_2 = xz' + yz$ (*d*) $E_4 = xy' + zt' + y'zt$

8.18 $Y_1 = AB + BC$, $Y_2 = AB + \bar{A}\bar{B} + \bar{B}C = AB + \bar{A}\bar{B} + AC$, $Y_3 = A\bar{B} + \bar{A}B + \bar{B}C$

8.19 $Y = A\bar{B} + A\bar{C}$

8.20 $Y = AB + B\bar{C}$

Chapter 9

Vectors, Matrices, Subscripted Variables

9.1 INTRODUCTION

Suppose that the weights (in pounds) of eight students are listed as follows:

$$134 \quad 156 \quad 127 \quad 145 \quad 203 \quad 186 \quad 145 \quad 138$$

One can denote all the values in the list using only one symbol, say w, but with different subscripts:

$$w_1 \quad w_2 \quad w_3 \quad w_4 \quad w_5 \quad w_6 \quad w_7 \quad w_8$$

Observe that the subscript denotes the position of the value in the list, i.e.

$$w_1 = 134, \text{ the first number} \qquad w_2 = 156, \text{ the second number} \qquad \ldots$$

Such a linear list of values is called a *vector* or a *linear array*.

Using such subscript notation, one can write the sum and average of the weights as

$$\sum_{i=1}^{8} w_i = w_1 + w_2 + \cdots + w_8 \qquad \text{and} \qquad \left(\sum_{i=1}^{8} w_i \right) \Big/ 8$$

(see Section 9.5). Subscript notation is indispensable in developing concise expressions for arithmetic manipulations.

Analogously, a chain of 28 stores, each store having 4 departments, could list its weekly sales (to the nearest dollar) as in Table 9-1. Again, one need use only one symbol, say s, but now with two subscripts, to denote the entries in the table as

$$s_{1,1} \quad s_{1,2} \quad s_{1,3} \quad s_{1,4} \quad s_{2,1} \quad s_{2,2} \quad \ldots \quad s_{28,4}$$

where $s_{i,j}$ denotes the sales in the ith store, jth department. (We shall simply write s_{ij} instead of $s_{i,j}$ when there is no possibility of misunderstanding.) Thus,

$$s_{11} = \$2872 \qquad s_{12} = \$805 \qquad s_{13} = \$3211 \qquad \ldots$$

Such a rectangular array of numbers is called a *matrix* or a *two-dimensional array*.

This chapter will investigate vectors and matrices, and certain algebraic operations involving them. In such a context, the numbers themselves are called *scalars*. We close the chapter with a discussion of *subscripted variables*, the mode of handling vectors and matrices in computer programs.

Table 9-1

Dept. / Store	1	2	3	4
1	2872	805	3211	1560
2	2196	1223	2525	1744
3	3257	1017	3686	1951
...	...	...	...	...
28	2618	931	2333	982

9.2 VECTORS

By a *vector*, *u*, we simply mean a list (or *n-tuple*) of numbers:

$$u = (u_1, u_2, \ldots, u_n)$$

The numbers u_i are called the *components* of *u*. If all the $u_i = 0$, then *u* is called the *zero vector*. Two vectors, *u* and *v*, are *equal*, written $u = v$, if they have the same number of components and if corresponding components are equal.

EXAMPLE 9.1

(*a*) The following are vectors:

$$(3, -4) \qquad (6, 8) \qquad (0, 0, 0) \qquad (2, 3, 4)$$

The first two vectors have two components, whereas the last two vectors have three components. The third vector is the zero vector with three components.

(*b*) Although the vectors $(1, 2, 3)$ and $(2, 3, 1)$ contain the same numbers, they are not equal, since corresponding components are not equal.

If two vectors, *u* and *v*, have the same number of components, their *sum*, written $u + v$, is the vector obtained by adding corresponding components from *u* and *v*:

$$u + v = (u_1, u_2, \ldots, u_n) + (v_1, v_2, \ldots, v_n)$$
$$= (u_1 + v_1, u_2 + v_2, \ldots, u_n + v_n)$$

The *product* of a scalar *k* and a vector *u*, written *ku*, is the vector obtained by multiplying each component of *u* by *k*:

$$ku = k(u_1, u_2, \ldots, u_n) = (ku_1, ku_2, \ldots, ku_n)$$

We also define:

$$-u = (-1)u$$

and

$$u - v = u + (-v)$$

and we let 0 denote the zero vector.

EXAMPLE 9.2   Let $u = (2, 3, -4)$ and $v = (1, -5, 8)$. Then

$$u + v = (2 + 1, 3 - 5, -4 + 8) = (3, -2, 4)$$
$$5u = (5 \cdot 2, 5 \cdot 3, 5 \cdot (-4)) = (10, 15, -20)$$
$$-v = (-1)(1, -5, 8) = (-1, 5, -8)$$
$$2u - 3v = (4, 6, -8) + (-3, 15, -24) = (1, 21, -32)$$

Under the operations of vector addition and scalar multiplication, vectors have various properties, e.g.

$$k(u + v) = ku + kv$$

where *k* is a scalar and *u* and *v* are vectors. Since vectors may be viewed as a special case of matrices, Theorem 9.1 (see Section 9.4) contains a list of such properties.

9.3 MATRICES

By a *matrix*, A, we mean a rectangular array of numbers:

$$A = \begin{pmatrix} a_{11} & a_{12} & \cdots & a_{1n} \\ a_{21} & a_{22} & \cdots & a_{2n} \\ \cdots\cdots\cdots\cdots\cdots\cdots\cdots \\ a_{m1} & a_{m2} & \cdots & a_{mn} \end{pmatrix}$$

The m horizontal n-tuples

$$(a_{11}, a_{12}, \ldots, a_{1n}) \qquad (a_{21}, a_{22}, \ldots, a_{2n}) \qquad \cdots \qquad (a_{m1}, a_{m2}, \ldots, a_{mn})$$

are called the *rows* of A, and the n vertical m-tuples,

$$\begin{pmatrix} a_{11} \\ a_{21} \\ \cdots \\ a_{m1} \end{pmatrix} \qquad \begin{pmatrix} a_{12} \\ a_{22} \\ \cdots \\ a_{m2} \end{pmatrix} \qquad \cdots \qquad \begin{pmatrix} a_{1n} \\ a_{2n} \\ \cdots \\ a_{mn} \end{pmatrix}$$

its *columns*. Note that the element a_{ij}, called the *ij-entry*, appears in the ith row and the jth column. We frequently denote such a matrix simply by $A = (a_{ij})$.

A matrix with m rows and n columns is said to be an *m by n* matrix, written $m \times n$. The pair of numbers m and n is called the *size* of the matrix. Two matrices, A and B, are *equal*, written $A = B$, if they have the same size and if corresponding elements are equal.

A matrix with only one row is sometimes called a *row vector*, and a matrix with only one column is called a *column vector*. A matrix whose entries are all zero is called a *zero matrix* and will usually be denoted by 0.

EXAMPLE 9.3

(a) The rectangular array

$$\begin{pmatrix} 1 & -3 & 4 \\ 0 & 5 & -2 \end{pmatrix}$$

is a 2×3 matrix. Its rows are $(1, -3, 4)$ and $(0, 5, -2)$, and its columns are

$$\begin{pmatrix} 1 \\ 0 \end{pmatrix} \qquad \begin{pmatrix} -3 \\ 5 \end{pmatrix} \qquad \text{and} \qquad \begin{pmatrix} 4 \\ -2 \end{pmatrix}$$

(b) The 2×4 zero matrix is

$$\begin{pmatrix} 0 & 0 & 0 & 0 \\ 0 & 0 & 0 & 0 \end{pmatrix}$$

(c) The statement

$$\begin{pmatrix} x+y & 2z+w \\ x-y & z-w \end{pmatrix} = \begin{pmatrix} 3 & 5 \\ 1 & 4 \end{pmatrix}$$

is equivalent to the system of equations

$$\begin{cases} x+y = 3 \\ x-y = 1 \\ 2z+w = 5 \\ z-w = 4 \end{cases}$$

(The solution of the system of equations is $x = 2$, $y = 1$, $z = 3$, $w = -1$.)

9.4 MATRIX ADDITION AND SCALAR MULTIPLICATION

Let A and B be two matrices of the same size. The *sum* of A and B, written $A + B$, is the matrix obtained by adding corresponding elements from A and B:

$$\begin{pmatrix} a_{11} & a_{12} & \dots & a_{1n} \\ a_{21} & a_{22} & \dots & a_{2n} \\ \dots & \dots & \dots & \dots \\ a_{m1} & a_{m2} & \dots & a_{mn} \end{pmatrix} + \begin{pmatrix} b_{11} & b_{12} & \dots & b_{1n} \\ b_{21} & b_{22} & \dots & b_{2n} \\ \dots & \dots & \dots & \dots \\ b_{m1} & b_{m2} & \dots & b_{mn} \end{pmatrix} = \begin{pmatrix} a_{11}+b_{11} & a_{12}+b_{12} & \dots & a_{1n}+b_{1n} \\ a_{21}+b_{21} & a_{22}+b_{22} & \dots & a_{2n}+b_{2n} \\ \dots & \dots & \dots & \dots \\ a_{m1}+b_{m1} & a_{m2}+b_{m2} & \dots & a_{mn}+b_{mn} \end{pmatrix}$$

The *product* of a scalar k and a matrix A, written kA or Ak, is the matrix obtained by multiplying each element of A by k:

$$k\begin{pmatrix} a_{11} & a_{12} & \dots & a_{1n} \\ a_{21} & a_{22} & \dots & a_{2n} \\ \dots & \dots & \dots & \dots \\ a_{m1} & a_{m2} & \dots & a_{mn} \end{pmatrix} = \begin{pmatrix} ka_{11} & ka_{12} & \dots & ka_{1n} \\ ka_{21} & ka_{22} & \dots & ka_{2n} \\ \dots & \dots & \dots & \dots \\ ka_{m1} & ka_{m2} & \dots & ka_{mn} \end{pmatrix}$$

We also define

$$-A = (-1)A \qquad \text{and} \qquad A - B = A + (-B)$$

The matrix $-A$ is called the *negative* of the matrix A.

EXAMPLE 9.4

(a) $\begin{pmatrix} 1 & -2 & 3 \\ 0 & 4 & 5 \end{pmatrix} + \begin{pmatrix} 3 & 0 & -6 \\ 2 & -3 & 1 \end{pmatrix} = \begin{pmatrix} 1+3 & -2+0 & 3+(-6) \\ 0+2 & 4+(-3) & 5+1 \end{pmatrix} = \begin{pmatrix} 4 & -2 & -3 \\ 2 & 1 & 6 \end{pmatrix}$

(b) $3\begin{pmatrix} 1 & -2 & 0 \\ 4 & 3 & -5 \end{pmatrix} = \begin{pmatrix} 3\cdot1 & 3\cdot(-2) & 3\cdot0 \\ 3\cdot4 & 3\cdot3 & 3\cdot(-5) \end{pmatrix} = \begin{pmatrix} 3 & -6 & 0 \\ 12 & 9 & -15 \end{pmatrix}$

(c) $2\begin{pmatrix} 3 & -1 \\ 4 & 6 \end{pmatrix} - 5\begin{pmatrix} 0 & 2 \\ 1 & -3 \end{pmatrix} = \begin{pmatrix} 6 & -2 \\ 8 & 12 \end{pmatrix} + \begin{pmatrix} 0 & -10 \\ -5 & 15 \end{pmatrix} = \begin{pmatrix} 6 & -12 \\ 3 & 27 \end{pmatrix}$

Under matrix addition and scalar multiplication, matrices have the following properties:

Theorem 9.1: Let A, B, and C be matrices of the same size and let k and k' be scalars. Then

(i) $(A+B)+C = A+(B+C)$, i.e. addition is associative.
(ii) $A+B = B+A$, i.e. addition is commutative.
(iii) $A+0 = 0+A = A$
(iv) $A+(-A) = (-A)+A = 0$
(v) $k(A+B) = kA + kB$
(vi) $(k+k')A = kA + k'A$
(vii) $(kk')A = k(k'A)$
(viii) $1A = A$

Since n-component vectors may be identified with either $1 \times n$ matrices or $n \times 1$ matrices, this theorem also holds for vectors under vector addition and scalar multiplication.

9.5 SUMMATION SYMBOL

Before we define matrix multiplication, it will be convenient to introduce the *summation symbol* Σ (the Greek letter sigma).

Suppose $f(k)$ is an algebraic expression involving the variable k. Then the expression

$$\sum_{k=1}^{n} f(k) \qquad \text{or equivalently} \qquad \sum_{k=1}^{n} f(k)$$

has the following meaning. First we let $k = 1$ in $f(k)$, obtaining

$$f(1)$$

Then we let $k = 2$ in $f(k)$, obtaining $f(2)$, and add this to $f(1)$, obtaining

$$f(1) + f(2)$$

Next we let $k = 3$ in $f(k)$, obtaining $f(3)$, and add this to the previous sum, obtaining

$$f(1) + f(2) + f(3)$$

We continue this process until we obtain the sum

$$f(1) + f(2) + f(3) + \cdots + f(n-1) + f(n)$$

Observe that at each step we increase the value of k by 1 until k is equal to n. Naturally we may use another variable besides k.

We also generalize our definition by allowing the sum to range from any integer n_1 to any integer n_2 such that $n_1 \le n_2$; that is, we define

$$\sum_{k=n_1}^{n_2} f(k) = f(n_1) + f(n_1 + 1) + f(n_1 + 2) + \cdots + f(n_2)$$

Thus we have, for example,

$$\sum_{k=1}^{5} x_k = x_1 + x_2 + x_3 + x_4 + x_5$$

$$\sum_{i=1}^{n} a_i b_i = a_1 b_1 + a_2 b_2 + \cdots + a_n b_n$$

$$\sum_{j=2}^{5} j^2 = 2^2 + 3^2 + 4^2 + 5^2 = 4 + 9 + 16 + 25 = 54$$

$$\sum_{i=0}^{n} a_i x^i = a_0 + a_1 x + a_2 x^2 + \cdots + a_n x^n$$

$$\sum_{k=1}^{p} a_{ik} b_{kj} = a_{i1} b_{1j} + a_{i2} b_{2j} + a_{i3} b_{3j} + \cdots + a_{ip} b_{pj}$$

9.6 MATRIX MULTIPLICATION

Now suppose A and B are two matrices such that the number of columns of A is equal to the number of rows of B, say A is an $m \times p$ matrix and B is a $p \times n$ matrix. Then the *product* of A and B, written AB, is the $m \times n$ matrix whose *ij*-entry is obtained by multiplying the elements of the *i*th row of A by the corresponding elements of the *j*th column of B and then adding:

$$
\begin{pmatrix}
a_{11} & \cdots & a_{1p} \\
\cdot & \cdots & \cdot \\
a_{i1} & \cdots & a_{ip} \\
\cdot & \cdots & \cdot \\
a_{m1} & \cdots & a_{mp}
\end{pmatrix}
\begin{pmatrix}
b_{11} & \cdots & b_{1j} & \cdots & b_{1n} \\
\cdot & \cdots & \cdot & \cdots & \cdot \\
\cdot & \cdots & \cdot & \cdots & \cdot \\
\cdot & \cdots & \cdot & \cdots & \cdot \\
b_{p1} & \cdots & b_{pj} & \cdots & b_{pn}
\end{pmatrix}
=
\begin{pmatrix}
c_{11} & \cdots & & & c_{1n} \\
\cdot & \cdots & & & \cdot \\
\cdot & & c_{ij} & & \cdot \\
\cdot & & & & \cdot \\
c_{m1} & \cdots & & & c_{mn}
\end{pmatrix}
$$

where

$$c_{ij} = a_{i1} b_{1j} + a_{i2} b_{2j} + \cdots + a_{ip} b_{pj} = \sum_{k=1}^{p} a_{ik} b_{kj}$$

If the number of columns of A is not equal to the number of rows of B, then the product AB is not defined.

EXAMPLE 9.5

(a) Given

$$A = \begin{pmatrix} 1 & 3 \\ 2 & -1 \end{pmatrix} \qquad B = \begin{pmatrix} 2 & 0 & -4 \\ 3 & -2 & 6 \end{pmatrix}$$

find AB.

Since A is 2×2 and B is 2×3, the product matrix AB is defined and is a 2×3 matrix.   To obtain the elements in the first row of the product matrix AB, multiply the first row $(1, 3)$ of A by the columns

$$\begin{pmatrix} 2 \\ 3 \end{pmatrix} \qquad \begin{pmatrix} 0 \\ -2 \end{pmatrix} \qquad \text{and} \qquad \begin{pmatrix} -4 \\ 6 \end{pmatrix}$$

of B, respectively:

$$\begin{pmatrix} 1 & 3 \\ 2 & -1 \end{pmatrix}\begin{pmatrix} 2 & 0 & -4 \\ 3 & -2 & 6 \end{pmatrix} = \begin{pmatrix} 1\cdot 2 + 3\cdot 3 & 1\cdot 0 + 3\cdot(-2) & 1\cdot(-4) + 3\cdot 6 \end{pmatrix} = \begin{pmatrix} 11 & -6 & 14 \end{pmatrix}$$

To obtain the elements in the second row of the product matrix AB, multiply the second row $(2, -1)$ of A by the columns of B, respectively:

$$\begin{pmatrix} 1 & 3 \\ 2 & -1 \end{pmatrix}\begin{pmatrix} 2 & 0 & -4 \\ 3 & -2 & 6 \end{pmatrix}$$

$$= \begin{pmatrix} 11 & -6 & 14 \\ 2\cdot 2 + (-1)\cdot 3 & 2\cdot 0 + (-1)\cdot(-2) & 2\cdot(-4) + (-1)\cdot 6 \end{pmatrix} = \begin{pmatrix} 11 & -6 & 14 \\ 1 & 2 & -14 \end{pmatrix}$$

Thus

$$AB = \begin{pmatrix} 11 & -6 & 14 \\ 1 & 2 & -14 \end{pmatrix}$$

$$\begin{pmatrix} 1 & 2 \\ 3 & 4 \end{pmatrix}\begin{pmatrix} 1 & 1 \\ 0 & 2 \end{pmatrix} = \begin{pmatrix} 1\cdot 1 + 2\cdot 0 & 1\cdot 1 + 2\cdot 2 \\ 3\cdot 1 + 4\cdot 0 & 3\cdot 1 + 4\cdot 2 \end{pmatrix} = \begin{pmatrix} 1 & 5 \\ 2 & 11 \end{pmatrix}$$

$$\begin{pmatrix} 1 & 1 \\ 0 & 2 \end{pmatrix}\begin{pmatrix} 1 & 2 \\ 3 & 4 \end{pmatrix} = \begin{pmatrix} 1\cdot 1 + 1\cdot 3 & 1\cdot 2 + 1\cdot 4 \\ 0\cdot 1 + 2\cdot 3 & 0\cdot 2 + 2\cdot 4 \end{pmatrix} = \begin{pmatrix} 4 & 6 \\ 6 & 8 \end{pmatrix}$$

We see by Example 9.5(b) that matrix multiplication does not obey the commutative law, i.e. the products AB and BA of matrices need not be equal.   Matrix multiplication does, however, possess the following properties:

Theorem 9.2:   Whenever the sums and products are defined,
    (a) $(AB)C = A(BC)$
    (b) $A(B + C) = AB + AC$
    (c) $(B + C)A = BA + CA$
    (d) $k(AB) = (kA)B = A(kB)$, where k is a scalar.

Remark: Systems of linear equations are closely related to matrix equations.   For example, the system

$$\begin{cases} x + 2y - 3z = 4 \\ 5x - 6y + 8z = 8 \end{cases}$$

is equivalent to the matrix equation

$$\begin{pmatrix} 1 & 2 & -3 \\ 5 & -6 & 8 \end{pmatrix}\begin{pmatrix} x \\ y \\ z \end{pmatrix} = \begin{pmatrix} 4 \\ 8 \end{pmatrix}$$

That is, any solution to the system of equations is also a solution to the matrix equation, and vice versa.   Systems of linear equations are treated in Chapter 10.

9.7 SQUARE MATRICES

A matrix with the same number of rows as columns is called a *square* matrix. A square matrix with n rows and n columns is said to be of *order n*, and is called an *n-square matrix*. The *main diagonal*, or simply *diagonal*, of an *n*-square matrix $A = (a_{ij})$ consists of the elements $a_{11}, a_{22}, \ldots, a_{nn}$.

EXAMPLE 9.6 The matrix

$$\begin{pmatrix} 1 & -2 & 0 \\ 0 & -4 & -1 \\ 5 & 3 & 2 \end{pmatrix}.$$

is a square matrix of order 3. The numbers along the main diagonal are 1, −4 and 2.

The *n*-square matrix with 1s along the main diagonal and 0s elsewhere, e.g.,

$$\begin{pmatrix} 1 & 0 & 0 & 0 \\ 0 & 1 & 0 & 0 \\ 0 & 0 & 1 & 0 \\ 0 & 0 & 0 & 1 \end{pmatrix}$$

is called the *unit matrix* and will be denoted by I. The unit matrix I plays the same role in matrix multiplication as the number 1 does in the usual multiplication of numbers. Specifically,

$$AI = IA = A$$

for any square matrix A.

We can form powers of a square matrix X by defining

$$X^2 = XX, \quad X^3 = X^2X, \quad \ldots \quad \text{and} \quad X^0 = I$$

We can also form polynomials in X. That is, for any polynomial

$$f(x) = a_0 + a_1x + a_2x^2 + \cdots + a_nx^n$$

we define $f(X)$ to be the matrix

$$f(X) = a_0I + a_1X + a_2X^2 + \cdots + a_nX^n$$

In the case that $f(A)$ is the zero matrix, then matrix A is said to be a *zero* of the polynomial $f(X)$, or a *root* of the polynomial equation $f(X) = 0$.

EXAMPLE 9.7 Given

$$A = \begin{pmatrix} 1 & 2 \\ 3 & -4 \end{pmatrix}$$

$$A^2 = \begin{pmatrix} 7 & -6 \\ -9 & 22 \end{pmatrix}$$

If $f(x) = 2x^2 - 3x + 5$, then

$$f(A) = 2\begin{pmatrix} 7 & -6 \\ -9 & 22 \end{pmatrix} - 3\begin{pmatrix} 1 & 2 \\ 3 & -4 \end{pmatrix} + 5\begin{pmatrix} 1 & 0 \\ 0 & 1 \end{pmatrix} = \begin{pmatrix} 16 & -18 \\ -27 & 61 \end{pmatrix}$$

On the other hand, if $g(x) = x^2 + 3x - 10$ then

$$g(A) = \begin{pmatrix} 7 & -6 \\ -9 & 22 \end{pmatrix} + 3\begin{pmatrix} 1 & 2 \\ 3 & -4 \end{pmatrix} - 10\begin{pmatrix} 1 & 0 \\ 0 & 1 \end{pmatrix} = \begin{pmatrix} 0 & 0 \\ 0 & 0 \end{pmatrix}$$

Thus A is a zero of the polynomial $g(X) = X^2 + 3X - 10I$.

9.8 INVERTIBLE MATRICES

A square matrix A is said to be *invertible* if there exists a matrix B with the property

$$AB = BA = I, \text{ the identity matrix}$$

Such a matrix B is unique; it is called the *inverse* of A and is denoted by A^{-1}.   Observe that B is the inverse of A if and only if A is the inverse of B.   For example, suppose

$$A = \begin{pmatrix} 2 & 5 \\ 1 & 3 \end{pmatrix} \quad \text{and} \quad B = \begin{pmatrix} 3 & -5 \\ -1 & 2 \end{pmatrix}$$

Then

$$AB = \begin{pmatrix} 6-5 & -10+10 \\ 3-3 & -5+6 \end{pmatrix} = \begin{pmatrix} 1 & 0 \\ 0 & 1 \end{pmatrix} \quad \text{and} \quad BA = \begin{pmatrix} 6-5 & 15-15 \\ -2+2 & -5+6 \end{pmatrix} = \begin{pmatrix} 1 & 0 \\ 0 & 1 \end{pmatrix}$$

Thus A and B are inverses.

It is known that $AB = I$ if and only if $BA = I$; hence it is necessary to test only one product to determine whether two given matrices are inverses.

EXAMPLE 9.8

$$\begin{pmatrix} 1 & 0 & 2 \\ 2 & -1 & 3 \\ 4 & 1 & 8 \end{pmatrix} \begin{pmatrix} -11 & 2 & 2 \\ -4 & 0 & 1 \\ 6 & -1 & -1 \end{pmatrix} = \begin{pmatrix} -11+0+12 & 2+0-2 & 2+0-2 \\ -22+4+18 & 4+0-3 & 4-1-3 \\ -44-4+48 & 8+0-8 & 8+1-8 \end{pmatrix} = \begin{pmatrix} 1 & 0 & 0 \\ 0 & 1 & 0 \\ 0 & 0 & 1 \end{pmatrix}$$

Thus the two matrices are invertible and are inverses of each other.

9.9 DETERMINANTS

To each n-square matrix $A = (a_{ij})$ we assign a specific number called the *determinant* of A, denoted det (A) or $|A|$ or

$$\begin{vmatrix} a_{11} & a_{12} & \cdots & a_{1n} \\ a_{21} & a_{22} & \cdots & a_{2n} \\ \cdots & \cdots & \cdots & \cdots \\ a_{n1} & a_{n2} & \cdots & a_{nn} \end{vmatrix}$$

We emphasize that an $n \times n$ array of numbers enclosed by straight lines, called a *determinant of order n*, is not a matrix but denotes the number that the determinant function assigns to the enclosed array of numbers, i.e. the enclosed square matrix.

The determinants of orders one, two, and three are defined as follows:

$$|a_{11}| = a_{11}$$

$$\begin{vmatrix} a_{11} & a_{12} \\ a_{21} & a_{22} \end{vmatrix} = a_{11}a_{22} - a_{12}a_{21}$$

$$\begin{vmatrix} a_{11} & a_{12} & a_{13} \\ a_{21} & a_{22} & a_{23} \\ a_{31} & a_{32} & a_{33} \end{vmatrix} = a_{11}a_{22}a_{33} + a_{12}a_{23}a_{31} + a_{13}a_{21}a_{32} - a_{13}a_{22}a_{31} - a_{12}a_{21}a_{33} - a_{11}a_{23}a_{32}$$

The following diagram may help the reader remember how to evaluate the determinant of order two:

That is, the determinant equals the product of the elements along the plus-labeled arrow minus the product of the elements along the minus-labeled arrow.   There is an analogous scheme for a determinant of order three.   For notational convenience we have separated the plus-labeled and minus-labeled arrows:

EXAMPLE 9.9

(a) $\begin{vmatrix} 5 & 4 \\ 2 & 3 \end{vmatrix} = 5 \cdot 3 - 4 \cdot 2 = 15 - 8 = 7$ $\begin{vmatrix} 2 & 1 \\ -4 & 6 \end{vmatrix} = 2 \cdot 6 - 1 \cdot (-4) = 12 + 4 = 16$

(b) $\begin{vmatrix} 2 & 1 & 3 \\ 4 & 6 & -1 \\ 5 & 1 & 0 \end{vmatrix} = 2 \cdot 6 \cdot 0 + 1 \cdot (-1) \cdot 5 + 3 \cdot 4 \cdot 1 - 3 \cdot 6 \cdot 5 - 1 \cdot 4 \cdot 0 - 2 \cdot (-1) \cdot 1 = -81$

The general definition of a determinant of order n is as follows:

$$\det(A) = \sum \operatorname{sgn}(\sigma)\, a_{1j_1} a_{2j_2} \cdots a_{nj_n}$$

where the sum is over all possible permutations $\sigma = (j_1, j_2, \ldots, j_n)$ of $(1, 2, \ldots, n)$.   Here $\operatorname{sgn}(\sigma)$ equals plus or minus one according as an even or an odd number of interchanges are required to change σ so that its numbers are in the usual order.   We have included the general definition of the determinant function for completeness.   The reader is referred to texts in matrix theory or linear algebra for techniques for computing determinants of order greater than three.   Permutations are studied in Chapter 11.

An important property of the determinant function is that it is multiplicative; that is,

Theorem 9.3:   For any two n-square matrices A and B, $\det(AB) = \det(A) \cdot \det(B)$.

9.10   INVERTIBLE MATRICES AND DETERMINANTS

Theorem 9.4:   A square matrix is invertible if and only if it has a nonzero determinant.

We now show how to calculate the inverse of a 2×2 matrix

$$A = \begin{pmatrix} a & b \\ c & d \end{pmatrix}$$

whose determinant is nonzero.   We seek scalars x, y, z, and w such that

$$\begin{pmatrix} a & b \\ c & d \end{pmatrix}\begin{pmatrix} x & y \\ z & w \end{pmatrix} = \begin{pmatrix} 1 & 0 \\ 0 & 1 \end{pmatrix} \quad \text{or} \quad \begin{pmatrix} ax + bz & ay + bw \\ cx + dz & cy + dw \end{pmatrix} = \begin{pmatrix} 1 & 0 \\ 0 & 1 \end{pmatrix}$$

which reduces to solving the following two systems of linear equations in two unknowns:

$$\begin{cases} ax + bz = 1 \\ cx + dz = 0 \end{cases} \qquad \begin{cases} ay + bw = 0 \\ cy + dw = 1 \end{cases}$$

Since $|A| = ad - bc$ is not zero, we can uniquely solve for the unknowns x, y, z, and w, obtaining:

$$x = \frac{d}{ad - bc} = \frac{d}{|A|}, \quad y = \frac{-b}{ad - bc} = \frac{-b}{|A|}, \quad z = \frac{-c}{ad - bc} = \frac{-c}{|A|}, \quad w = \frac{a}{ad - bc} = \frac{a}{|A|}$$

Accordingly,

$$A^{-1} = \begin{pmatrix} a & b \\ c & d \end{pmatrix}^{-1} = \begin{pmatrix} d/|A| & -b/|A| \\ -c/|A| & a/|A| \end{pmatrix} = \frac{1}{|A|}\begin{pmatrix} d & -b \\ -c & a \end{pmatrix}$$

In other words, we can obtain the inverse of a 2×2 matrix, with determinant nonzero, by (i) inter-changing the elements on the main diagonal, (ii) taking the negative of the other elements, and (iii) dividing each element by the determinant of the original matrix.   For example, if

$$A = \begin{pmatrix} 2 & 3 \\ 4 & 5 \end{pmatrix}$$

then $|A| = -2$, and so

$$A^{-1} = \frac{1}{-2}\begin{pmatrix} 5 & -3 \\ -4 & 2 \end{pmatrix} = \begin{pmatrix} -\frac{5}{2} & \frac{3}{2} \\ 2 & -1 \end{pmatrix}$$

An analogous, though more complicated, determinantal formula may be given for the inverse of an invertible matrix of order greater than 2.

9.11 SUBSCRIPTED VARIABLES

The data-names or variables in computer programs, discussed in Section 5.2, are also called *unsubscripted variables* or *scalar variables*, since each such variable represents a memory cell in which a single value is stored.   Frequently, one wishes to use the same data-name to refer to an array of values rather than to a single value.   This can be done by the use of *subscripted variables*.   How-ever, since computer statements must be punched (or typed) on one line, the "subscripts" appear in parentheses, as, e.g.,

$$W(1), \; W(2), \; W(3), \ldots \qquad \text{instead of} \qquad w_1, \; w_2, \; w_3, \ldots$$

or

$$S(1, 1), S(1, 2), S(1, 3), \ldots \qquad \text{instead of} \qquad s_{11}, s_{12}, s_{13}, \ldots$$

We will use both notations in our flowcharts.

The number of subscripts is called the *dimension* of the array or subscripted variable.   Thus the variable W above is a one-dimensional array, also called a *linear array* or *vector array*, and the variable S is a two-dimensional array, also called a *matrix array*.   Most computers can handle arrays with one, two, or three subscripts, and some of the larger computers accept as many as seven subscripts.   If the dimension is two or three, the first subscript of an array is called its *row*, the second subscript its *column*, and the third subscript (if it has one) its *page*.

EXAMPLE 9.10   Figure 9-1(a) is the flowchart of an algorithm for calculating the average weight of 35 students.   As shown by the input box

WEIGHT is a subscripted variable that represents the addresses

$$\text{WEIGHT}(1), \; \text{WEIGHT}(2), \; \text{WEIGHT}(3), \ldots, \; \text{WEIGHT}(35)$$

of the input data items.   Figure 9-1(b) gives an equivalent pseudocode program.

Generally speaking, when using a subscripted variable in a computer program, one must provide the compiler with the following information before the program is actually run:

1. The name of the subscripted variable.

2. The number of subscripts (i.e. the dimension of the array).

3. The ranges of the subscripts, which jointly determine the total number of storage locations allocated to the variable.

Each programming language has its own syntax for doing this.

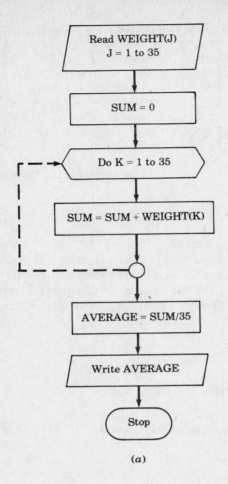

(a)

```
read WEIGHT(J), J = 1 to 35
SUM = 0
DO K = 1 to 35
      SUM = SUM + WEIGHT(K)
ENDDO
AVERAGE = SUM/35
write AVERAGE
END
```

(b)

Fig. 9-1

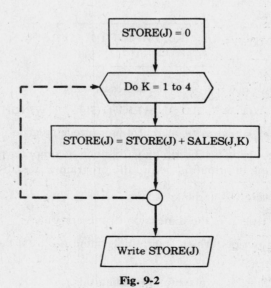

Fig. 9-2

EXAMPLE 9.11 Suppose that the sales data of Table 9-1 are read into the computer as the 28×4 matrix array SALES. Thus,

$$SALES(J, 1) \qquad SALES(J, 2) \qquad SALES(J, 3) \qquad SALES(J, 4)$$

are the weekly sales in the four departments of the Jth store. Let STORE(J) denote the total weekly sales in the Jth store, and let CHAIN denote the total weekly sales in the chain. We want a flowchart which finds the entire linear array STORE and also the value of CHAIN. Figure 9-2 gives a flowchart, involving a loop, that calculates and outputs STORE(J). To find the entire linear array STORE, we then require nested loops, where the outer loop runs through each of the 28 stores. Such a flowchart appears in Fig. 9-3(a), which also uses the

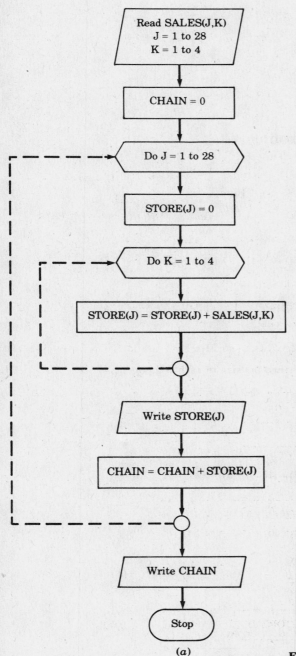

```
read SALES(J,K), J = 1 to 28, K = 1 to 4
CHAIN = 0
DO J = 1 to 28
      STORE(J) = 0
      DO K = 1 to 4
            STORE(J) = STORE(J) + SALES(J,K)
      ENDDO
      write STORE(J)
      CHAIN = CHAIN + STORE(J)
ENDDO
write CHAIN
END
```

(b)

(a)   **Fig. 9-3**

outer loop to calculate the value of CHAIN by summing up all the values in STORE. Observe that the input box

$$\begin{array}{c} \text{Read SALES(J, K)} \\ \text{J = 1 to 28} \\ \text{K = 1 to 4} \end{array}$$

indicates that SALES is a 28×4 matrix array. Figure 9-3(b) gives an equivalent pseudocode program.

Solved Problems

VECTORS

9.1 Let $u = (2, -7, 1)$, $v = (-3, 0, 4)$ and $w = (0, 5, -8)$. Find:

$$(a) \quad u + v \qquad (b) \quad v + w \qquad (c) \quad -3u \qquad (d) \quad -w$$

(a) Add corresponding components:

$$u + v = (2, -7, 1) + (-3, 0, 4) = (2 - 3, -7 + 0, 1 + 4) = (-1, -7, 5)$$

(b) Add corresponding components:

$$v + w = (-3, 0, 4) + (0, 5, -8) = (-3 + 0, 0 + 5, 4 - 8) = (-3, 5, -4)$$

(c) Multiply each component of u by the scalar -3:

$$-3u = -3(2, -7, 1) = (-6, 21, -3)$$

(d) Multiply each component of w by -1, i.e. change the sign of each component:

$$-w = -(0, 5, -8) = (0, -5, 8)$$

9.2 For the vectors of Problem 9.1, calculate (a) $3u - 4v$, (b) $2u + 3v - 5w$.

First perform the scalar multiplication and then the vector addition.

(a) $3u - 4v = 3(2, -7, 1) - 4(-3, 0, 4) = (6, -21, 3) + (12, 0, -16) = (18, -21, -13)$

(b) $2u + 3v - 5w = 2(2, -7, 1) + 3(-3, 0, 4) - 5(0, 5, -8)$

$$= (4, -14, 2) + (-9, 0, 12) + (0, -25, 40)$$
$$= (4 - 9 + 0, -14 + 0 - 25, 2 + 12 + 40) = (-5, -39, 54)$$

9.3 The *dot product* or *inner product* of vectors $u = (u_1, u_2, \ldots, u_n)$ and $v = (v_1, v_2, \ldots, v_n)$ is denoted and defined by

$$u \cdot v = u_1 v_1 + u_2 v_2 + \cdots + u_n v_n$$

and the *norm* (or *length*) of u is denoted and defined by

$$\|u\| = \sqrt{u \cdot u} = \sqrt{u_1^2 + u_2^2 + \cdots + u_n^2}$$

(Observe that $\|u\| \geq 0$, with equality if and only if $u = 0$.) For the vectors of Problem 9.1, find:

$$(a) \quad u \cdot v, \ u \cdot w, \ v \cdot w \qquad (b) \quad \|u\|, \|v\|, \|w\|$$

(a) Multiply corresponding components and then add:

$$u \cdot v = 2 \cdot (-3) + (-7) \cdot 0 + 1 \cdot 4 = -6 + 0 + 4 = -2$$
$$u \cdot w = 0 - 35 - 8 = -43$$
$$v \cdot w = 0 + 0 - 32 = -32$$

(b) Take the square root of the sum of the squares of the components:

$$\|u\| = \sqrt{2^2 + (-7)^2 + 1^2} = \sqrt{4 + 49 + 1} = \sqrt{54} = 3\sqrt{6}$$
$$\|v\| = \sqrt{9 + 0 + 16} = \sqrt{25} = 5$$
$$\|w\| = \sqrt{0 + 25 + 64} = \sqrt{89}$$

9.4 Find x and y if $x(1, 1) + y(2, -1) = (1, 4)$.

First multiply by the scalars x and y and then add:

$$x(1, 1) + y(2, -1) = (x, x) + (2y, -y) = (x + 2y, x - y) = (1, 4)$$

Two vectors are equal only if their corresponding components are equal; hence

$$x + 2y = 1$$
$$x - \ y = 4$$

Solve the system of equations to find $x = 3$ and $y = -1$.

MATRIX ADDITION AND SCALAR MULTIPLICATION

9.5 Compute:

$$(a) \ \begin{pmatrix} 1 & 2 & 3 \\ 4 & 5 & 6 \end{pmatrix} + \begin{pmatrix} 1 & -1 & 2 \\ 0 & 3 & -5 \end{pmatrix} \qquad (b) \ -2 \begin{pmatrix} 1 & 7 \\ 2 & -3 \\ 0 & -1 \end{pmatrix} \qquad (c) \ - \begin{pmatrix} 2 & -3 & 8 \\ 1 & -2 & -6 \end{pmatrix}$$

(a) Add corresponding elements:

$$\begin{pmatrix} 1 & 2 & 3 \\ 4 & 5 & 6 \end{pmatrix} + \begin{pmatrix} 1 & -1 & 2 \\ 0 & 3 & -5 \end{pmatrix} = \begin{pmatrix} 1+1 & 2+(-1) & 3+2 \\ 4+0 & 5+3 & 6+(-5) \end{pmatrix} = \begin{pmatrix} 2 & 1 & 5 \\ 4 & 8 & 1 \end{pmatrix}$$

(b) Multiply each element of the matrix by the scalar -2:

$$-2 \begin{pmatrix} 1 & 7 \\ 2 & -3 \\ 0 & -1 \end{pmatrix} = \begin{pmatrix} (-2) \cdot 1 & (-2) \cdot 7 \\ (-2) \cdot 2 & (-2) \cdot (-3) \\ (-2) \cdot 0 & (-2) \cdot (-1) \end{pmatrix} = \begin{pmatrix} -2 & -14 \\ -4 & 6 \\ 0 & 2 \end{pmatrix}$$

(c) Multiply each element of the matrix by -1, or, equivalently, change the sign of each element in the matrix:

$$- \begin{pmatrix} 2 & -3 & 8 \\ 1 & -2 & -6 \end{pmatrix} = \begin{pmatrix} -2 & 3 & -8 \\ -1 & 2 & 6 \end{pmatrix}$$

9.6 Compute:

$$3 \begin{pmatrix} 2 & -5 & 1 \\ 3 & 0 & -4 \end{pmatrix} - 2 \begin{pmatrix} 1 & -2 & -3 \\ 0 & -1 & 5 \end{pmatrix} + 4 \begin{pmatrix} 0 & 1 & -2 \\ 1 & -1 & -1 \end{pmatrix}$$

First perform the scalar multiplications and then the matrix additions:

$$3\begin{pmatrix} 2 & -5 & 1 \\ 3 & 0 & -4 \end{pmatrix} - 2\begin{pmatrix} 1 & -2 & -3 \\ 0 & -1 & 5 \end{pmatrix} + 4\begin{pmatrix} 0 & 1 & -2 \\ 0 & -1 & -1 \end{pmatrix}$$

$$= \begin{pmatrix} 6 & -15 & 3 \\ 9 & 0 & -12 \end{pmatrix} + \begin{pmatrix} -2 & 4 & 6 \\ 0 & 2 & -10 \end{pmatrix} + \begin{pmatrix} 0 & 4 & -8 \\ 4 & -4 & -4 \end{pmatrix}$$

$$= \begin{pmatrix} 6+(-2)+0 & -15+4+4 & 3+6+(-8) \\ 9+0+4 & 0+2+(-4) & -12+(-10)+(-4) \end{pmatrix} = \begin{pmatrix} 4 & -7 & 1 \\ 13 & -2 & -26 \end{pmatrix}$$

MATRIX MULTIPLICATION

9.7 Let $(r \times s)$ denote an $r \times s$ matrix. Find the sizes of those matrix products which are defined:

     (a) $(2 \times 3)(3 \times 4)$ (c) $(1 \times 2)(3 \times 1)$ (e) $(4 \times 4)(3 \times 3)$
     (b) $(4 \times 1)(1 \times 2)$ (d) $(5 \times 2)(2 \times 3)$ (f) $(2 \times 2)(2 \times 4)$

In each case the product is defined if the inner numbers are equal, and then the product will have the size of the outer numbers in the given order.

     (a) 2×4 (c)   not defined   (e)   not defined
     (b) 4×2 (d) 5×3 (f) 2×4

9.8 Let

$$A = \begin{pmatrix} 2 & -1 \\ 1 & 0 \\ -3 & 4 \end{pmatrix} \qquad B = \begin{pmatrix} 1 & -2 & -5 \\ 3 & 4 & 0 \end{pmatrix}$$

Find AB.

Now A is 3×2 and B is 2×3, so the product AB is defined and is a 3×3 matrix. To obtain the first row of the product matrix AB, multiply the first row of A by the columns of B, in sequence:

$$\begin{pmatrix} 2 & -1 \\ 1 & 0 \\ -3 & 4 \end{pmatrix}\begin{pmatrix} 1 & -2 & -5 \\ 3 & 4 & 0 \end{pmatrix} = \begin{pmatrix} 2-3 & -4-4 & -10+0 \\ & & \end{pmatrix} = \begin{pmatrix} -1 & -8 & -10 \\ & & \end{pmatrix}$$

To obtain the second row of the product matrix AB, multiply the second row of A by the columns of B, in sequence:

$$\begin{pmatrix} 2 & -1 \\ 1 & 0 \\ -3 & 4 \end{pmatrix}\begin{pmatrix} 1 & -2 & -5 \\ 3 & 4 & 0 \end{pmatrix} = \begin{pmatrix} -1 & -8 & -10 \\ 1+0 & -2+0 & -5+0 \\ & & \end{pmatrix} = \begin{pmatrix} -1 & -8 & -10 \\ 1 & -2 & -5 \\ & & \end{pmatrix}$$

To obtain the third row of the product matrix AB, multiply the third row of A by the columns of B, in sequence:

$$\begin{pmatrix} 2 & -1 \\ 1 & 0 \\ -3 & 4 \end{pmatrix}\begin{pmatrix} 1 & -2 & -5 \\ 3 & 4 & 0 \end{pmatrix} = \begin{pmatrix} -1 & -8 & -10 \\ 1 & -2 & -5 \\ -3+12 & 6+16 & 15+0 \end{pmatrix} = \begin{pmatrix} -1 & -8 & -10 \\ 1 & -2 & -5 \\ 9 & 22 & 15 \end{pmatrix}$$

Thus

$$AB = \begin{pmatrix} -1 & -8 & -10 \\ 1 & -2 & -5 \\ 9 & 22 & 15 \end{pmatrix}$$

9.9 Let

$$A = \begin{pmatrix} 2 & -1 & 0 \\ 1 & 0 & -3 \end{pmatrix} \qquad B = \begin{pmatrix} 1 & -4 & 0 & 1 \\ 2 & -1 & 3 & -1 \\ 4 & 0 & -2 & 0 \end{pmatrix}$$

(a) Determine the size of AB.   (b) Let c_{ij} denote the element in the ith row and jth column of the product matrix AB, that is, $AB = (c_{ij})$.   Find: c_{23}, c_{14}, c_{21} and c_{12}.

(a) Since A is 2×3 and B is 3×4, the product AB is a 2×4 matrix.
(b) Now c_{ij} is defined as the product of the ith row of A by the jth column of B.   Hence:

$$c_{23} = (1, 0, -3)\begin{pmatrix} 0 \\ 3 \\ -2 \end{pmatrix} = 1 \cdot 0 + 0 \cdot 3 + (-3) \cdot (-2) = 0 + 0 + 6 = 6$$

$$c_{14} = (2, -1, 0)\begin{pmatrix} 1 \\ -1 \\ 0 \end{pmatrix} = 2 \cdot 1 + (-1) \cdot (-1) + 0 \cdot 0 = 2 + 1 + 0 = 3$$

$$c_{21} = (1, 0, -3)\begin{pmatrix} 1 \\ 2 \\ 4 \end{pmatrix} = 1 \cdot 1 + 0 \cdot 2 + (-3) \cdot 4 = 1 + 0 - 12 = -11$$

$$c_{12} = (2, -1, 0)\begin{pmatrix} -4 \\ -1 \\ 0 \end{pmatrix} = 2 \cdot (-4) + (-1) \cdot (-1) + 0 \cdot 0 = -8 + 1 + 0 = -7$$

9.10   Compute:

(a) $\begin{pmatrix} 1 & 6 \\ -3 & 5 \end{pmatrix}\begin{pmatrix} 4 & 0 \\ 2 & -1 \end{pmatrix}$ (c) $\begin{pmatrix} 1 \\ -6 \end{pmatrix}\begin{pmatrix} 1 & 6 \\ -3 & 5 \end{pmatrix}$ (e) $(2, -1)\begin{pmatrix} 1 \\ -6 \end{pmatrix}$

(b) $\begin{pmatrix} 1 & 6 \\ -3 & 5 \end{pmatrix}\begin{pmatrix} 2 \\ -7 \end{pmatrix}$ (d) $\begin{pmatrix} 1 \\ 6 \end{pmatrix}(3, 2)$

(a) The first factor is 2×2 and the second is 2×2, so the product is defined and is a 2×2 matrix:

$$\begin{pmatrix} 1 & 6 \\ -3 & 5 \end{pmatrix}\begin{pmatrix} 4 & 0 \\ 2 & -1 \end{pmatrix} = \begin{pmatrix} 1 \cdot 4 + 6 \cdot 2 & 1 \cdot 0 + 6 \cdot (-1) \\ (-3) \cdot 4 + 5 \cdot 2 & (-3) \cdot 0 + 5 \cdot (-1) \end{pmatrix} = \begin{pmatrix} 16 & -6 \\ -2 & -5 \end{pmatrix}$$

(b) The first factor is 2×2 and the second is 2×1, so the product is defined and is a 2×1 matrix:

$$\begin{pmatrix} 1 & 6 \\ -3 & 5 \end{pmatrix}\begin{pmatrix} 2 \\ -7 \end{pmatrix} = \begin{pmatrix} 1 \cdot 2 + 6 \cdot (-7) \\ (-3) \cdot 2 + 5 \cdot (-7) \end{pmatrix} = \begin{pmatrix} -40 \\ -41 \end{pmatrix}$$

(c) Now the first factor is 2×1 and the second is 2×2.   Since the inner numbers 1 and 2 are distinct, the product is not defined.

(d) Here the first factor is 2×1 and the second is 1×2, so the product is defined and is a 2×2 matrix:

$$\begin{pmatrix} 1 \\ 6 \end{pmatrix}(3, 2) = \begin{pmatrix} 1 \cdot 3 & 1 \cdot 2 \\ 6 \cdot 3 & 6 \cdot 2 \end{pmatrix} = \begin{pmatrix} 3 & 2 \\ 18 & 12 \end{pmatrix}$$

(e) The first factor is 1×2 and the second is 2×1, so the product is defined and is a 1×1 matrix, which we frequently write as a scalar.

$$(2, -1)\begin{pmatrix} 1 \\ -6 \end{pmatrix} = (2 \cdot 1) + (-1) \cdot (-6) = (8) = 8$$

9.11   The *transpose* of a matrix A, written A^T, is the matrix obtained by writing the rows of A, in order, as columns.   That is, if $A = (a_{ij})$, then $B = (b_{ij})$ is the transpose of A if $b_{ij} = a_{ji}$ for all i and j.   (Note that if A is an $m \times n$ matrix, then A^T is an $n \times m$ matrix.)   If

$$A = \begin{pmatrix} 1 & 2 & 0 \\ 3 & -1 & 4 \end{pmatrix}$$

find: (a) A^T, (b) AA^T, (c) A^TA.

(a) To obtain A^T, rewrite the rows of A as columns:

$$A^T = \begin{pmatrix} 1 & 3 \\ 2 & -1 \\ 0 & 4 \end{pmatrix}$$

(b) $$AA^T = \begin{pmatrix} 1 & 2 & 0 \\ 3 & -1 & 4 \end{pmatrix} \begin{pmatrix} 1 & 3 \\ 2 & -1 \\ 0 & 4 \end{pmatrix} = \begin{pmatrix} 1+4+0 & 3-2+0 \\ 3-2+0 & 9+1+16 \end{pmatrix} = \begin{pmatrix} 5 & 1 \\ 1 & 26 \end{pmatrix}$$

(c) $$A^T A = \begin{pmatrix} 1 & 3 \\ 2 & -1 \\ 0 & 4 \end{pmatrix} \begin{pmatrix} 1 & 2 & 0 \\ 3 & -1 & 4 \end{pmatrix} = \begin{pmatrix} 1+9 & 2-3 & 0+12 \\ 2-3 & 4+1 & 0-4 \\ 0+12 & 0-4 & 0+16 \end{pmatrix} = \begin{pmatrix} 10 & -1 & 12 \\ -1 & 5 & -4 \\ 12 & -4 & 16 \end{pmatrix}$$

SQUARE MATRICES

9.12   Let

$$A = \begin{pmatrix} 1 & 2 \\ 4 & -3 \end{pmatrix}$$

Find: (a) A^2; (b) A^3; (c) $f(A)$, where $f(x) = 2x^3 - 4x + 5$.   (d) Show that A is a zero of the polynomial $g(X) = X^2 + 2X - 11I$.

(a) $$A^2 = AA = \begin{pmatrix} 1 & 2 \\ 4 & -3 \end{pmatrix} \begin{pmatrix} 1 & 2 \\ 4 & -3 \end{pmatrix}$$

$$= \begin{pmatrix} 1 \cdot 1 + 2 \cdot 4 & 1 \cdot 2 + 2 \cdot (-3) \\ 4 \cdot 1 + (-3) \cdot 4 & 4 \cdot 2 + (-3) \cdot (-3) \end{pmatrix} = \begin{pmatrix} 9 & -4 \\ -8 & 17 \end{pmatrix}$$

(b) $$A^3 = AA^2 = \begin{pmatrix} 1 & 2 \\ 4 & -3 \end{pmatrix} \begin{pmatrix} 9 & -4 \\ -8 & 17 \end{pmatrix}$$

$$= \begin{pmatrix} 1 \cdot 9 + 2 \cdot (-8) & 1 \cdot (-4) + 2 \cdot 17 \\ 4 \cdot 9 + (-3) \cdot (-8) & 4 \cdot (-4) + (-3) \cdot 17 \end{pmatrix} = \begin{pmatrix} -7 & 30 \\ 60 & -67 \end{pmatrix}$$

(c) To find $f(A)$, first substitute A^3 for x^3, A for x, and $5I$ for the constant 5 in the given polynomial $f(x) = 2x^3 - 4x + 5$:

$$f(A) = 2A^3 - 4A + 5I = 2\begin{pmatrix} -7 & 30 \\ 60 & -67 \end{pmatrix} - 4\begin{pmatrix} 1 & 2 \\ 4 & -3 \end{pmatrix} + 5\begin{pmatrix} 1 & 0 \\ 0 & 1 \end{pmatrix}$$

Then multiply each matrix by its respective scalar:

$$f(A) = \begin{pmatrix} -14 & 60 \\ 120 & -134 \end{pmatrix} + \begin{pmatrix} -4 & -8 \\ -16 & 12 \end{pmatrix} + \begin{pmatrix} 5 & 0 \\ 0 & 5 \end{pmatrix}$$

Lastly, add the corresponding elements in the matrices:

$$f(A) = \begin{pmatrix} -14-4+5 & 60-8+0 \\ 120-16+0 & -134+12+5 \end{pmatrix} = \begin{pmatrix} -13 & 52 \\ 104 & -117 \end{pmatrix}$$

(d) Matrix A is a zero of $g(X)$ if the matrix $g(A)$ is the zero matrix.   We have:

$$g(A) = A^2 + 2A - 11I = \begin{pmatrix} 9 & -4 \\ -8 & 17 \end{pmatrix} + 2\begin{pmatrix} 1 & 2 \\ 4 & -3 \end{pmatrix} - 11\begin{pmatrix} 1 & 0 \\ 0 & 1 \end{pmatrix}$$

Now multiply each matrix by the scalar preceding it:

$$g(A) = \begin{pmatrix} 9 & -4 \\ -8 & 17 \end{pmatrix} + \begin{pmatrix} 2 & 4 \\ 8 & -6 \end{pmatrix} + \begin{pmatrix} -11 & 0 \\ 0 & -11 \end{pmatrix}$$

Lastly, add the corresponding elements in the matrices:

$$g(A) = \begin{pmatrix} 9+2-11 & -4+4+0 \\ -8+8+0 & 17-6-11 \end{pmatrix} = \begin{pmatrix} 0 & 0 \\ 0 & 0 \end{pmatrix}$$

Since $g(A) = 0$, A is a zero of the polynomial $g(X)$.

9.13 Compute the determinant of each matrix:

(a) $\begin{pmatrix} 3 & -2 \\ 4 & 5 \end{pmatrix}$ (b) $\begin{pmatrix} -1 & 6 \\ 0 & 4 \end{pmatrix}$ (c) $\begin{pmatrix} a-b & b \\ b & a+b \end{pmatrix}$ (d) $\begin{pmatrix} a-b & a \\ a & a+b \end{pmatrix}$

(a)
$$\begin{vmatrix} 3 & -2 \\ 4 & 5 \end{vmatrix} = 3 \cdot 5 - (-2) \cdot 4 = 15 + 8 = 23$$

(b)
$$\begin{vmatrix} -1 & 6 \\ 0 & 4 \end{vmatrix} = -1 \cdot 4 - 6 \cdot 0 = -4$$

(c)
$$\begin{vmatrix} a-b & b \\ b & a+b \end{vmatrix} = (a-b)(a+b) - b \cdot b = a^2 - b^2 - b^2 = a^2 - 2b^2$$

(d)
$$\begin{vmatrix} a-b & a \\ a & a+b \end{vmatrix} = (a-b)(a+b) - a \cdot a = a^2 - b^2 - a^2 = -b^2$$

9.14 Find the determinant of each matrix:

(a) $\begin{pmatrix} 1 & 2 & 3 \\ 4 & -2 & 3 \\ 0 & 5 & -1 \end{pmatrix}$ (b) $\begin{pmatrix} 4 & -1 & -2 \\ 0 & 2 & -3 \\ 5 & 2 & 1 \end{pmatrix}$ (c) $\begin{pmatrix} 2 & -3 & 4 \\ 1 & 2 & -3 \\ -1 & -2 & 5 \end{pmatrix}$

(*Hint*: Use diagram in Section 9.9.)

(a)
$$\begin{vmatrix} 1 & 2 & 3 \\ 4 & -2 & 3 \\ 0 & 5 & -1 \end{vmatrix} = 2 + 0 + 60 - 0 - 15 + 8 = 55$$

(b)
$$\begin{vmatrix} 4 & -1 & -2 \\ 0 & 2 & -3 \\ 5 & 2 & 1 \end{vmatrix} = 8 + 15 + 0 + 20 + 24 + 0 = 67$$

(c)
$$\begin{vmatrix} 2 & -3 & 4 \\ 1 & 2 & -3 \\ -1 & -2 & 5 \end{vmatrix} = 20 - 9 - 8 + 8 - 12 + 15 = 14$$

9.15 Find the inverse of $\begin{pmatrix} 3 & 5 \\ 2 & 3 \end{pmatrix}$

Method 1.
We seek scalars x, y, z and w for which

$$\begin{pmatrix} 3 & 5 \\ 2 & 3 \end{pmatrix}\begin{pmatrix} x & y \\ z & w \end{pmatrix} = \begin{pmatrix} 1 & 0 \\ 0 & 1 \end{pmatrix} \quad \text{or} \quad \begin{pmatrix} 3x+5z & 3y+5w \\ 2x+3z & 2y+3w \end{pmatrix} = \begin{pmatrix} 1 & 0 \\ 0 & 1 \end{pmatrix}$$

or which satisfy

$$\begin{cases} 3x+5z = 1 \\ 2x+3z = 0 \end{cases} \quad \text{and} \quad \begin{cases} 3y+5w = 0 \\ 2y+3w = 1 \end{cases}$$

To solve the first system, multiply the first equation by 2 and the second equation by −3 and then add:

$$\begin{array}{lr} 2 \times \text{first:} & 6x + 10z = 2 \\ -3 \times \text{second:} & -6x - 9z = 0 \\ \hline \text{Addition:} & z = 2 \end{array}$$

Substitute $z = 2$ into the first equation to obtain

$$3x + 5 \cdot 2 = 1 \quad \text{or} \quad 3x + 10 = 1 \quad \text{or} \quad 3x = -9 \quad \text{or} \quad x = -3$$

To solve the second system, multiply the first equation by 2 and the second equation by -3 and then add:

$$\begin{array}{rl} 2 \times \text{first:} & 6y + 10w = 0 \\ -3 \times \text{second:} & -6y - 9w = -3 \\ \hline \text{Addition:} & w = -3 \end{array}$$

Substitute $w = -3$ in the first equation to obtain

$$3y + 5 \cdot (-3) = 0 \quad \text{or} \quad 3y - 15 = 0 \quad \text{or} \quad 3y = 15 \quad \text{or} \quad y = 5$$

Thus the inverse of the given matrix is

$$\begin{pmatrix} -3 & 5 \\ 2 & -3 \end{pmatrix}$$

Method 2.

We use the general formula for the inverse of a 2×2 matrix. First find the determinant of the given matrix:

$$\begin{vmatrix} 3 & 5 \\ 2 & 3 \end{vmatrix} = 3 \cdot 3 - 2 \cdot 5 = 9 - 10 = -1$$

Now interchange the elements on the main diagonal of the given matrix and take the negative of the other elements to obtain

$$\begin{pmatrix} 3 & -5 \\ -2 & 3 \end{pmatrix}$$

Lastly, divide each element of this matrix by the determinant of the given matrix, that is, by -1:

$$\begin{pmatrix} -3 & 5 \\ 2 & -3 \end{pmatrix}$$

The above is the required inverse.

SUBSCRIPTED VARIABLES

9.16 Determine the dimensions of and the numbers of elements in whatever arrays are defined by the following input boxes:

(a) Read TEST(J) J = 1 to 86

(b) Read B(J,K,L) J = 1 to 6 K = 1 to 8 L = 1 to 4

(c) Read NAME, RATE, HOURS

(d) Read N, $A_{J,K}$ J = 1 to N K = 1 to N + 1

(a) TEST is a linear (one-dimensional) array with 86 elements.
(b) B is a $6 \times 8 \times 4$, three-dimensional array. Hence B contains $(6)(8)(4) = 192$ elements.
(c) The variables NAME, RATE, and HOURS are not arrays.
(d) Here A is an $N \times (N + 1)$, two-dimensional array. Hence it has $N(N + 1) = N^2 + N$ elements. Note that a value for N is read before values for A are read.

9.17   Determine the output of each flowchart in Fig. 9-4.

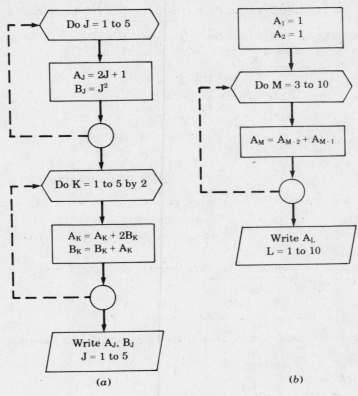

(a)   (b)

Fig. 9-4

(a)   The first DO loop assigns:

$$A_1 = 2 \cdot 1 + 1 = 3, \quad B_1 = 1^2 = 1 \qquad A_4 = 2 \cdot 4 + 1 = 9, \quad B_4 = 4^2 = 16$$
$$A_2 = 2 \cdot 2 + 1 = 5, \quad B_2 = 2^2 = 4 \qquad A_5 = 2 \cdot 5 + 1 = 11, \quad B_5 = 5^2 = 25$$
$$A_3 = 2 \cdot 3 + 1 = 7, \quad B_3 = 3^2 = 9$$

The second DO loop assigns:

$$A_1 = A_1 + 2B_1 = 3 + 2 \cdot 1 = 5 \qquad B_1 = B_1 + A_1 = 1 + 5 = 6$$
$$A_3 = A_3 + 2B_3 = 7 + 2 \cdot 9 = 25 \qquad B_3 = B_3 + A_3 = 9 + 25 = 34$$
$$A_5 = A_5 + 2B_5 = 11 + 2 \cdot 25 = 61 \qquad B_5 = B_5 + A_5 = 25 + 61 = 86$$

The output is: $A_1, B_1, A_2, B_2, \ldots, A_5, B_5$; that is,

5,   6,   5,   4,   25,   34,   9,   16,   61,   86

(b)   Here $A_1 = 1$ and $A_2 = 1$, and each subsequent term is the sum of the two preceding terms; that is,

$$A_3 = A_1 + A_2 = 1 + 1 = 2 \qquad A_4 = A_2 + A_3 = 1 + 2 = 3 \qquad A_5 = A_3 + A_4 = 2 + 3 = 5 \qquad \ldots$$

Thus the output is: 1, 1, 2, 3, 5, 8, 13, 21, 34, 55 (the first ten terms of the *Fibonacci sequence*).

9.18   Thirty-five students take an exam where scores range between 0 and 100.   (a) Draw a flow-chart which reads the scores into an array S and which finds the number FAIL of scores less than 60 and the number PERFECT of scores 100.   (b) Write an equivalent pseudocode program.

(a) The flowchart appears in Fig. 9-5. Observe that the input box

$$\boxed{\begin{array}{c} \text{Read } S_K \\ K = 1 \text{ to } 35 \end{array}}$$

stores the scores in the array S. The loop tests each score to see if it is less than 60, in which case FAIL is increased by one, or if it is equal to 100, in which case PERFECT is increased by one.

(b)

```
read S_K, K = 1 to 35
FAIL = 0
PERFECT = 0
DO K = 1 to 35
    IF S_K < 60 THEN
        FAIL = FAIL + 1
    ELSEIF S_K = 100 THEN
        PERFECT = PERFECT + 1
    ENDIF
ENDDO
write FAIL, PERFECT
END
```

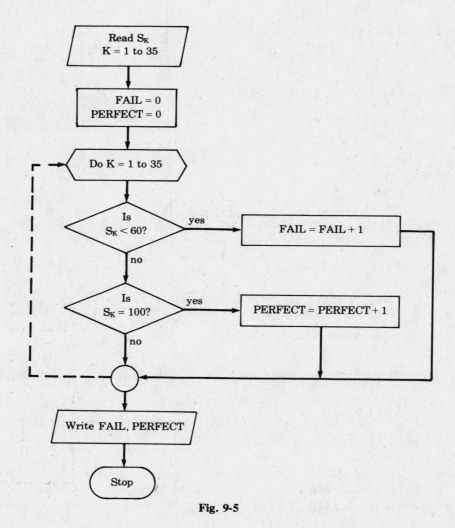

Fig. 9-5

9.19   Consider the list of 10 distinct numbers:

$$8, \ 6, \ 11, \ 14, \ 5, \ 10, \ 3, \ 2, \ 12, \ 7$$

The smallest number, 2, appears in the position $L = 8$.   Interchanging this number with the first number gives the new list

$$2, \ 6, \ 11, \ 14, \ 5, \ 10, \ 3, \ 8, \ 12, \ 7$$

(a) Draw a flowchart of an algorithm which, for an arbitrary linear array $A_1, A_2, \ldots, A_N$, finds the position L of the smallest number in the array, and then exchanges this number, A_L, with the first number, A_1.   (b) How would this algorithm work if applied to the above list of 10 numbers?

(a)   Figure 9-6 gives the flowchart.   Observe that we first read the value of N and then read the N numbers into the array A.   We let SMALL denote the current smallest number and L its position.   Hence we first assign $SMALL = A_1$ and $L = 1$.   Using a loop for $K = 2$ to N, we examine each A_K and compare it to SMALL.   If any A_K is less than SMALL, we reassign $SMALL = A_K$ and $L = K$.   After the last step of the loop, SMALL will contain the smallest element and L will be its position.   We then interchange A_1 and A_L.

(b)   The algorithm first assigns $SMALL = 8$ and $L = 1$.   Figure 9-7 shows the values of SMALL and L after executing the loop for the given value of K and A_K.   Observe that the value of SMALL decreases for

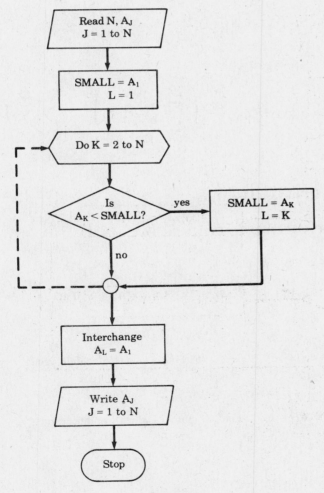

Fig. 9-6

K	A_K	SMALL	L
Initial		8	1
2	6	6	2
3	11	6	2
4	14	6	2
5	5	5	5
6	10	5	5
7	3	3	7
8	2	2	8
9	12	2	8
10	7	2	8

Fig. 9-7

K = 2, 5, 7, 8, i.e. when A_K is less than the preceding value of SMALL.   The final value, SMALL = 2, is the smallest number in the list, and it is in the 8th position.   The algorithm then interchanges the first and eighth elements to obtain the desired list.

Supplementary Problems

VECTORS

9.20 Let $u = (2, -1, 0, -3)$, $v = (1, -1, -1, 3)$ and $w = (1, 3, -2, 2)$. Find (a) $3u$, (b) $u + v$, (c) $2u - 3v$, (d) $5u - 3v - 4w$, (e) $-u + 2v - 2w$.

9.21 Compute the following, where u, v, and w are the vectors in Problem 9.20.   (See Problem 9.3.)

   (a) $v \cdot v$ (b) $u \cdot w$ (c) $w \cdot v$ (d) $\|u\|$ (e) $\|v\|$ (f) $\|w\|$

9.22 Find x and y if (a) $(x, x + y) = (y - 2, 6)$, (b) $x(3, 2) = 2(y, -1)$.

MATRIX OPERATIONS

In Problems 9.23, 9.24, and 9.25,

$$A = \begin{pmatrix} 1 & -1 & 2 \\ 0 & 3 & 4 \end{pmatrix} \qquad B = \begin{pmatrix} 4 & 0 & -3 \\ -1 & -2 & 3 \end{pmatrix} \qquad C = \begin{pmatrix} 2 & -3 & 0 & 1 \\ 5 & -1 & -4 & 2 \\ -1 & 0 & 0 & 3 \end{pmatrix} \qquad D = \begin{pmatrix} 2 \\ -1 \\ 3 \end{pmatrix}$$

9.23 Find (a) $A + B$, (b) $A + C$, (c) $3A - 4B$.

9.24 Find (a) AB, (b) AC, (c) AD, (d) BC, (e) BD, (f) CD.

9.25 Find (a) A^T, (b) $A^T C$, (c) $D^T A^T$, (d) $B^T A$, (e) $D^T D$, (f) DD^T.   (See Problem 9.11.)

SQUARE MATRICES

9.26 Let $A = \begin{pmatrix} 2 & 2 \\ 3 & -1 \end{pmatrix}$.   (a) Find A^2 and A^3.   (b) If $f(x) = x^3 - 3x^2 - 2x + 4$, find $f(A)$.   (c) If $g(x) = x^2 - x - 8$, find $g(A)$.

9.27 Let $B = \begin{pmatrix} 1 & 3 \\ 5 & 3 \end{pmatrix}$.   (a) Find $f(B)$ where $f(x) = 2x^2 - 4x + 3$.   (b) Find a nonzero column vector u such that $Bu = 6u$.

9.28 Matrices A and B are said *to commute* if $AB = BA$. Find all matrices which commute with

$$\begin{pmatrix} 1 & 1 \\ 0 & 1 \end{pmatrix}$$

9.29 Let $A = \begin{pmatrix} 1 & 2 \\ 0 & 1 \end{pmatrix}$. Find A^n.

9.30   Find the determinant of each matrix:

(a) $\begin{pmatrix} 2 & 5 \\ 4 & 1 \end{pmatrix}$, (b) $\begin{pmatrix} 6 & 1 \\ 3 & -2 \end{pmatrix}$, (c) $\begin{pmatrix} 4 & -5 \\ 0 & 2 \end{pmatrix}$,

(d) $\begin{pmatrix} 1 & 0 \\ 0 & 1 \end{pmatrix}$, (e) $\begin{pmatrix} -2 & 8 \\ -5 & -2 \end{pmatrix}$, (f) $\begin{pmatrix} 4 & 9 \\ 5 & -3 \end{pmatrix}$

9.31   Find the determinant of each matrix:

(a) $\begin{pmatrix} 2 & 1 & 1 \\ 0 & 5 & -2 \\ 1 & -3 & 4 \end{pmatrix}$, (b) $\begin{pmatrix} 3 & -2 & -4 \\ 2 & 5 & -1 \\ 0 & 6 & 1 \end{pmatrix}$, (c) $\begin{pmatrix} -2 & -1 & 4 \\ 6 & -3 & -2 \\ 4 & 1 & 2 \end{pmatrix}$, (d) $\begin{pmatrix} 7 & 6 & 5 \\ 1 & 2 & 1 \\ 3 & -2 & 1 \end{pmatrix}$

9.32   Find the inverse of each matrix:

(a) $\begin{pmatrix} 3 & 2 \\ 7 & 5 \end{pmatrix}$, (b) $\begin{pmatrix} 2 & -3 \\ 1 & 3 \end{pmatrix}$

SUBSCRIPTED VARIABLES

9.33   Determine the dimensions of and the numbers of elements in whatever arrays are defined by the following input boxes:

(a) Read HOURS(J,K,L) / J = 1 to 24 / K = 1 to 5 / L = 1 to 52

(c) Read N, NAME(K) / K = 1 to N

(b) Read CITY, DOCTORS, LAWYERS

(d) Read STUDENT(K), TEST(K, L) / K = 1 to 36, L = 1 to 4

9.34   Determine the output of each flowchart in Fig. 9-8.

9.35   Draw a flowchart or write a pseudocode program which reads 25 positive integers and determines if any two of them add up to 25.

9.36   Draw a flowchart or write a pseudocode program which reads 25 distinct positive integers and finds the second-largest of them and its position.

9.37   Write a pseudocode program which continues the algorithm of Problem 9.19 until all the numbers A_1, $A_2, \ldots, A_N$ are arranged in increasing order   (This is called *selection sort.*)

9.38   A department store chain has 12 stores (numbered from 1 to 12), and each store has the same 14 departments (numbered from 1 to 14). The daily sales of each department in each store are sent each week to the main office of the chain, and the data are stored in a $12 \times 14 \times 7$ array, SALES. Draw a flowchart or write a pseudocode program which finds (a) the daily sales of the chain, (b) the weekly sales of each store, (c) the weekly sales of each department, (d) the total weekly sales of the chain.

9.39   A student writes 6 papers (graded from 0 to 100). His final grade is the average of the 5 best papers. Draw a flowchart or write a pseudocode program which reads the grades on the 6 papers and determines his final grade.

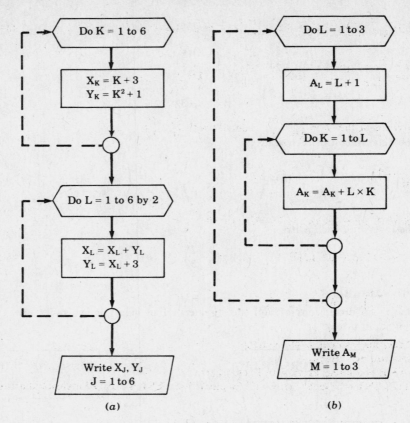

Fig. 9-8

Answers to Supplementary Problems

9.20   (a) $3u = (6, -3, 0, -9)$ (c) $2u - 3v = (1, 1, 3, -15)$ (e) $-u + 2v - 2w = (-2, -7, 2, 5)$
       (b) $u + v = (3, -2, -1, 0)$ (d) $5u - 3v - 4w = (3, -14, 11, -32)$

9.21   (a) -6, (b) -7, (c) 6, (d) $\sqrt{14}$, (e) $\sqrt{12} = 2\sqrt{3}$, (f) $\sqrt{18} = 3\sqrt{2}$

9.22   (a) $x = 2$, $y = 4$; (b) $x = -1$, $y = -3/2$

9.23   (a) $\begin{pmatrix} 5 & -1 & -1 \\ -1 & 1 & 7 \end{pmatrix}$, (b) Not defined, (c) $\begin{pmatrix} -13 & -3 & 18 \\ 4 & 17 & 0 \end{pmatrix}$

9.24   (a) Not defined   (c) $\begin{pmatrix} 9 \\ 9 \end{pmatrix}$ (e) $\begin{pmatrix} -1 \\ 9 \end{pmatrix}$

       (b) $\begin{pmatrix} -5 & -2 & 4 & 5 \\ 11 & -3 & -12 & 18 \end{pmatrix}$ (d) $\begin{pmatrix} 11 & -12 & 0 & -5 \\ -15 & 5 & 8 & 4 \end{pmatrix}$ (f) Not defined

9.25   (a) $\begin{pmatrix} 1 & 0 \\ -1 & 3 \\ 2 & 4 \end{pmatrix}$ (c) $(9, 9)$ (e) 14

       (b) Not defined   (d) $\begin{pmatrix} 4 & -7 & 4 \\ 0 & -6 & -8 \\ -3 & 12 & 6 \end{pmatrix}$ (f) $\begin{pmatrix} 4 & -2 & 6 \\ -2 & 1 & -3 \\ 6 & -3 & 9 \end{pmatrix}$

9.26 (a) $A^2 = \begin{pmatrix} 10 & 2 \\ 3 & 7 \end{pmatrix}$, $A^3 = \begin{pmatrix} 26 & 18 \\ 27 & -1 \end{pmatrix}$;   (b) $f(A) = \begin{pmatrix} -4 & 8 \\ 12 & -16 \end{pmatrix}$,   (c) $g(A) = \begin{pmatrix} 0 & 0 \\ 0 & 0 \end{pmatrix}$

9.27 (a) $f(B) = \begin{pmatrix} 31 & 12 \\ 20 & 39 \end{pmatrix}$,   (b) $u = \begin{pmatrix} 3k \\ 5k \end{pmatrix}$, $k \neq 0$

9.28 Only matrices of the form $\begin{pmatrix} a & b \\ 0 & a \end{pmatrix}$ commute with $\begin{pmatrix} 1 & 1 \\ 0 & 1 \end{pmatrix}$.

9.29 $A^n = \begin{pmatrix} 1 & 2n \\ 0 & 1 \end{pmatrix}$

9.30 (a) -18, (b) -15, (c) 8, (d) 1, (e) 44, (f) -57

9.31 (a) 21, (b) -11, (c) 102, (d) 0

9.32 (a) $\begin{pmatrix} 5 & -2 \\ -7 & 3 \end{pmatrix}$,   (b) $\begin{pmatrix} \frac{1}{3} & \frac{1}{3} \\ -\frac{1}{9} & \frac{2}{9} \end{pmatrix}$

9.33 (a) Three-dimensional, 6240 elements.
   (b) They are not arrays.
   (c) NAME is a linear array with N elements.
   (d) STUDENT is a linear array with 36 elements; TEST is a two-dimensional array with 144 elements.

9.34 (a) 6, 9, 5, 5, 16, 19, 7, 17, 34, 37, 9, 36; (b) 8, 13, 13

9.35
```
read N_K, K = 1 to 25
DO J = 1 to 24
     DO L = J + 1 to 25
          IF N_J + N_L = 25 THEN
               write N_J, N_L
          ENDIF
     ENDDO
ENDDO
END
```

9.36
```
read N_K, K = 1 to 25
IF N_2 < N_1 THEN
     FIRST = N_1
     L1 = 1
     SECOND = N_2
     L2 = 2
ELSE
     FIRST = N_2
     L1 = 2
     SECOND = N_1
     L2 = 1
ENDIF
DO J = 3 to 25
     IF FIRST < N_J THEN
          SECOND = FIRST
          L2 = L1
          FIRST = N_J
          L1 = J
     ELSEIF SECOND < N_J THEN
          SECOND = N_J
          L2 = J
     ENDIF
ENDDO
write SECOND, L2
END
```

9.37   read N, A_J, J = 1 to N
DO M = 1 to N − 1
    SMALL = A_M
    L = M
    DO K = M + 1 to N
        IF A_K < SMALL THEN
            SMALL = A_K
            L = K
        ENDIF
    ENDDO
    Interchange A_L and A_M
ENDDO
write A_J, J = 1 to N
END

(Note that line 11 is a macroinstruction.)

9.38   read SALES(I,J,K), I = 1 to 12, J = 1 to 14, K = 1 to 7
(*a*) DO M = 1 to 7
    DAILY(M) = 0
    DO I = 1 to 12
        DO J = 1 to 14
            DAILY(M) = DAILY(M) + SALES(I,J,M)
        ENDDO
    ENDDO
ENDDO
write DAILY(M), M = 1 to 7
(*b*) DO M = 1 to 12
    STORE(M) = 0
    DO J = 1 to 14
      DO K = 1 to 7
          STORE(M) = STORE(M) + SALES(M,J,K)
      ENDDO
    ENDDO
ENDDO
write STORE(M), M = 1 to 14
(*c*) DO M = 1 to 14
    DEPT(M) = 0
    DO I = 1 to 12
      DO K = 1 to 14
          DEPT(M) = DEPT(M) + SALES(I,M,K)
      ENDDO
    ENDDO
ENDDO
write DEPT(M), M = 1 to 14
(*d*) CHAIN = 0
DO M = 1 to 7
    CHAIN = CHAIN + DAILY(M)
ENDDO
write CHAIN
END

9.39   read $GRADE_J$, J = 1 to 6
Find SUM of GRADES
Find SMALL, the lowest grade
FINAL = (SUM − SMALL)/5
write FINAL
END   (Note the two macroinstructions.)

Chapter 10

Linear Equations

10.1 LINEAR EQUATIONS IN ONE UNKNOWN

A linear equation in one unknown, x, can always be simplified into the standard form $ax = b$, where a and b are constants. If $a \neq 0$, this equation has the unique solution

$$x = \frac{b}{a}$$

EXAMPLE 10.1

(a) Consider the linear equations

        (i) $5x = 10$ (ii) $3x = 27$ (iii) $2x = 9$ (iv) $4x = 22$

The solutions are as follows:

      (i) $x = \dfrac{10}{5}$, or $x = 2$ (iii) $x = \dfrac{9}{2}$

      (ii) $x = \dfrac{27}{3}$, or $x = 9$ (iv) $x = \dfrac{22}{4}$, or $x = \dfrac{11}{2}$

(b) Consider the linear equation

$$9x - 4 - 2x = 5x + 15 - 3$$

Simplifying both sides gives

$$7x - 4 = 5x + 12$$

Adding 4 to both sides gives

$$7x = 5x + 16$$

Subtracting $5x$ from both sides gives

$$2x = 16$$

which now is in standard form. The solution is

$$x = \frac{16}{2} = 8$$

10.2 LINEAR EQUATIONS IN TWO UNKNOWNS

A linear equation in two unknowns, x and y, can be put into the form

$$ax + by = c$$

where a, b, c are real numbers. We also assume that a and b are not both zero. A solution of the equation consists of a pair of numbers, $u = (k_1, k_2)$, which satisfies the equation; i.e. is such that

$$ak_1 + bk_2 = c$$

Solutions of the equation can be found by assigning arbitrary values to x and solving for y (or vice versa).

EXAMPLE 10.2   Consider the equation

$$2x + y = 4$$

If we substitute $x = -2$ in the equation, we obtain

$$2 \cdot (-2) + y = 4 \quad \text{or} \quad -4 + y = 4 \quad \text{or} \quad y = 8$$

Hence $(-2, 8)$ is a solution.   If we substitute $x = 3$ in the equation, we obtain

$$2 \cdot 3 + y = 4 \quad \text{or} \quad 6 + y = 4 \quad \text{or} \quad y = -2$$

Hence $(3, -2)$ is a solution.   Figure 10-1(a) lists six possible values for x and the corresponding values for y; they give six solutions of the equation.

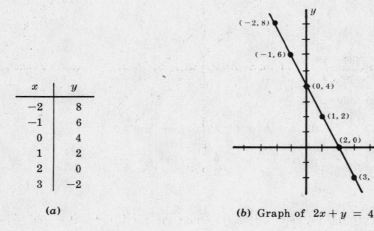

x	y
-2	8
-1	6
0	4
1	2
2	0
3	-2

(a)

(b) Graph of $2x + y = 4$

Fig. 10-1

Recall from Example 6.9(a) that any solution $u = (k_1, k_2)$ of the linear equation

$$ax + by = c$$

determines a point in the cartesian plane $\mathbf{R}^2$.   Since a and b are not both zero, the solutions u correspond precisely to the points on a straight line (whence the name "linear equation").   For example, the six solutions of the equation $2x + y = 4$, appearing in Fig. 10-1(a), have been plotted in Fig. 10-1(b).   Observe that they all lie on the same straight line, the *graph* of the equation.

10.3   SYSTEM OF TWO LINEAR EQUATIONS IN TWO UNKNOWNS

Consider now a system of two linear equations in the two unknowns x and y:

$$a_1 x + b_1 y = c_1$$
$$a_2 x + b_2 y = c_2 \tag{10.1}$$

Again we assume that a_1 and b_1 are not both zero, and that a_2 and b_2 are not both zero.   A pair of numbers which satisfies both equations is called a *simultaneous solution* of the given equations or a *solution of the system* of equations.   There are three cases, which can be described geometrically.

(1)   *The system has exactly one solution.*   Here the graphs of the linear equations intersect in one point, as in Fig. 10-2(a).

(2) *The system has no solutions.*   Here the graphs of the linear equations are parallel, as in Fig. 10-2(*b*).

(3) *The system has an infinite number of solutions.*   Here the graphs of the linear equations coincide, as in Fig. 10-2(*c*).

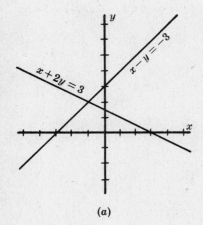
(*a*)

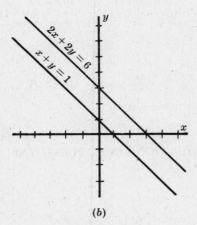
(*b*)

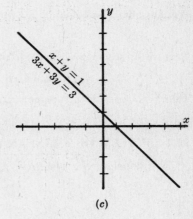
(*c*)

Fig. 10-2

The special cases (2) and (3) can only occur when the coefficients of x and y in the two linear equations are proportional:

$$\frac{a_1}{a_2} = \frac{b_1}{b_2} \qquad \text{i.e.} \qquad \begin{vmatrix} a_1 & b_1 \\ a_2 & b_2 \end{vmatrix} = 0$$

Specifically, the system has an infinite number of solutions when the coefficients and constant terms are proportional,

$$\frac{a_1}{a_2} = \frac{b_1}{b_2} = \frac{c_1}{c_2}$$

but has no solution when

$$\frac{a_1}{a_2} = \frac{b_1}{b_2} \neq \frac{c_1}{c_2}$$

Unless it is otherwise stated or implied, we shall assume that the determinant of the coefficients does not vanish, so that we are dealing with case (1).

The solution to system (*10.1*) can be obtained by the process known as *elimination*, whereby we reduce the system to a single equation in only one unknown.   This is accomplished by the following two steps:

STEP 1:   Multiply the two equations by two numbers which are such that the resulting coefficients of one of the unknowns (usually x) are negatives of each other.

STEP 2:   Add the equations obtained in Step 1.

The result of Step 2 is a linear equation in one unknown (usually y).   This equation may be solved for that unknown, and the solution may be substituted in one of the original equations to yield the value of the other unknown.

EXAMPLE 10.3   Consider the system

$$(1) \quad 3x + 2y = 8$$
$$(2) \quad 2x - 5y = -1$$

We multiply (1) by 2 and (2) by −3 and then add:

$$
\begin{array}{ll}
2 \times (1): & 6x + 4y = 16 \\
-3 \times (2): & -6x + 15y = 3 \\
\hline
\text{Addition:} & 19y = 19
\end{array}
$$

We thus obtain an equation involving only y.   We solve for y to obtain

$$y = 1$$

Next we substitute $y = 1$ in (1) to obtain

$$3x + 2 \cdot 1 = 8 \quad \text{or} \quad 3x + 2 = 8 \quad \text{or} \quad 3x = 6 \quad \text{or} \quad x = 2$$

Thus $x = 2$, $y = 1$ is the unique solution to the system.

10.4   SYSTEM OF n LINEAR EQUATIONS IN n UNKNOWNS

We consider the system

$$
\begin{aligned}
a_{11}x_1 + a_{12}x_2 + \cdots + a_{1n}x_n &= b_1 \\
a_{21}x_1 + a_{22}x_2 + \cdots + a_{2n}x_n &= b_2 \\
&\cdots\cdots\cdots\cdots\cdots\cdots \\
a_{n1}x_1 + a_{n2}x_2 + \cdots + a_{nn}x_n &= b_n
\end{aligned}
\qquad (10.2)
$$

where the a_{ij}, b_i are real numbers.   The number a_{ij} is called the *coefficient* of x_j in the ith equation, and the number b_i is called the *constant* of the ith equation.   A list of values for the unknowns,

$$x_1 = k_1, \; x_2 = k_2, \ldots, \; x_n = k_n$$

or, equivalently, a list of n numbers, $u = (k_1, k_2, \ldots, k_n)$, is called a *solution* of the system if, with k_j substituted for x_j, the left-hand side of each equation in fact equals the right-hand side.

System (10.2) is equivalent to the matrix equation

$$
\begin{pmatrix}
a_{11} & a_{12} & \cdots & a_{1n} \\
a_{21} & a_{22} & \cdots & a_{2n} \\
\cdots & \cdots & \cdots & \cdots \\
a_{n1} & a_{n2} & \cdots & a_{nn}
\end{pmatrix}
\begin{pmatrix}
x_1 \\ x_2 \\ \cdots \\ x_n
\end{pmatrix}
=
\begin{pmatrix}
b_1 \\ b_2 \\ \cdots \\ b_n
\end{pmatrix}
\qquad (10.3)
$$

or, simply, $AX = B$, where $A = (a_{ij})$, $X = (x_i)$, and $B = (b_i)$.   That is, any solution of (10.2) is a solution of (10.3), and vice versa.   The matrix A is called the *coefficient matrix* of the system (10.2), and the matrix

$$
\begin{pmatrix}
a_{11} & a_{12} & \cdots & a_{1n} & b_1 \\
a_{21} & a_{22} & \cdots & a_{2n} & b_2 \\
\cdots & \cdots & \cdots & \cdots & \cdots \\
a_{n1} & a_{n2} & \cdots & a_{nn} & b_n
\end{pmatrix}
$$

is called the *augmented matrix* of the system (10.2).   Observe that the system (10.2) is completely determined by its augmented matrix.

EXAMPLE 10.4   Consider the system

$$
\begin{aligned}
x + 2y - z &= 8 \\
2x + 3y + z &= 5 \\
x - 2y - 3z &= 6
\end{aligned}
\quad \text{or} \quad
\begin{pmatrix}
1 & 2 & -1 \\
2 & 3 & 1 \\
1 & -2 & -3
\end{pmatrix}
\begin{pmatrix}
x \\ y \\ z
\end{pmatrix}
=
\begin{pmatrix}
8 \\ 5 \\ 6
\end{pmatrix}
$$

The augmented matrix of the system is

$$\begin{pmatrix} 1 & 2 & -1 & 8 \\ 2 & 3 & 1 & 5 \\ 1 & -2 & -3 & 6 \end{pmatrix}$$

We note that one stores a system in the computer as its augmented matrix.

We claim that $x = 1$, $y = 2$, $z = -3$, or $u = (1, 2, -3)$, is a solution of this system. This is verified by substituting the values of the unknowns in each of the equations,

$$1 + 2\cdot 2 - (-3) = 8 \qquad 2\cdot 1 + 3\cdot 2 + (-3) = 5 \qquad 1 - 2\cdot 2 - 3(-3) = 6$$

and obtaining true statements. On the other hand, $x = 6$, $y = -1$, $z = -4$, or $v = (6, -1, -4)$, is not a solution of the system, because a test of the third equation gives

$$6 - 2(-2) - 3(-4) \ne 6$$

As with (*10.1*), (*10.2*) has a unique solution provided $|A| \ne 0$. In Section 10.5, we show how to find this solution for an important special case of (*10.2*). Then, in Section 10.6, we show how to reduce the general system (*10.2*) to the already-solved special case.

10.5 SOLUTION OF A TRIANGULAR SYSTEM

In the special case that $a_{ij} = 0$ for $i > j$, (*10.2*) assumes the *triangular form*

$$a_{11}x_1 + a_{12}x_2 + \cdots\cdots\cdots\cdots + a_{1,n-1}x_{n-1} + a_{1n}x_n = b_1$$
$$a_{22}x_2 + \cdots\cdots\cdots\cdots + a_{2,n-1}x_{n-1} + a_{2n}x_n = b_2$$
$$\cdots\cdots\cdots\cdots\cdots\cdots\cdots\cdots\cdots\cdots$$
$$a_{n-2,n-2}x_{n-2} + a_{n-2,n-1}x_{n-1} + a_{n-2,n}x_n = b_{n-2}$$
$$a_{n-1,n-1}x_{n-1} + a_{n-1,n}x_n = b_{n-1}$$
$$a_{nn}x_n = b_n$$

$$(10.4)$$

Here, $|A| = a_{11}a_{22}\cdots a_{nn}$; consequently, if none of the diagonal entries $a_{11}, a_{22}, \ldots, a_{nn}$ is zero, the system has a unique solution.

We obtain this solution by the technique of *back-substitution*. First, we solve the last equation for the last unknown, x_n:

$$x_n = \frac{b_n}{a_{nn}}$$

Second, we substitute this value for x_n in the next-to-last equation and solve it for the next-to-last unknown, x_{n-1}:

$$x_{n-1} = \frac{b_{n-1} - a_{n-1,n}(b_n/a_{nn})}{a_{n-1,n-1}}$$

Third, we substitute these values for x_n and x_{n-1} in the third-from-last equation and solve it for the third-from-last unknown, x_{n-2}:

$$x_{n-2} = \frac{b_{n-2} - (a_{n-2,n-1}/a_{n-1,n-1})[b_{n-1} - a_{n-1,n}(b_n/a_{nn})] - (a_{n-2,n}/a_{nn})b_n}{a_{n-2,n-2}}$$

In general, we determine x_k by substituting the previously obtained values of $x_n, x_{n-1}, \ldots, x_{k+1}$ in the kth equation:

$$x_k = \frac{b_k - \sum_{m=k+1}^{n} a_{km}x_m}{a_{kk}}$$

$$(10.5)$$

The process ceases when we have determined the first unknown, x_1.

A flowchart of the back-substitution algorithm is given in Fig. 10-3.   System (10.4) is stored in the computer as an $N \times (N + 1)$ matrix array A, the augmented matrix of the system.   Thus, the constants $b_1, b_2, \ldots, b_n$ are respectively stored as $A_{1,N+1}, A_{2,N+1}, \ldots, A_{N,N+1}$.   Observe that the flowchart has one loop inside another.   The inner loop, which we picture as a DO loop, is used to calculate the summation expression appearing in (10.5).   The outer loop then evaluates X_K in terms of this sum.

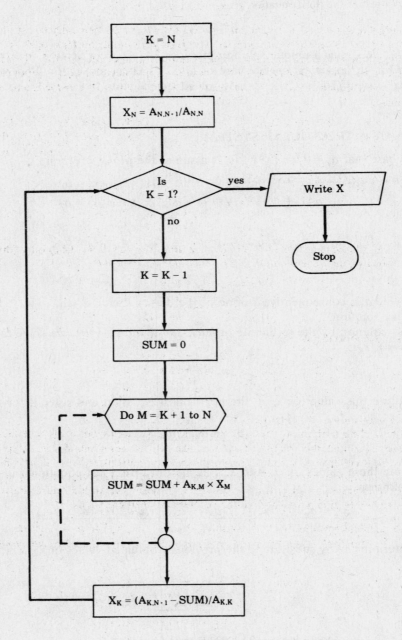

Fig. 10-3

10.6   GAUSSIAN ELIMINATION

Gaussian elimination, one of the oldest and most frequently used methods for finding the solution of a system of linear equations, is made up of two parts. The first part consists in a step-by-step triangularization of the system, so that it is finally put into triangular form; the second part consists in solving the triangular system by back-substitution, as discussed in Section 10.5. We illustrate this method with a specific example, and then we discuss the method in general.

EXAMPLE 10.5   Consider the following system of three equations in three unknowns:

$$\begin{aligned}
\mathscr{E}_1: & \quad x - 3y - 2z = 6 \\
\mathscr{E}_2: & \quad 2x - 4y + 2z = 18 \\
\mathscr{E}_3: & \quad -3x + 8y + 9z = -9
\end{aligned}$$

The first stage in reducing the system to triangular form is to eliminate the first unknown, x, from the second and third equations, $\mathscr{E}_2$ and $\mathscr{E}_3$. This can be accomplished by adding a suitable multiple of $\mathscr{E}_1$ to $\mathscr{E}_2$, thereby obtaining a new second equation without x, and then adding a suitable multiple of $\mathscr{E}_1$ to $\mathscr{E}_3$, thereby obtaining a new third equation without x. The multiplier in the first case is -2, the negative of the quotient of the coefficients of x in $\mathscr{E}_2$ and $\mathscr{E}_1$; in the second case it is 3, the negative of the quotient of the coefficients of x in $\mathscr{E}_3$ and $\mathscr{E}_1$. Accordingly, we have:

$$\begin{array}{lrcccr}
-2 \times \mathscr{E}_1: & -2x + 6y + 4z & = & -12 \\
\mathscr{E}_2: & \underline{2x - 4y + 2z} & = & \underline{18} \\
\text{new } \mathscr{E}_2: & 2y + 6z & = & 6
\end{array}
\qquad
\begin{array}{lrcccr}
3 \times \mathscr{E}_1: & 3x - 9y - 6z & = & 18 \\
\mathscr{E}_3: & \underline{-3x + 8y + 9z} & = & \underline{-9} \\
\text{new } \mathscr{E}_3: & -y + 3z & = & 9
\end{array}$$

and the original system is reduced to the system

$$\begin{aligned}
\mathscr{E}_1: & \quad x - 3y - 2z = 6 \\
\mathscr{E}_2: & \quad 2y + 6z = 6 \\
\mathscr{E}_3: & \quad -y + 3z = 9
\end{aligned}$$

The second stage is to eliminate the second unknown, y, from the (new) $\mathscr{E}_3$, using only (the new) $\mathscr{E}_2$ and $\mathscr{E}_3$. Specifically, we multiply $\mathscr{E}_2$ by 1/2 and add it to $\mathscr{E}_3$:

$$\begin{array}{lrcr}
\tfrac{1}{2} \times \mathscr{E}_2: & y + 3z & = & 3 \\
\mathscr{E}_3: & \underline{-y + 3z} & = & \underline{9} \\
\text{new } \mathscr{E}_3: & 6z & = & 12
\end{array}$$

Thus our system is reduced to the system

$$\begin{aligned}
\mathscr{E}_1: & \quad x - 3y - 2z = 6 \\
\mathscr{E}_2: & \quad 2y + 6z = 6 \\
\mathscr{E}_3: & \quad 6z = 12
\end{aligned}$$

which is in triangular form.

Solving by back-substitution, we obtain $x = 1$, $y = -3$, $z = 2$.

In the first stage of the algorithm, the coefficient of x in the first equation, $\mathscr{E}_1$, is called the *pivot*, and in the second stage of the algorithm, the coefficient of y in the second equation, $\mathscr{E}_2$, is the pivot. Clearly, the algorithm cannot work if either pivot is zero. In such a case one must interchange equations so that a pivot is not zero. In fact, if one uses a computer to solve the system, then the greatest accuracy is attained when the pivot is as large in absolute value as possible. Thus, for instance, we would interchange $\mathscr{E}_1$ and $\mathscr{E}_3$ in the original system above before eliminating x from the second and third equations.

Consider the general system of n equations, $\mathscr{E}_1, \mathscr{E}_2, \ldots, \mathscr{E}_n$, in n unknowns, $x_1, x_2, \ldots, x_n$, as given in (10.2). In the first stage of the Gaussian elimination algorithm we want to reduce the system to the form

$$\begin{aligned}
a_{11}x_1 + a_{12}x_2 + a_{13}x_3 + \cdots + a_{1n}x_n &= b_1 \\
a_{22}x_2 + a_{23}x_3 + \cdots + a_{2n}x_n &= b_2 \\
\cdots\cdots\cdots\cdots\cdots\cdots\cdots\cdots\cdots\cdots \\
a_{n2}x_2 + a_{n3}x_3 + \cdots + a_{nn}x_n &= b_n
\end{aligned}$$

$$(10.6)$$

i.e. we want to eliminate x_1 from the equations $\mathscr{E}_2, \mathscr{E}_3, \ldots, \mathscr{E}_n$. (Clearly, the a_{ij}, b_i in (10.6) need not be the same as those in (10.2).) This first stage can be accomplished by the following two steps:

STEP 1. Find the equation $\mathscr{E}_\ell$ of (10.2) such that the coefficient of x_1 has the largest absolute value among all the coefficients of x_1, and then interchange $\mathscr{E}_1$ and $\mathscr{E}_\ell$. (In particular, since the determinant of the coefficients is presumed nonzero, this guarantees that the new $a_{11} \neq 0$.)

STEP 2. For each $k > 1$, eliminate x_1 from $\mathscr{E}_k$ by multiplying $\mathscr{E}_1$ by

$$m_{k1} = -\frac{a_{k1}}{a_{11}}$$

and adding this multiple of $\mathscr{E}_1$ to $\mathscr{E}_k$. (Here a_{11} is the pivot.)

We repeat the above process with the subsystem $\mathscr{E}_2, \mathscr{E}_3, \ldots, \mathscr{E}_n$; that is, in the second stage of our algorithm we reduce the system (10.6) to the form

$$\begin{aligned}
a_{11}x_1 + a_{12}x_2 + a_{13}x_3 + a_{14}x_4 + \cdots + a_{1n}x_n &= b_1 \\
a_{22}x_2 + a_{23}x_3 + a_{24}x_4 + \cdots + a_{2n}x_n &= b_2 \\
a_{33}x_3 + a_{34}x_4 + \cdots + a_{3n}x_n &= b_3 \\
&\cdots \cdots \\
a_{n3}x_3 + a_{n4}x_4 + \cdots + a_{nn}x_n &= b_n
\end{aligned} \tag{10.7}$$

where x_2 is eliminated from $\mathscr{E}_3, \mathscr{E}_4, \ldots, \mathscr{E}_n$. This second stage is accomplished by an analogous two steps:

STEP 1′. Interchange $\mathscr{E}_2$ with an equation in the subsystem (i.e. excluding $\mathscr{E}_1$) so that the new pivot element, which becomes a_{22}, is as large in absolute value as possible.

STEP 2′. For each $k > 2$, multiply $\mathscr{E}_2$ by

$$m_{k2} = -\frac{a_{k2}}{a_{22}}$$

and add this multiple of $\mathscr{E}_2$ to $\mathscr{E}_k$.

After $n-1$ repetitions of the process, the original system (10.2) is reduced to the triangular system (10.4), which is then solved by back-substitution. Figure 10-4 is a flowchart for the triangularization part of the Gaussian method. Again we assume that system (10.2) is stored in the computer as its augmented matrix, A.

Remark 1: The pivot a_{jj} in the jth stage of the triangularization procedure appears in the denominator of each multiplier m_{kj} ($k = j+1, j+2, \ldots, n$). We maximize the pivot since division in a computer is most accurate when the divisor is as large in absolute value as possible.

Remark 2: Suppose the first two equations in a system are

$$\begin{aligned}
\mathscr{E}_1: \quad 5x - 7y + 3z &= 8 \\
\mathscr{E}_2: \quad 2x + y - 9z &= 7
\end{aligned}$$

The above algorithm eliminates x from $\mathscr{E}_2$ by multiplying $\mathscr{E}_1$ by $-2/5$ and adding this multiple of $\mathscr{E}_1$ to $\mathscr{E}_2$. Observe that this will result in fractional coefficients. One can avoid such fractional coefficients by multiplying $\mathscr{E}_1$ by -2 and $\mathscr{E}_2$ by 5, and then adding to obtain a new $\mathscr{E}_2$ without x:

$$\begin{array}{ll}
-2 \times \mathscr{E}_1: & -10x + 14y - 6z = -16 \\
5 \times \mathscr{E}_2: & \underline{10x + 5y - 45z = 35} \\
\text{new } \mathscr{E}_2: & 19y - 51z = 19
\end{array}$$

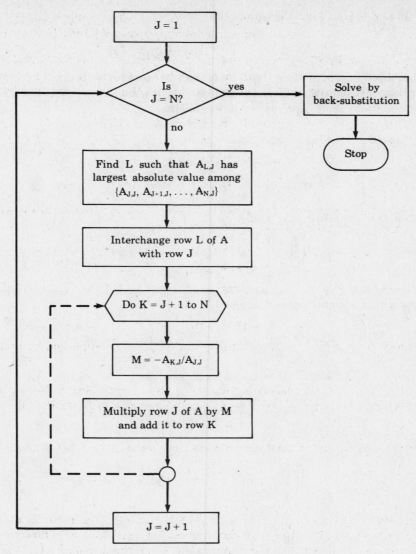

Fig. 10-4

Remark 3:   Generally, one applies the triangularization process to (*10.2*) without calculating in advance the determinant of coefficients, |A|.   It may happen, then, that at some stage of the process a *degenerate equation*, of the form

$$0x_1 + 0x_2 + \cdots + 0x_n = b$$

is obtained, signaling that |A| = 0.   In such case, the system either has no solution, and is said to be *inconsistent*, or has an infinite number of solutions.   (See Problem 10.11.)   We note that a system is inconsistent only if some degenerate equation has a nonzero constant.

10.7   DETERMINANTS AND SYSTEMS OF LINEAR EQUATIONS

For the system (*10.2*), let D denote the determinant of the matrix $A = (a_{ij})$ of coefficients; that is, let $D = |A|$.   Also, let N_i denote the determinant of the matrix obtained by replacing the *i*th column of A by the column of constant terms.

Theorem 10.1:   If $D \neq 0$, (10.2) has the unique solution

$$u = \left(\frac{N_1}{D}, \frac{N_2}{D}, \ldots, \frac{N_n}{D} \right)$$

We emphasize that this theorem (*Cramer's rule*) is of interest more for theoretical reasons and historical reasons than for practical reasons.   Gaussian elimination is usually much more efficient for solving systems of linear equations than is the use of determinants.

EXAMPLE 10.6   Solve the following system by determinants:

$$2x + y - z = 3$$
$$x + y + z = 1$$
$$x - 2y - 3z = 4$$

The determinant D of the matrix of coefficients is

$$D = \begin{vmatrix} 2 & 1 & -1 \\ 1 & 1 & 1 \\ 1 & -2 & -3 \end{vmatrix} = -6 + 1 + 2 + 1 + 4 + 3 = 5$$

Since $D \neq 0$, the system has a unique solution.   To obtain the numerator N_x replace, in the matrix of coefficients, the coefficients of x by the constant terms:

$$N_x = \begin{vmatrix} 3 & 1 & -1 \\ 1 & 1 & 1 \\ 4 & -2 & -3 \end{vmatrix} = -9 + 4 + 2 + 4 + 6 + 3 = 10$$

To obtain the numerator N_y replace, in the matrix of coefficients, the coefficients of y by the constant terms:

$$N_y = \begin{vmatrix} 2 & 3 & -1 \\ 1 & 1 & 1 \\ 1 & 4 & -3 \end{vmatrix} = -6 + 3 - 4 + 1 - 8 + 9 = -5$$

To obtain the numerator N_z replace, in the matrix of coefficients, the coefficients of z by the constant terms:

$$N_z = \begin{vmatrix} 2 & 1 & 3 \\ 1 & 1 & 1 \\ 1 & -2 & 4 \end{vmatrix} = 8 + 1 - 6 - 3 + 4 - 4 = 0$$

Thus the unique solution is

$$x = \frac{N_x}{D} = \frac{10}{5} = 2, \; y = \frac{N_y}{D} = \frac{-5}{5} = -1, \; z = \frac{N_z}{D} = \frac{0}{5} = 0$$

Solved Problems

LINEAR EQUATIONS IN ONE UNKNOWN

10.1 Solve:

$$(a) \quad 7x = 21 \qquad (c) \quad 5x = 17 \qquad (e) \quad 0x = 5$$
$$(b) \quad 2x = -12 \qquad (d) \quad -3x = 11 \qquad (f) \quad 0x = 0$$

If $a \neq 0$, then $ax = b$ has the unique solution $x = b/a$. Hence:

$$(a) \quad x = \frac{21}{7} = 3 \qquad (c) \quad x = \frac{17}{5}$$
$$(b) \quad x = \frac{-12}{2} = -6 \qquad (d) \quad x = -\frac{11}{3}$$

Note that the coefficient of x is zero in (e) and (f).

(e) The equation has no solution.

(f) Every number is a solution of this equation.

10.2 Solve:

$$(a) \quad x + 3 = 9 - 2x \qquad\qquad (c) \quad 2x - \frac{2}{3} = \frac{1}{2}x + 4$$
$$(b) \quad 6x - 8 + x + 4 = 2x + 11 - 5x \qquad (d) \quad 3x - 4 - x = 5x + 8 - 3x - 3$$

First put each equation in standard form, $ax = b$.

(a) Add $2x$ to both sides:

$$3x + 3 = 9$$

Add -3 to both sides:

$$3x = 6$$

Hence, $x = 2$ is the solution.

(b) Simplify each side:

$$7x - 4 = -3x + 11 \quad \text{or} \quad 10x = 15 \quad \text{or} \quad x = \frac{15}{10} \quad \text{or} \quad x = \frac{3}{2}$$

(c) Multiply both sides by 6 to remove fractions:

$$12x - 8 = 3x + 24 \quad \text{or} \quad 9x = 32 \quad \text{or} \quad x = \frac{32}{9}$$

(d) Simplify each side:

$$2x - 4 = 2x + 5 \quad \text{or} \quad 0x = 9$$

The equation has no solution.

LINEAR EQUATIONS IN TWO UNKNOWNS

10.3 Determine three distinct solutions of $2x - 3y = 14$ and plot its graph.

Choose any value for either unknown, say $x = -2$. Substitute $x = -2$ into the equation to obtain

$$2 \cdot (-2) - 3y = 14 \quad \text{or} \quad -4 - 3y = 14 \quad \text{or} \quad -3y = 18 \quad \text{or} \quad y = -6$$

Thus $x = -2$ and $y = -6$ or, in other words, the pair $(-2, -6)$ is a solution.
Now substitute, say, $x = 0$ into the equation to obtain

$$2 \cdot 0 - 3y = 14 \quad \text{or} \quad -3y = 14 \quad \text{or} \quad y = -14/3$$

Thus $(0, -14/3)$ is another solution.

     Lastly, substitute, say, $y = 0$ into the equation to obtain

$$2x - 3 \cdot 0 = 14 \quad \text{or} \quad 2x = 14 \quad \text{or} \quad x = 7$$

Hence $(7, 0)$ is still another solution.

     Now plot the three solutions on the cartesian plane $\mathbf{R}^2$, as shown in Fig. 10-5. The line passing through these three points is the graph of the equation.

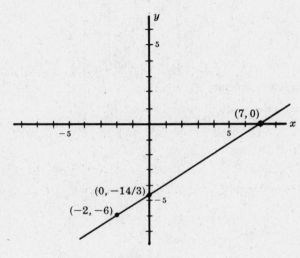

Fig. 10-5

10.4   Solve the system

$$3x - 2y = 7$$
$$x + 2y = 1$$

     Now the coefficients of y are already the negatives of each other; hence add the equations:

$$3x - 2y = 7$$
$$\underline{x + 2y = 1}$$

$$\text{Addition:} \quad 4x \quad = 8 \quad \text{or} \quad x = 2$$

Substitute $x = 2$ into the second equation to obtain

$$2 + 2y = 1 \quad \text{or} \quad 2y = -1 \quad \text{or} \quad y = -\tfrac{1}{2}$$

Thus $x = 2$, $y = -\tfrac{1}{2}$ or, in other words, the pair $(2, -\tfrac{1}{2})$, is the solution of the system.

     Check your answer by substituting the solution back into both original equations:

$$3 \cdot 2 - 2 \cdot (-\tfrac{1}{2}) = 7 \quad \text{or} \quad 6 + 1 = 7 \quad \text{or} \quad 7 = 7$$
$$2 + 2 \cdot (-\tfrac{1}{2}) = 1 \quad \text{or} \quad 2 - 1 = 1 \quad \text{or} \quad 1 = 1$$

10.5   Solve the system

$$(1) \quad 2x + 5y = 8$$
$$(2) \quad 3x - 2y = -7$$

To eliminate x, multiply (1) by 3 and (2) by -2 and then add:

$$3 \times (1): \quad 6x + 15y = 24$$
$$-2 \times (2): \quad -6x + 4y = 14$$
$$\text{Addition:} \qquad 19y = 38 \quad \text{or} \quad y = 2$$

Substitute $y = 2$ into one of the original equations, say (1), to obtain

$$2x + 5 \cdot 2 = 8 \quad \text{or} \quad 2x + 10 = 8 \quad \text{or} \quad 2x = -2 \quad \text{or} \quad x = -1$$

Hence $x = -1$, $y = 2$ or, in other words, the pair $(-1, 2)$, is the unique solution to the system.
Check your answer by substituting the solution back into both original equations:

$$(1): \quad 2 \cdot (-1) + 5 \cdot 2 = 8 \quad \text{or} \quad -2 + 10 = 8 \quad \text{or} \quad 8 = 8$$
$$(2): \quad 3 \cdot (-1) - 2 \cdot 2 = -7 \quad \text{or} \quad -3 - 4 = -7 \quad \text{or} \quad -7 = -7$$

We could also solve the system by first eliminating y as follows. Multiply (1) by 2 and (2) by 5 and then add:

$$2 \times (1): \quad 4x + 10y = 16$$
$$5 \times (2): \quad 15x - 10y = -35$$
$$\text{Addition:} \quad 19x = -19 \quad \text{or} \quad x = -1$$

Substitute $x = -1$ in (1) to obtain

$$2 \cdot (-1) + 5y = 8 \quad \text{or} \quad -2 + 5y = 8 \quad \text{or} \quad 5y = 10 \quad \text{or} \quad y = 2$$

Again we get $(-1, 2)$ as the solution.

10.6 Solve the system

$$(1) \quad 5x - 2y = 8$$
$$(2) \quad 3x + 4y = 10$$

To eliminate y, multiply (1) by 2 and add it to (2):

$$2 \times (1): \quad 10x - 4y = 16$$
$$(2): \quad 3x + 4y = 10$$
$$\text{Addition:} \quad 13x = 26 \quad \text{or} \quad x = 2$$

Substitute $x = 2$ in either original equation, say (2), to obtain

$$3 \cdot 2 + 4y = 10 \quad \text{or} \quad 6 + 4y = 10 \quad \text{or} \quad 4y = 4 \quad \text{or} \quad y = 1$$

Thus the pair $(2, 1)$ is the unique solution to the system.
Check the answer by substituting the solution back into both original equations:

$$(1): \quad 5 \cdot 2 - 2 \cdot 1 = 8 \quad \text{or} \quad 10 - 2 = 8 \quad \text{or} \quad 8 = 8$$
$$(2): \quad 3 \cdot 2 + 4 \cdot 1 = 10 \quad \text{or} \quad 6 + 4 = 10 \quad \text{or} \quad 10 = 10$$

10.7 Solve the system

$$(1) \quad x - 2y = 5$$
$$(2) \quad -3x + 6y = -10$$

Observe that the coefficients of x and y are proportional, i.e. $1/-3 = -2/6$. Hence the system does not have a unique solution. Observe also that the equations are not multiples of each other:

$$\frac{1}{-3} = \frac{-2}{6} \neq \frac{5}{-10}$$

Hence the system has no solution.

10.8   Solve the system

$$(1) \quad x - 2y = 5$$
$$(2) \quad -3x + 6y = -15$$

Observe that the coefficients and the constants of the two equations are proportional to each other:

$$\frac{1}{-3} = \frac{-2}{6} = \frac{5}{-15}$$

Accordingly, the equations are multiples of each other, and the system has an infinite number of solutions, which are the solutions of either equation.

Particular solutions can be found as follows: Let $y = 1$ and substitute in (1) to obtain

$$x - 2 \cdot 1 = 5 \quad \text{or} \quad x - 2 = 5 \quad \text{or} \quad x = 7$$

Hence $(7, 1)$ is a particular solution.   Let $y = 2$ and substitute in (1) to obtain

$$x - 2 \cdot 2 = 5 \quad \text{or} \quad x - 4 = 5 \quad \text{or} \quad x = 9$$

Then $(9, 2)$ is another specific solution of the system.

Let $y = 0$ and substitute in (1) to obtain $x = 5$.   Thus $(5, 0)$ is a third solution of the system.   And so forth.

One can graph the solution of the system as one graphs the solution of a single equation in two unknowns.   That is, we plot the three solutions in the plane $\mathbf{R}^2$, as shown in Fig. 10-6.   The line passing through these three points is the solution of the system.

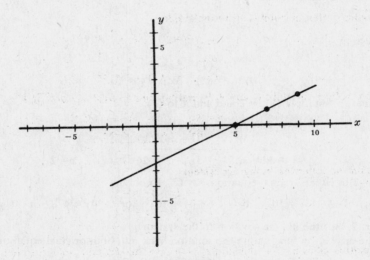

Fig. 10-6

n LINEAR EQUATIONS IN n UNKNOWNS

10.9   Find the solution of the following system:

$$2x - 3y + 5z - 2t = 9$$
$$5y - z + 3t = 1$$
$$7z - t = 3$$
$$2t = 8$$

The system is in triangular form; hence we solve by back-substitution.   The last equation gives $t = 4$.   Substituting in the third equation gives

$$7z - 4 = 3 \qquad \text{or} \qquad 7z = 7 \qquad \text{or} \qquad z = 1$$

Substituting $z = 1$ and $t = 4$ in the second equation gives

$$5y - 1 + 3 \cdot 4 = 1 \quad \text{or} \quad 5y - 1 + 12 = 1 \quad \text{or} \quad 5y = -10 \quad \text{or} \quad y = -2$$

Finally, substituting $y = -2$, $z = 1$, $t = 4$ in the first equation gives

$$2x - 3(-2) + 5(1) - 2(4) = 9 \quad \text{or} \quad 2x + 6 + 5 - 8 = 9 \quad \text{or} \quad 2x = 6 \quad \text{or} \quad x = 3$$

Thus, $x = 3$, $y = -2$, $z = 1$, $t = 4$, or, equivalently, the list $(3, -2, 1, 4)$, is the unique solution of the system.

10.10 Solve the system

$$\begin{aligned} \mathscr{E}_1: & \quad x + 2y - 4z = -4 \\ \mathscr{E}_2: & \quad 5x - 3y - 7z = 6 \\ \mathscr{E}_3: & \quad 3x - 2y + 3z = 11 \end{aligned}$$

First multiply $\mathscr{E}_1$ by -5 and add to $\mathscr{E}_2$ to obtain a new second equation without x; then multiply $\mathscr{E}_1$ by -3 and add to $\mathscr{E}_3$ to obtain a new third equation without x:

$$\begin{array}{rl} -5 \times \mathscr{E}_1: & -5x - 10y + 20z = 20 \\ \mathscr{E}_2: & \underline{5x - 3y - 7z = 6} \\ \text{new } \mathscr{E}_2: & -13y + 13z = 26 \\ \text{or:} & y - z = -2 \end{array} \qquad \begin{array}{rl} -3 \times \mathscr{E}_1: & -3x - 6y + 12z = 12 \\ \mathscr{E}_3: & \underline{3x - 2y + 3z = 11} \\ \text{new } \mathscr{E}_3: & -8y + 15z = 23 \end{array}$$

Thus the original system is equivalent to the system

$$\begin{aligned} \mathscr{E}_1: & \quad x + 2y - 4z = -4 \\ \mathscr{E}_2: & \quad y - z = -2 \\ \mathscr{E}_3: & \quad -8y + 15z = 23 \end{aligned}$$

Now eliminate y from the third equation by multiplying $\mathscr{E}_2$ by 8 and adding it to $\mathscr{E}_3$:

$$\begin{array}{rl} 8 \times \mathscr{E}_2: & 8y - 8z = -16 \\ \mathscr{E}_3: & \underline{-8y + 15z = 23} \\ \text{new } \mathscr{E}_3: & 7z = 7 \end{array}$$

Thus we obtain the following triangular system:

$$\begin{aligned} x + 2y - 4z &= -4 \\ y - z &= -2 \\ 7z &= 7 \end{aligned}$$

The third equation gives $z = 1$. Substitute $z = 1$ into the second equation to obtain

$$y - 1 = -2 \quad \text{or} \quad y = -1$$

Now substitute $z = 1$ and $y = -1$ into the first equation to obtain

$$x + 2 \cdot (-1) - 4 \cdot 1 = -4 \quad \text{or} \quad x - 2 - 4 = -4 \quad \text{or} \quad x - 6 = -4 \quad \text{or} \quad x = 2$$

Thus $x = 2$, $y = -1$, $z = 1$, or, in other words, the ordered triple $(2, -1, 1)$, is the unique solution to the system.

10.11 Solve each system:

$$(a) \quad \begin{aligned} x + 2y - 3z &= -1 \\ -3x + y - 2z &= -7 \\ 5x + 3y - 4z &= 2 \end{aligned} \qquad (b) \quad \begin{aligned} x + 2y - 3z &= 6 \\ 2x - y + 4z &= 2 \\ -4x - 5y + 6z &= -18 \end{aligned}$$

(*a*) Eliminate x from the first and second equations:

$$3 \times \mathcal{E}_1: \quad 3x + 6y - 9z = -3 \qquad\qquad -5 \times \mathcal{E}_1: \quad -5x - 10y + 15z = 6$$
$$\mathcal{E}_2: \quad -3x + y - 2z = -7 \qquad\qquad \mathcal{E}_3: \quad 5x + 3y - 4z = 2$$

new $\mathcal{E}_2$: $\qquad\qquad 7y - 11z = -10 \qquad$ new $\mathcal{E}_3$: $\qquad -7y + 11z = 7$

This gives the new system

$$x + 2y - 3z = -1$$
$$7y - 11z = -10$$
$$-7y + 11z = 7$$

Adding the second equation to the third equation to eliminate y from the third equation, we obtain

$$0x + 0y + 0z = -3$$

This equation, and hence the system, has no solution.

(*b*) Eliminate x from the second and third equations:

$$-2 \times \mathcal{E}_1: \quad -2x - 4y + 6z = -12 \qquad\qquad 4 \times \mathcal{E}_1: \quad 4x + 8y - 12z = 24$$
$$\mathcal{E}_2: \quad 2x - y + 4z = 2 \qquad\qquad \mathcal{E}_3: \quad -4x - 5y + 6z = -18$$

new $\mathcal{E}_2$: $\qquad -5y + 10z = -10 \qquad$ new $\mathcal{E}_3$: $\qquad 3y - 6z = 6$

This gives the new system

$$x + 2y - 3z = 6$$
$$-5y + 10z = -10$$
$$3y - 6z = 6$$

Multiply the second equation by 3/5 and add it to the third equation to eliminate y from the third equation. This gives

$$0x + 0y + 0z = 0$$

which shows that the second and third equations are proportional. Thus the third equation can be dropped, and the system has been reduced to

$$x + 2y - 3z = 6$$
$$-5y + 10z = -10$$

Observe that any value substituted for z gives a triangular system involving x and y. Hence this system has an infinite number of solutions, one for each value for z.

10.12 Solve the system

$$2x + y - 3z = 1$$
$$5x + 2y - 6z = 5$$
$$3x - y - 4z = 7$$

Eliminate x from the second and third equations. To avoid fractions, multiply both $\mathcal{E}_1$ and $\mathcal{E}_2$ to eliminate x from the second equation, and multiply both $\mathcal{E}_1$ and $\mathcal{E}_3$ to eliminate x from the third equation:

$$-5 \times \mathcal{E}_1: \quad -10x - 5y + 15z = -5 \qquad\qquad -3 \times \mathcal{E}_1: \quad -6x - 3y + 9z = -3$$
$$2 \times \mathcal{E}_2: \quad 10x + 4y - 12z = 10 \qquad\qquad 2 \times \mathcal{E}_3: \quad 6x - 2y - 8z = 14$$

new $\mathcal{E}_2$: $\qquad -y + 3z = 5 \qquad$ new $\mathcal{E}_3$: $\qquad -5y + z = 11$

This gives the system

$$2x + y - 3z = 1$$
$$-y + 3z = 5$$
$$-5y + z = 11$$

Eliminate y from the third equation:

$$
\begin{array}{rl}
-5 \times \mathscr{E}_2: & 5y - 15z = -25 \\
\mathscr{E}_3: & -5y + z = 11 \\
\hline
\text{new } \mathscr{E}_3: & -14z = -14
\end{array}
$$

Thus we obtain the triangular system

$$
\begin{array}{rl}
2x + y - 3z = & 1 \\
-y + 3z = & 5 \\
-14z = & -14
\end{array}
$$

Back-substitution gives the solution $x = 3$, $y = -2$, $z = 1$.

10.13 A computer, programmed according to the flowchart in Fig. 10-4, solves the system defined by the augmented matrix

$$
A = \begin{pmatrix}
1 & 7 & 1 & -10 \\
1 & 2 & -2 & 7 \\
2 & 8 & -1 & -2
\end{pmatrix}
$$

Describe the steps the computer performs.

First the computer finds the row of A such that its first-column entry has the largest absolute value among all the first-column entries. This is the third row; so the computer interchanges the first and third rows:

$$
A = \begin{pmatrix}
2 & 8 & -1 & -2 \\
1 & 2 & -2 & 7 \\
1 & 7 & 1 & -10
\end{pmatrix}
$$

Now the computer uses the multiplier $m_{12} = -1/2$ to produce a 0 in the (2, 1)-position, and the multiplier $m_{13} = -1/2$ to produce a 0 in the (3, 1)-position:

$$
A = \begin{pmatrix}
2 & 8 & -1 & -2 \\
1 - \frac{1}{2}(2) & 2 - \frac{1}{2}(8) & -2 - \frac{1}{2}(-1) & 7 - \frac{1}{2}(-2) \\
1 - \frac{1}{2}(2) & 7 - \frac{1}{2}(8) & 1 - \frac{1}{2}(-1) & -10 - \frac{1}{2}(-2)
\end{pmatrix} = \begin{pmatrix}
2 & 8 & -1 & -2 \\
0 & -2 & -1.5 & 8 \\
0 & 3 & 1.5 & -9
\end{pmatrix}
$$

Next the computer examines the second-column entries below the first row; it interchanges the second and third rows, since 3 in the third row has larger absolute value than -2 in the second row. Thus the array becomes

$$
A = \begin{pmatrix}
2 & 8 & -1 & -2 \\
0 & 3 & 1.5 & -9 \\
0 & -2 & -1.5 & 8
\end{pmatrix}
$$

The computer now uses the multiplier $m_{23} = -(-2)/3 = 2/3$ to produce a 0 in the (3, 2)-position, thus bringing the first three columns of the array into triangular form:

$$
A = \begin{pmatrix}
2 & 8 & -1 & -2 \\
0 & 3 & 1.5 & -9 \\
0 + \frac{2}{3}(0) & -2 + \frac{2}{3}(3) & -1.5 + \frac{2}{3}(1.5) & 8 + \frac{2}{3}(-9)
\end{pmatrix} = \begin{pmatrix}
2 & 8 & -1 & -2 \\
0 & 3 & 1.5 & -9 \\
0 & 0 & -0.5 & 2
\end{pmatrix}
$$

Control now passes to the back-substitution part of the program (see Fig. 10-3), which yields the solution $(1, -1, -4)$.

DETERMINANTS AND LINEAR EQUATIONS

10.14 Solve, using determinants:

$$2x + y = 7$$
$$3x - 5y = 4$$

First, compute the determinant D of the matrix of coefficients:

$$D = \begin{vmatrix} 2 & 1 \\ 3 & -5 \end{vmatrix} = -10 - 3 = -13$$

Since $D \neq 0$, the system has a unique solution.   Replace the first column in the matrix of coefficients by the column of constant terms and calculate its determinant, N_x:

$$N_x = \begin{vmatrix} 7 & 1 \\ 4 & -5 \end{vmatrix} = -35 - 4 = -39$$

Replace the second column in the matrix of coefficients by the column of constant terms and calculate its determinant, N_y:

$$N_y = \begin{vmatrix} 2 & 7 \\ 3 & 4 \end{vmatrix} = 8 - 21 = -13$$

The solution of the system is then

$$x = \frac{N_x}{D} = \frac{-39}{-13} = 3 \qquad y = \frac{N_y}{D} = \frac{-13}{-13} = 1$$

10.15 Solve, using determinants:

$$3y + 2x = z + 1$$
$$3x + 2z = 8 - 5y$$
$$3z - 1 = x - 2y$$

First arrange the equations in standard form with the unknowns appearing in columns:

$$2x + 3y - z = 1$$
$$3x + 5y + 2z = 8$$
$$x - 2y - 3z = -1$$

Compute the determinant D of the matrix of coefficients:

$$D = \begin{vmatrix} 2 & 3 & -1 \\ 3 & 5 & 2 \\ 1 & -2 & -3 \end{vmatrix} = -30 + 6 + 6 + 5 + 8 + 27 = 22$$

Since $D \neq 0$, the system has a unique solution.   To compute N_x, N_y and N_z, replace the coefficients of the unknown in the matrix of coefficients by the constant terms:

$$N_x = \begin{vmatrix} 1 & 3 & -1 \\ 8 & 5 & 2 \\ -1 & -2 & -3 \end{vmatrix} = -15 - 6 + 16 - 5 + 72 + 4 = 66$$

$$N_y = \begin{vmatrix} 2 & 1 & -1 \\ 3 & 8 & 2 \\ 1 & -1 & -3 \end{vmatrix} = -48 + 2 + 3 + 8 + 4 + 9 = -22$$

$$N_z = \begin{vmatrix} 2 & 3 & 1 \\ 3 & 5 & 8 \\ 1 & -2 & -1 \end{vmatrix} = -10 + 24 - 6 - 5 + 32 + 9 = 44$$

Hence the unique solution of the system is

$$x = \frac{N_x}{D} = \frac{66}{22} = 3 \qquad y = \frac{N_y}{D} = \frac{-22}{22} = -1 \qquad z = \frac{N_z}{D} = \frac{44}{22} = 2$$

Supplementary Problems

LINEAR EQUATIONS IN ONE AND TWO UNKNOWNS

10.16 Solve each equation:

(a) $4x = 28$ (c) $0x = -2$ (e) $-2x = 9$ (g) $-7x = -21$
(b) $3x = -18$ (d) $7x = 11$ (f) $0x = 0$ (h) $0x = 6$

10.17 Solve: (a) $3x + 2 = 10 - x$, (b) $5x - 4 = 2x + 11$.

10.18 Solve:

(a) $2x + \dfrac{1}{2} = \dfrac{1}{3}x + 5$ (b) $\dfrac{x-2}{2x+3} = \dfrac{3}{7}$

10.19 Plot each equation in the plane:

(a) $3x - 2y = 6$ (b) $x + 2y = 4$ (c) $2x = 6$ (d) $y = -3$

10.20 Solve each system:

(a) $2x - 5y = 1$ (b) $2x + 3y = 1$ (c) $3x - 2y = 7$
     $3x + 2y = 11$ $5x + 7y = 3$ $x + 3y = -16$

10.21 Solve each system:

(a) $2x + 4y = 10$ (b) $4x - 2y = 5$ (c) $2x - 4 = 3y$
     $3x + 6y = 15$ $-6x + 3y = 1$ $5y - x = 5$

10.22 Solve each system:

(a) $\begin{aligned} \dfrac{2x}{3} + \dfrac{y}{2} &= 8 \\ \dfrac{x}{6} - \dfrac{y}{4} &= -1 \end{aligned}$ (b) $\begin{aligned} \dfrac{2x-1}{3} + \dfrac{y+2}{4} &= 4 \\ \dfrac{x+3}{2} - \dfrac{x-y}{3} &= 3 \end{aligned}$

SQUARE SYSTEMS OF LINEAR EQUATIONS

10.23 Solve each system:

(a) $x + y - 3z = 5$ (b) $r - 2s + t = 2$
     $2y + z = 8$ $3s - 4t = 9$
     $3z = 6$ $2t = 6$

10.24 Solve each system:

(a) $x + 3y - z - 2t = 13$ (b) $2x + 4y - 3z + 2t = 6$
     $2y + 4z - t = 10$ $3y - 5z + 3t = 2$
     $5z - 2t = 7$ $4z - t = 5$
     $3t = 12$ $5t = 15$

10.25 Solve:

(a) $x + 2y - z = 3$ (b) $2x + y - 2z = 1$
     $2x + 5y - 4z = 5$ $3x + 2y - 4z = 1$
     $3x + 4y + 2z = 12$ $5x + 4y - z = 8$

10.26 Solve:

(a) $2x + 3y - 2z = 5$ (b) $x + 2y + 3z = 3$
     $x - 2y + 3z = 2$ $2x + 3y + 8z = 4$
     $4x - y + 4z = 1$ $3x + 2y + 17z = 1$

MATRICES AND LINEAR EQUATIONS

10.27   Consider the system

$$3x - 5y = 1$$
$$6x + 7y = 19$$

(a) Write as a matrix equation.   (b) Find the augmented matrix of the system.   (c) Solve the system.

10.28   Consider the following system:

$$x + y + 2z = 3$$
$$2x + 3y + 5z = 7$$
$$3x + 5y + 7z = 12$$

(a) Write as a matrix equation.   (b) Find the augmented matrix of the system.   (c) Solve the system.

DETERMINANTS AND LINEAR EQUATIONS

10.29   Solve each system, using determinants:

    (a) $3x + 4y = 11$ (b) $2x - 3y = 2$ (c) $4x - 3y = 2$
       $2x + 5y = 10$ $3x + 4y = 10$ $2y - 5 = 7x$

10.30   Solve, using determinants:

    (a) $x + 2y - z = 3$ (b) $x + y - z = 2$
       $2x - 5y + 2z = 1$ $2x + 3y + z = 1$
       $3x - 4y - z = 2$ $5x + 7y + z = 4$

Answers to Supplementary Problems

10.16   (a) 4, (b) −6, (c) no solution, (d) 11/7, (e) −9/2, (f) every number, (g) 3, (h) no solution

10.17   (a) 2, (b) 5

10.18   (a) 27/10, (b) 23

10.19   See Fig. 10-7.

10.20   (a) $x = 3$, $y = 1$; (b) $x = 2$, $y = -1$; (c) $x = -1$, $y = -5$

10.21   (a)   There are an infinite number of solutions, since the two equations coincide.
      (b)   No solution; the graphs of the two equations are parallel.
      (c) $x = 5$, $y = 2$

10.22   (a) $x = 6$, $y = 8$; (b) $x = 5$, $y = 2$

10.23   (a) $x = 8$, $y = 3$, $z = 2$; (b) $r = 13$, $s = 7$, $t = 3$

10.24   (a) $x = 21$, $y = 1$, $z = 3$, $t = 4$; (b) $x = 1$, $y = 1$, $z = 2$, $t = 3$

10.25   (a) $x = 2$, $y = 1$, $z = 1$; (b) $x = 1$, $y = 1$, $z = 1$

10.26   (a) No solution.   (b) An infinite number of solutions; solving for x and y in terms of z gives
    $x = -7z - 1$, $y = 2z + 2$.

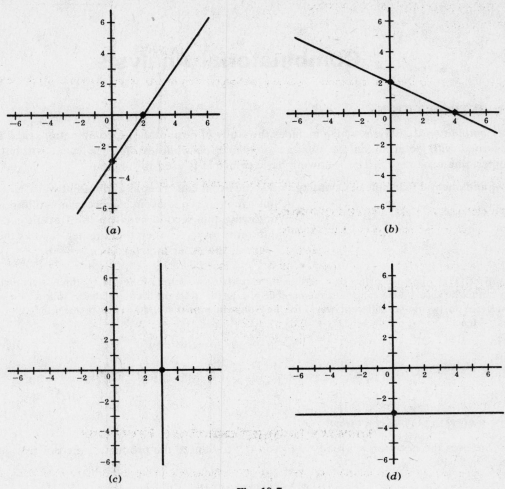

(a)　　　　　　　　　　　　　　(b)

(c)　　　　　　　　　　　　　　(d)

Fig. 10-7

10.27 (a) $\begin{pmatrix} 3 & -5 \\ 6 & 7 \end{pmatrix}\begin{pmatrix} x \\ y \end{pmatrix} = \begin{pmatrix} 1 \\ 19 \end{pmatrix}$; (b) $\begin{pmatrix} 3 & -5 & 1 \\ 6 & 7 & 19 \end{pmatrix}$; (c) $x = 1$, $y = 1$

10.28 (a) $\begin{pmatrix} 1 & 1 & 2 \\ 2 & 3 & 5 \\ 3 & 5 & 7 \end{pmatrix}\begin{pmatrix} x \\ y \\ z \end{pmatrix} = \begin{pmatrix} 3 \\ 7 \\ 12 \end{pmatrix}$; (b) $\begin{pmatrix} 1 & 1 & 2 & 3 \\ 2 & 3 & 5 & 7 \\ 3 & 5 & 7 & 12 \end{pmatrix}$; (c) $x = 3$, $y = 2$, $z = -1$

10.29 (a) $x = \dfrac{15}{7}$, $y = \dfrac{8}{7}$; (b) $x = \dfrac{38}{17}$, $y = \dfrac{14}{17}$; (c) $x = -\dfrac{19}{13}$, $y = -\dfrac{34}{13}$

10.30 (a) $x = \dfrac{43}{22}$, $y = \dfrac{18}{22}$, $z = \dfrac{13}{22}$

(b) Because the determinant of the matrix of coefficients is zero, the (nonunique) solution of the system cannot be found by Cramer's rule.

Chapter 11

Combinatorial Analysis

11.1 INTRODUCTION

Combinatorial analysis, which includes the study of permutations, combinations, and partitions, is concerned with determining the number of logical possibilities of some event without necessarily enumerating each case.   The following basic principle is used throughout.

> **Fundamental Principle of Counting:** If some event can occur in n_1 different ways, and if, following this event, a second event can occur in n_2 different ways, and, following this second event, a third event can occur in n_3 different ways, . . . , then the number of ways the events can occur in the order indicated is $n_1 \cdot n_2 \cdot n_3 \cdots$.

EXAMPLE 11.1   License plates of a certain state contain two letters followed by three digits, where the first digit cannot be zero.   One calculates the number of possible different license plates as follows: Each letter can be selected in twenty-six different ways, the first digit in nine ways, and each of the other two digits in ten ways.   Hence

$$26 \cdot 26 \cdot 9 \cdot 10 \cdot 10 = 608\,400$$

different plates are possible.

11.2 FACTORIAL NOTATION

One uses the notation $n!$, read "n factorial," to denote the product of the positive integers from 1 to n, inclusive:

$$n! = 1 \cdot 2 \cdot 3 \cdot \cdots \cdot (n-2)(n-1)n$$

Equivalently, $n!$ is defined by

$$1! = 1 \qquad \text{and} \qquad n! = n \cdot (n-1)!$$

It is also convenient to define $0! = 1$.

EXAMPLE 11.2

(a) $2! = 1 \cdot 2 = 2 \qquad 3! = 1 \cdot 2 \cdot 3 = 6 \qquad 4! = 1 \cdot 2 \cdot 3 \cdot 4 = 24$
$5! = 5 \cdot 4! = 5 \cdot 24 = 120 \qquad 6! = 6 \cdot 5! = 6 \cdot 120 = 720$

(b) $\dfrac{8!}{6!} = \dfrac{8 \cdot 7 \cdot 6!}{6!} = 8 \cdot 7 = 56 \qquad 12 \cdot 11 \cdot 10 = \dfrac{12 \cdot 11 \cdot 10 \cdot 9!}{9!} = \dfrac{12!}{9!}$

$\dfrac{12 \cdot 11 \cdot 10}{1 \cdot 2 \cdot 3} = 12 \cdot 11 \cdot 10 \cdot \dfrac{1}{3!} = \dfrac{12!}{3!\,9!}$

(c) $n(n-1) \cdots (n-r+1) = \dfrac{n(n-1) \cdots (n-r+1)(n-r)(n-r-1) \cdots 3 \cdot 2 \cdot 1}{(n-r)(n-r-1) \cdots 3 \cdot 2 \cdot 1} = \dfrac{n!}{(n-r)!}$

$\dfrac{n(n-1) \cdots (n-r+1)}{1 \cdot 2 \cdot 3 \cdots (r-1)r} = n(n-1) \cdots (n-r+1) \cdot \dfrac{1}{r!} = \dfrac{n!}{(n-r)!} \cdot \dfrac{1}{r!} = \dfrac{n!}{r!(n-r)!}$

Figure 11-1 shows a flowchart which reads a positive integer N and calculates the output N!, represented by the variable NFACT.

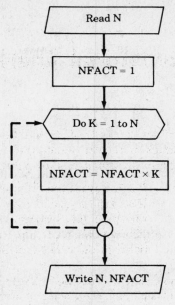

Fig. 11-1

11.3   BINOMIAL COEFFICIENTS

One defines the symbol

$$\binom{n}{r}$$

read "en-see-are," where r and n are positive integers with $r \le n$, by

$$\binom{n}{r} = \frac{n(n-1)(n-2) \cdots (n-r+1)}{1 \cdot 2 \cdot 3 \cdots (r-1)r}$$

or, from Example 11.2(*c*), by

$$\binom{n}{r} = \frac{n!}{r!(n-r)!}$$

which also extends to the case $r = 0$. But $n - (n - r) = r$; hence we have the following important relation:

$$\binom{n}{n-r} = \binom{n}{r}$$

In other words, if $a + b = n$, then

$$\binom{n}{a} = \binom{n}{b}$$

EXAMPLE 11.3

(*a*) $\binom{8}{2} = \frac{8 \cdot 7}{1 \cdot 2} = 28$ $\binom{9}{4} = \frac{9 \cdot 8 \cdot 7 \cdot 6}{1 \cdot 2 \cdot 3 \cdot 4} = 126$ $\binom{12}{5} = \frac{12 \cdot 11 \cdot 10 \cdot 9 \cdot 8}{1 \cdot 2 \cdot 3 \cdot 4 \cdot 5} = 792$

$\binom{10}{3} = \frac{10 \cdot 9 \cdot 8}{1 \cdot 2 \cdot 3} = 120$ $\binom{13}{1} = \frac{13}{1} = 13$

Note that $\binom{n}{r}$ has exactly r factors in both the numerator and the denominator.

(b)   Compute $\binom{10}{7}$.   By definition,

$$\binom{10}{7} = \frac{10 \cdot 9 \cdot 8 \cdot 7 \cdot 6 \cdot 5 \cdot 4}{1 \cdot 2 \cdot 3 \cdot 4 \cdot 5 \cdot 6 \cdot 7} = 120$$

On the other hand, $10 - 7 = 3$, and so

$$\binom{10}{7} = \binom{10}{3} = \frac{10 \cdot 9 \cdot 8}{1 \cdot 2 \cdot 3} = 120$$

Observe that the second computation saves space and time.

Figure 11-2 shows a flowchart which reads N and R and calculates and outputs

$$\binom{N}{R}$$

represented by variable NCR.   We use the fact that NCR has R factors in the numerator and denominator, and hence may be evaluated as

$$NCR = \frac{N}{1} \times \frac{N-1}{2} \times \frac{N-2}{3} \times \cdots \times \frac{N-R+1}{R}$$

Thus we initialize NCR at $(N/1) = N$ and use a DO loop with index K running from 1 to $R - 1$.

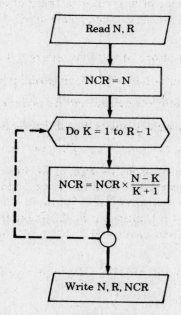

Fig. 11-2

The numbers $\binom{n}{r}$ are called the *binomial coefficients* since they appear as the coefficients in the expansion of $(a + b)^n$.   Specifically, one can prove that

$$(a + b)^n = \sum_{k=0}^{n} \binom{n}{k} a^{n-k} b^k$$

The coefficients in the expansions of the successive powers of $a + b$ can be arranged in a triangular array of numbers, called *Pascal's triangle*, as shown in Fig. 11-3.

$$(a+b)^0 = 1$$
$$(a+b)^1 = a+b$$
$$(a+b)^2 = a^2 + 2ab + b^2$$
$$(a+b)^3 = a^3 + 3a^2b + 3ab^2 + b^3$$
$$(a+b)^4 = a^4 + 4a^3b + 6a^2b^2 + 4ab^3 + b^4$$
$$(a+b)^5 = a^5 + 5a^4b + 10a^3b^2 + 10a^2b^3 + 5ab^4 + b^5$$
$$(a+b)^6 = a^6 + 6a^5b + 15a^4b^2 + 20a^3b^3 + 15a^2b^4 + 6ab^5 + b^6$$

```
        1
       1  1
      1  2  1
     1  3  3  1
    1  4  6  4  1
   1  5 10 10  5  1
  1  6 15 20 15  6  1
```

Fig. 11-3

Pascal's triangle has the following interesting properties.

(i)   Each row exhibits a central symmetry.

(ii) Any interior number in the array can be obtained by adding the two numbers appearing directly above it.   For example $10 = 4 + 6$, $15 = 5 + 10$, $20 = 10 + 10$.

Property (ii) of Pascal's triangle reflects the following theorem (proved in Problem 11.8):

Theorem 11.1: $\binom{n+1}{r} = \binom{n}{r-1} + \binom{n}{r}$

11.4   PERMUTATIONS

Any arrangement of a set of n objects in a given order is called a *permutation* of the objects (taken all at a time).   Any arrangement of any $r \le n$ of these objects in a given order is called an *r-permutation* or a *permutation of the n objects taken r at a time*.   Consider, for example, the set of letters a, b, c and d.   Then:

(i)   *bdca*, *dcba* and *acdb* are permutations of the four letters (taken all at a time);

(ii)   *bad*, *adb*, *cbd* and *bca* are permutations of the four letters taken three at a time;

(iii) *ad*, *cb*, *da* and *bd* are permutations of the four letters taken two at a time.

The number of permutations of n objects taken r at a time is denoted by

$$P(n,r), \quad {}_nP_r, \quad P_{n,r}, \quad P_r^n, \quad \text{or} \quad (n)_r$$

We shall use $P(n,r)$.   To find an expression for $P(n,r)$ we observe that the first element in an r-permutation of n objects can be chosen in n different ways; following this, the second element in the permutation can be chosen in $n-1$ ways; and, following this, the third element in the permutation can be chosen in $n-2$ ways.   Continuing in this manner, we have that the rth (last) element in the r-permutation can be chosen in $n-(r-1) = n-r+1$ ways.   Thus, by the fundamental principle of counting,

$$P(n,r) = n(n-1)(n-2) \cdots (n-r+1)$$

or, using Example 11.2(c),

Theorem 11.2: $P(n,r) = \dfrac{n!}{(n-r)!}$

In the special case in which $r = n$, we have

$$P(n,n) = n(n-1)(n-2) \cdots 3 \cdot 2 \cdot 1 = n!$$

Accordingly,

Corollary 11.3:   There are $n!$ permutations of n objects (taken all at a time).

For example, there are $3! = 1 \cdot 2 \cdot 3 = 6$ permutations of the three letters a, b and c.   These are *abc, acb, bac, bca, cab, cba*.

11.5   PERMUTATIONS WITH REPETITION; PARTITIONS

Frequently we want to find the number of permutations of objects some of which are alike.   The general formula is as follows:

Theorem 11.4:   The number of permutations of n objects of which n_1 are alike in one respect, n_2 are alike in another respect, . . . , n_r are alike in yet another respect, is

$$\frac{n!}{n_1! \, n_2! \cdots n_r!}$$

In the theorem, $n_1, n_2, \ldots, n_r$ are positive integers whose sum is n.

We indicate the proof of the above theorem by a particular example.   Suppose we want to form all possible five-letter "words" using the letters from the word "DADDY".   Now there are $5! = 120$ permutations of the objects D_1, A, D_2, D_3, Y, where the three D's are distinguished.   Observe that the following six permutations

$$D_1D_2D_3AY, \ D_2D_1D_3AY, \ D_3D_1D_2AY, \ D_1D_3D_2AY, \ D_2D_3D_1AY, \text{ and } D_3D_2D_1AY$$

produce the same word when the subscripts are removed.   The 6 comes from the fact that there are $3! = 3 \cdot 2 \cdot 1 = 6$ different ways of placing the three D's in the first three positions in the permutation.   This is true for each set of three positions in which the D's can appear.   Accordingly there are

$$\frac{5!}{3!} = \frac{120}{6} = 20$$

different five-letter words that can be formed using the letters from the word "DADDY".   (The complete expression as given by Theorem 11.4 would be $5!/(3! \, 1! \, 1!)$, where one 1! arose from the unique A and the other from the unique Y.   Since $1! = 1$, we can omit $1!$.)

EXAMPLE 11.4   How many different signals, each consisting of eight flags hung in a vertical line, can be formed with four indistinguishable red flags, three indistinguishable white flags, and a blue flag?

We seek the number of permutations of eight objects of which four are alike and three are alike and one is alike.   There are

$$\frac{8!}{4! \, 3! \, 1!} = \frac{8 \cdot 7 \cdot 6 \cdot 5 \cdot 4 \cdot 3 \cdot 2 \cdot 1}{4 \cdot 3 \cdot 2 \cdot 1 \cdot 3 \cdot 2 \cdot 1 \cdot 1} = 280$$

different signals.

Ordered Partitions

Consider a set, A, of n distinct elements.   Theorem 11.4 also counts the *ordered partitions*,

$$[A_1, A_2, \ldots, A_r]$$

of A into a cell A_1 containing n_1 elements, a cell A_2 containing n_2 elements, . . . , and a cell A_r containing n_r elements.   In fact, if we view the permutations in Theorem 11.4 as fillings of n fixed positions (e.g. $\underline{Y} \ \underline{D} \ \underline{D} \ \underline{A} \ \underline{D}$), then it is clear that each permutation determines an ordered partition *of the positions* into a cell of n_1 positions filled by objects of the first kind, a cell of n_2 positions filled by objects of the second kind, etc.   (For instance, the cells determined by the permutation YDDAD are a D-cell consisting of positions 2, 3, and 5; an A-cell consisting of position 4; and a Y-cell consisting of position 1.)   Moreover, distinct permutations determine distinct ordered partitions.   Thus we have proved

Corollary 11.5:   The number of ordered partitions of n distinct elements into cells having sizes $n_1, n_2, \ldots, n_r$ is

$$\frac{n!}{n_1! \, n_2! \cdots n_r!}$$

EXAMPLE 11.5 In how many ways may nine toys be divided among four children, if the youngest child is to receive three toys and each of the other children two toys?

We wish to find the number of ordered partitions of the nine toys into four cells containing 3, 2, 2, and 2 toys, respectively. By Corollary 11.5, there are

$$\frac{9!}{3!\,2!\,2!\,2!} = 7560$$

such ordered partitions.

Unordered Partitions

These are partitions as defined in Section 6.10; here, only the composition of the cells matters, not their ordering. In counting the unordered partitions of n elements, it is convenient to characterize a partition not by the cell sizes, but—what amounts to the same thing—by the number i_1 of cells containing 1 element, the number i_2 of cells containing 2 elements, ..., and the number i_n of cells containing n elements. The same reasoning that led to Corollary 11.5 gives us

Corollary 11.6: The number of unordered partitions of n distinct elements into i_1 cells containing 1 element each, i_2 cells containing 2 elements each, ..., i_n cells containing n elements each, is

$$\frac{n!}{[(1!)^{i_1}i_1!][(2!)^{i_2}i_2!] \cdots [(n!)^{i_n}i_n!]}$$

In this corollary, $i_1, i_2, \ldots, i_n$ are nonnegative integers such that

$$1i_1 + 2i_2 + \cdots + ni_n = n$$

In particular, i_n can equal 1 only if all the other i_α are 0.

EXAMPLE 11.6. In how many ways may nine toys be sorted into a pile of three toys, a pile of two toys, a pile of two toys, and a pile of two toys? (Compare Example 11.5.)

We wish to find the number of unordered partitions of the nine toys such that 3 cells contain two toys and 1 cell contains three toys (and 0 cells contain one toy, etc.). By Corollary 11.6, there are

$$\frac{9!}{[(2!)^3 3!][(3!)^1 1!]} = 1260$$

such unordered partitions.

11.6 COMBINATIONS

Suppose we have a collection of n objects. A *combination* of these n objects taken r at a time is any selection of r of the objects where order doesn't count. In other words, an *r-combination* of a set of n objects is any subset containing r objects. For example, the combinations of the letters a, b, c, d taken three at a time are

$$\{a, b, c\}, \{a, b, d\}, \{a, c, d\}, \{b, c, d\} \qquad \text{or simply} \qquad abc, abd, acd, bcd$$

Observe that the following combinations are equal:

$$abc, acb, bac, bca, cab, cba$$

That is, each denotes the same set $\{a, b, c\}$.

The number of combinations of n objects taken r at a time is denoted by $C(n, r)$. The symbols $_nC_r$, $C_{n,r}$ and C_r^n also appear in various texts. To find the general formula for $C(n, r)$, we note that any combination of n objects taken r at a time determines $r!$ permutations of the objects in the combination. Consequently,

$$P(n, r) = r!C(n, r)$$

and we obtain

Theorem 11.7:
$$C(n, r) = \frac{P(n, r)}{r!} = \frac{n!}{r!(n - r)!} = \binom{n}{r}$$

We shall henceforth use $C(n, r)$ and $\binom{n}{r}$ interchangeably.

EXAMPLE 11.7

(a) How many committees of three can be formed from eight people? Each committee represents a combination of the eight people taken three at a time. Thus

$$C(8, 3) = \binom{8}{3} = \frac{8 \cdot 7 \cdot 6}{1 \cdot 2 \cdot 3} = 56$$

different committees can be formed.

(b) A farmer buys three cows, two pigs and four hens from a man who has six cows, five pigs and eight hens. How many choices does the farmer have?

The farmer can choose the cows in $C(6, 3)$ ways, the pigs in $C(5, 2)$ ways, and the hens in $C(8, 4)$ ways. Hence altogether he can choose the animals in

$$\binom{6}{3}\binom{5}{2}\binom{8}{4} = \frac{6 \cdot 5 \cdot 4}{1 \cdot 2 \cdot 3} \cdot \frac{5 \cdot 4}{1 \cdot 2} \cdot \frac{8 \cdot 7 \cdot 6 \cdot 5}{1 \cdot 2 \cdot 3 \cdot 4} = 20 \cdot 10 \cdot 70 = 14\,000 \text{ ways}$$

11.7 TREE DIAGRAMS

A (rooted) tree diagram augments the fundamental principle of counting by exhibiting all possible outcomes of a sequence of events where each event can occur in a finite number of ways. We illustrate this device with an example. (Tree diagrams are discussed in their own right in Chapter 14.)

EXAMPLE 11.8 Marc and Erik are to play a tennis tournament. The first person to win two games in a row or to win a total of three games wins the tournament.

Figure 11-4 gives a tree diagram which shows how the tournament can go. The tree is constructed from left to right. At each point (game) other than an endpoint, there originate two branches, which correspond to the two possible outcomes of that game, i.e. Marc winning or Erik winning. Observe that there are 10 endpoints, corresponding to the 10 possible courses of the tournament:

MM, MEMM, MEMEM, MEMEE, MEE, EMM, EMEMM, EMEME, EMEE, EE

The path from the beginning of the tree to a particular endpoint describes who won which game in that particular course.

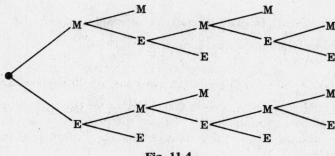

Fig. 11-4

Solved Problems

FACTORIALS, BINOMIAL COEFFICIENTS

11.1 Compute 4!, 5!, 6!, 7!, and 8!.

$$4! = 1 \cdot 2 \cdot 3 \cdot 4 = 24 \qquad\qquad 7! = 7 \cdot 6! = 7 \cdot 720 = 5040$$
$$5! = 1 \cdot 2 \cdot 3 \cdot 4 \cdot 5 = 5 \cdot 4! = 5 \cdot 24 = 120 \qquad 8! = 8 \cdot 7! = 8 \cdot 5040 = 40\,320$$
$$6! = 1 \cdot 2 \cdot 3 \cdot 4 \cdot 5 \cdot 6 = 6 \cdot 5! = 6 \cdot 120 = 720$$

11.2 Compute: (a) $\dfrac{13!}{11!}$ (b) $\dfrac{7!}{10!}$

(a) $\dfrac{13!}{11!} = \dfrac{13 \cdot 12 \cdot 11 \cdot 10 \cdot 9 \cdot 8 \cdot 7 \cdot 6 \cdot 5 \cdot 4 \cdot 3 \cdot 2 \cdot 1}{11 \cdot 10 \cdot 9 \cdot 8 \cdot 7 \cdot 6 \cdot 5 \cdot 4 \cdot 3 \cdot 2 \cdot 1} = 13 \cdot 12 = 156$

or $\dfrac{13!}{11!} = \dfrac{13 \cdot 12 \cdot 11!}{11!} = 13 \cdot 12 = 156$

(b) $\dfrac{7!}{10!} = \dfrac{7!}{10 \cdot 9 \cdot 8 \cdot 7!} = \dfrac{1}{10 \cdot 9 \cdot 8} = \dfrac{1}{720}$

11.3 Write in terms of factorials: (a) $27 \cdot 26$ (b) $\dfrac{1}{14 \cdot 13 \cdot 12}$

(a) $27 \cdot 26 = \dfrac{27 \cdot 26 \cdot 25!}{25!} = \dfrac{27!}{25!}$

(b) $\dfrac{1}{14 \cdot 13 \cdot 12} = \dfrac{11!}{14 \cdot 13 \cdot 12 \cdot 11!} = \dfrac{11!}{14!}$

11.4 Simplify: (a) $\dfrac{n!}{(n-1)!}$ (b) $\dfrac{(n+2)!}{n!}$

(a) $\dfrac{n!}{(n-1)!} = \dfrac{n(n-1)(n-2) \cdots 3 \cdot 2 \cdot 1}{(n-1)(n-2) \cdots 3 \cdot 2 \cdot 1} = n$ or simply $\dfrac{n!}{(n-1)!} = \dfrac{n(n-1)!}{(n-1)!} = n$

(b) $\dfrac{(n+2)!}{n!} = \dfrac{(n+2)(n+1)n(n-1)(n-2) \cdots 3 \cdot 2 \cdot 1}{n(n-1)(n-2) \cdots 3 \cdot 2 \cdot 1} = (n+2)(n+1) = n^2 + 3n + 2$

or simply

$$\dfrac{(n+2)!}{n!} = \dfrac{(n+2)(n+1) \cdot n!}{n!} = (n+2)(n+1) = n^2 + 3n + 2$$

11.5 Compute: (a) $\dbinom{16}{3}$ (b) $\dbinom{12}{4}$ (c) $\dbinom{15}{5}$

Recall that there are as many factors in the numerator as in the denominator.

(a) $\dbinom{16}{3} = \dfrac{16 \cdot 15 \cdot 14}{1 \cdot 2 \cdot 3} = 560$ (c) $\dbinom{15}{5} = \dfrac{15 \cdot 14 \cdot 13 \cdot 12 \cdot 11}{1 \cdot 2 \cdot 3 \cdot 4 \cdot 5} = 3003$

(b) $\dbinom{12}{4} = \dfrac{12 \cdot 11 \cdot 10 \cdot 9}{1 \cdot 2 \cdot 3 \cdot 4} = 495$

11.6 Compute: (a) $\binom{8}{5}$ (b) $\binom{9}{7}$

(a) $\binom{8}{5} = \dfrac{8 \cdot 7 \cdot 6 \cdot 5 \cdot 4}{1 \cdot 2 \cdot 3 \cdot 4 \cdot 5} = 56$

Note that $8 - 5 = 3$; hence we could also compute $\binom{8}{5}$ as follows:

$$\binom{8}{5} = \binom{8}{3} = \frac{8 \cdot 7 \cdot 6}{1 \cdot 2 \cdot 3} = 56$$

(b) Now $9 - 7 = 2$; hence $\binom{9}{7} = \binom{9}{2} = \dfrac{9 \cdot 8}{1 \cdot 2} = 36$

11.7 Prove: $\binom{17}{6} = \binom{16}{5} + \binom{16}{6}$

Now $\binom{16}{5} + \binom{16}{6} = \dfrac{16!}{5! \, 11!} + \dfrac{16!}{6! \, 10!}$

Multiply the first fraction by 6/6 and the second by 11/11 to obtain the same denominator in both fractions; and then add:

$$\binom{16}{5} + \binom{16}{6} = \frac{6 \cdot 16!}{6 \cdot 5! \cdot 11!} + \frac{11 \cdot 16!}{6! \cdot 11 \cdot 10!} = \frac{6 \cdot 16!}{6! \cdot 11!} + \frac{11 \cdot 16!}{6! \cdot 11!}$$

$$= \frac{6 \cdot 16! + 11 \cdot 16!}{6! \cdot 11!} = \frac{(6 + 11) \cdot 16!}{6! \cdot 11!} = \frac{17 \cdot 16!}{6! \cdot 11!} = \frac{17!}{6! \cdot 11!} = \binom{17}{6}$$

11.8 Prove Theorem 11.1: $\binom{n+1}{r} = \binom{n}{r-1} + \binom{n}{r}$

The technique in this proof is similar to that of Problem 11.7.   We have:

$$\binom{n}{r-1} + \binom{n}{r} = \frac{n!}{(r-1)! \cdot (n-r+1)!} + \frac{n!}{r! \cdot (n-r)!}$$

To obtain the same denominator in both fractions, multiply the first fraction by r/r and the second fraction by

$$\frac{n-r+1}{n-r+1}$$

Hence

$$\binom{n}{r-1} + \binom{n}{r} = \frac{r \cdot n!}{r \cdot (r-1)! \cdot (n-r+1)!} + \frac{(n-r+1) \cdot n!}{r! \cdot (n-r+1) \cdot (n-r)!}$$

$$= \frac{r \cdot n!}{r! \, (n-r+1)!} + \frac{(n-r+1) \cdot n!}{r! \, (n-r+1)!}$$

$$= \frac{r \cdot n! + (n-r+1) \cdot n!}{r! \, (n-r+1)!}$$

$$= \frac{[r + (n-r+1)] \cdot n!}{r! \, (n-r+1)!}$$

$$= \frac{(n+1)n!}{r! \, (n-r+1)!}$$

$$= \frac{(n+1)!}{r! \, (n-r+1)!}$$

$$= \binom{n+1}{r}$$

PERMUTATIONS, PARTITIONS

11.9 There are four bus lines between A and B; and three bus lines between B and C. In how many ways can a person travel   (a) by bus from A to C by way of B?   (b) roundtrip by bus from A to C by way of B?   (c) roundtrip by bus from A to C by way of B, without using any bus line more than once?

(a)   There are four ways to go from A to B and three ways to go from B to C; hence there are $4 \cdot 3 = 12$ ways to go from A to C by way of B.

(b)   There are twelve ways to go from A to C by way of B, and 12 ways to return.   Hence there are $12 \cdot 12 = 144$ ways to travel roundtrip.

(c)   The person will travel from A to B to C to B to A.   Enter these letters with connecting arrows as follows:

$$A \to B \to C \to B \to A$$

The person can travel four ways from A to B and three ways from B to C; this leaves available two ways from C to B and three ways from B to A.   Enter these numbers above the corresponding arrows as follows:

$$A \overset{4}{\to} B \overset{3}{\to} C \overset{2}{\to} B \overset{3}{\to} A$$

Thus there are $4 \cdot 3 \cdot 2 \cdot 3 = 72$ ways to travel roundtrip without using the same bus line more than once.

11.10 Suppose repetitions are not permitted.   (a) How many three-digit numbers can be formed from the six digits 2, 3, 5, 6, 7 and 9?   (b) How many of these numbers are less than 400?   (c) How many are even?   (d) How many are odd?   (e) How many are multiples of 5?

In each case draw three boxes,

☐ ☐ ☐

to represent an arbitrary number, and then write in each box the number of digits that can be placed there.

(a)   The box on the left can be filled in six ways; following this, the middle box can be filled in five ways; and, lastly, the box on the right can be filled in four ways:

6 5 4

Thus there are $6 \cdot 5 \cdot 4 = 120$ numbers.

(b)   The box on the left can be filled in only two ways, by 2 or 3, since each number must be less than 400; the middle box can be filled in five ways; and, lastly, the box on the right can be filled in four ways:

2 5 4

Thus there are $2 \cdot 5 \cdot 4 = 40$ numbers.

(c)   The box on the right can be filled in only two ways, by 2 or 6, since the numbers must be even; the box on the left can then be filled in five ways; and, lastly, the middle box can be filled in four ways:

5 4 2

Thus there are $5 \cdot 4 \cdot 2 = 40$ numbers.

(d)   The box on the right can be filled in only four ways, by 3, 5, 7 or 9, since the numbers must be odd; the box on the left can then be filled in five ways; and, lastly, the box in the middle can be filled in four ways:

$$\boxed{5} \quad \boxed{4} \quad \boxed{4}$$

Thus there are $5 \cdot 4 \cdot 4 = 80$ numbers.   (Alternatively, from (a) and (c), $120 - 40 = 80$.)

(e) The box on the right can be filled in only one way, by 5, since the numbers must be multiples of 5; the box on the left can then be filled in five ways; and, lastly, the box in the middle can be filled in four ways:

$$\boxed{5} \quad \boxed{4} \quad \boxed{1}$$

Thus there are $5 \cdot 4 \cdot 1 = 20$ numbers.

11.11 Find the number of distinct permutations that can be formed from all the letters of each word: (a) THEM, (b) THAT, (c) RADAR, (d) UNUSUAL, (e) SOCIOLOGICAL.

(a) $4! = 24$, since there are four letters and no repetition.
(b) $4!/2! = 12$, since there are four letters of which two are T.
(c) $5!/(2!\,2!) = 30$, since there are five letters of which two are R and two are A.
(d) $7!/3! = 840$, since there are seven letters of which three are U.
(e)
$$\frac{12!}{3!\,2!\,2!\,2!} = 9\,979\,200$$

since there are twelve letters of which three are O, two are C, two are I, and two are L.

11.12 In how many ways can four mathematics books, three history books, three chemistry books and two sociology books be arranged on a shelf so that all books of the same subject are together?

First the books must be arranged on the shelf in four units according to subject matter:

$$\boxed{} \quad \boxed{} \quad \boxed{} \quad \boxed{}$$

The box on the left can be filled by any of the four subjects; the next by any three subjects remaining; the next by any two subjects remaining; and the box on the right by the last subject:

$$\boxed{4} \quad \boxed{3} \quad \boxed{2} \quad \boxed{1}$$

Thus there are $4 \cdot 3 \cdot 2 \cdot 1 = 4!$ ways to arrange the books on the shelf according to subject matter.
In each of the above cases, the mathematics books can be arranged in 4! ways, the history books in 3! ways, the chemistry books in 3! ways, and the sociology books in 2! ways. Thus, altogether, there are $4!\,4!\,3!\,3!\,2! = 41\,472$ arrangements.

11.13 In how many ways can a party of seven persons arrange themselves (a) in a row of seven chairs? (b) around a circular table?

(a) The seven persons can arrange themselves in a row in $7 \cdot 6 \cdot 5 \cdot 4 \cdot 3 \cdot 2 \cdot 1 = 7!$ ways.
(b) One person can sit at any place around the circular table. The other six persons can then arrange themselves in $6 \cdot 5 \cdot 4 \cdot 3 \cdot 2 \cdot 1 = 6!$ ways around the table. This is an example of a *circular permutation*. In general, n objects can be arranged in a circle in $(n-1)!$ ways.

11.14 Find n if (a) $P(n, 2) = 72$, (b) $P(n, 4) = 42P(n, 2)$, (c) $2P(n, 2) + 50 = P(2n, 2)$.

(a) $P(n, 2) = n(n-1) = n^2 - n$; hence $n^2 - n = 72$ or $n^2 - n - 72 = 0$ or $(n-9)(n+8) = 0$.
Since n must be positive, the only answer is $n = 9$.
(b) $P(n, 4) = n(n-1)(n-2)(n-3)$ and $P(n, 2) = n(n-1)$.   Hence

$$n(n-1)(n-2)(n-3) = 42n(n-1)$$

or, if $n \neq 0$, $n \neq 1$,

$$(n-2)(n-3) = 42 \quad \text{or} \quad n^2 - 5n + 6 = 42 \quad \text{or} \quad n^2 - 5n - 36 = 0 \quad \text{or} \quad (n-9)(n+4) = 0$$

Since n must be positive, the only answer is $n = 9$.

(c) $P(n,2) = n(n-1) = n^2 - n$ and $P(2n,2) = 2n(2n-1) = 4n^2 - 2n$.   Hence

$$2(n^2 - n) + 50 = 4n^2 - 2n \quad \text{or} \quad 2n^2 - 2n + 50 = 4n^2 - 2n \quad \text{or} \quad 50 = 2n^2 \quad \text{or} \quad n^2 = 25$$

Since n must be positive, the only answer is $n = 5$.

11.15 There are twelve students in a class. In how many ways can the twelve students take three different tests if four students are to take each test?

We seek the number of ordered partitions of the twelve students into cells containing four students each.   By Corollary 11.5, there are

$$\frac{12!}{4!\,4!\,4!} = 34\,650$$

such partitions.

11.16 In how many ways can twelve students be partitioned into three teams, so that each team contains four students?

As the teams are defined only by their composition, we are dealing with unordered partitions here. For 3 cells of four elements each, Corollary 11.6 gives

$$\frac{12!}{(4!)^3 3!} = 5775$$

such partitions.

11.17 In how many ways can ten students be partitioned into four teams, such that two teams contain two students each and two teams contain three students each.

By Corollary 11.6, there are

$$\frac{10!}{[(2!)^2 2!][(3!)^2 2!]} = 6300$$

such (unordered) partitions.

11.18 Count the (unordered) partitions of a set of n distinct elements.

Method 1.
Sum the expression given by Corollary 11.6 over all allowable values of $i_1, i_2, \ldots, i_n$. This would lead to a rather complicated formula.

Method 2.
Apply combinatorial reasoning. Distinguish one element of the set as the *green element*, and, in each partition, distinguish the cell that contains the green element as the *green cell*. To find $\phi(n)$, the total number of partitions of n elements, we observe that if the green cell contains only the green element, the remaining $n-1$ elements can be partitioned in $\phi(n-1)$ ways; if the green cell contains one element besides the green element, which can be chosen in $C(n-1,1)$ ways, the remaining $n-2$ elements can be partitioned in $\phi(n-2)$ ways; ...; if the green cell contains $n-1$ elements besides the green element, which can be chosen in $C(n-1, n-1)$ ways, the remaining 0 elements can be partitioned in $\phi(0) = 1$ way. As these partitions account for all the possible partitions, we have

$$\phi(n) = \phi(n-1) + \binom{n-1}{1}\phi(n-2) + \binom{n-1}{2}\phi(n-3) + \cdots + \binom{n-1}{n-2}\phi(1) + \binom{n-1}{n-1}\phi(0)$$

Starting with $\phi(0) = 1$, we can solve successively for $\phi(1)$, $\phi(2)$, $\phi(3)$, ... (a triangular system of linear equations). Thus:

$$\phi(1) = \phi(0) = 1$$

$$\phi(2) = \phi(1) + \binom{1}{1}\phi(0) = 1 + 1 = 2$$

$$\phi(3) = \phi(2) + \binom{2}{1}\phi(1) + \binom{2}{2}\phi(0) = 2 + 2 \cdot 1 + 1 = 5$$

$$\phi(4) = \phi(3) + \binom{3}{1}\phi(2) + \binom{3}{2}\phi(1) + \binom{3}{3}\phi(0)$$

$$= 5 + 3 \cdot 2 + 3 \cdot 1 + 1 = 15$$

$$\phi(5) = \phi(4) + \binom{4}{1}\phi(3) + \binom{4}{2}\phi(2) + \binom{4}{3}\phi(1) + \binom{4}{4}\phi(0)$$

$$= 15 + 4 \cdot 5 + 6 \cdot 2 + 4 \cdot 1 + 1 = 52$$

....

Method 3.

See Problem 11.45.

COMBINATIONS

11.19 In how many ways can a committee consisting of three men and two women be chosen from seven men and five women?

The three men can be chosen from the seven men in $C(7, 3)$ ways, and the two women can be chosen from the five women in $C(5, 2)$ ways. Hence the committee can be chosen in

$$\binom{7}{3}\binom{5}{2} = \frac{7 \cdot 6 \cdot 5}{1 \cdot 2 \cdot 3} \cdot \frac{5 \cdot 4}{1 \cdot 2} = 350 \text{ ways}$$

11.20 How many committees of five with a given chairman can be selected from twelve persons?

The chairman can be chosen in twelve ways and, following this, the other four on the committee can be chosen from the eleven remaining in $C(11, 4)$ ways. Thus there are

$$12 \cdot \binom{11}{4} = 12 \cdot 330 = 3960$$

such committees.

11.21 A bag contains six white marbles and five black marbles. Find the number of ways four marbles can be drawn from the bag if (*a*) they can be of either color, (*b*) two must be white and two black, (*c*) they must all be of the same color.

(*a*) The four marbles (of either color) can be chosen from the eleven marbles in

$$\binom{11}{4} = \frac{11 \cdot 10 \cdot 9 \cdot 8}{1 \cdot 2 \cdot 3 \cdot 4} = 330 \text{ ways}$$

(*b*) The two white marbles can be chosen in $C(6, 2)$ ways, and the two black marbles can be chosen in $C(5, 2)$ ways. Thus there are

$$\binom{6}{2}\binom{5}{2} = \frac{6 \cdot 5}{1 \cdot 2} \cdot \frac{5 \cdot 4}{1 \cdot 2} = 150$$

ways of drawing two white marbles and two black marbles.

(*c*) There are $C(6, 4) = 15$ ways of drawing four white marbles, and $C(5, 4) = 5$ ways of drawing four black marbles. Thus there are $15 + 5 = 20$ ways of drawing four marbles of the same color.

11.22 A student is to answer eight out of ten questions on an exam. (*a*) How many choices has he? (*b*) How many if he must answer the first three questions? (*c*) How many if he must answer at least four of the first five questions?

(*a*) The eight questions can be selected in

$$\binom{10}{8} = \binom{10}{2} = \frac{10 \cdot 9}{1 \cdot 2} = 45 \text{ ways}$$

(*b*) If he answers the first three questions, then we can choose the other five questions from the last seven questions in

$$\binom{7}{5} = \binom{7}{2} = \frac{7 \cdot 6}{1 \cdot 2} = 21 \text{ ways}$$

(*c*) If he answers all the first five questions, then he can choose the other three questions from the last five in

$$\binom{5}{3} = 10 \text{ ways}$$

On the other hand, if he answers only four of the first five questions, then he can choose these four in

$$\binom{5}{4} = \binom{5}{1} = 5 \text{ ways}$$

and he can choose the other four questions from the last five in

$$\binom{5}{4} = \binom{5}{1} = 5 \text{ ways}$$

hence he can choose the eight questions in $5 \cdot 5 = 25$ ways. Thus he has a total of thirty-five choices.

TREE DIAGRAMS

11.23 Find the product set $A \times B \times C$ where $A = \{1, 2\}$, $B = \{a, b, c\}$ and $C = \{3, 4\}$.

The product set is obtained by constructing a tree diagram as shown in Fig. 11-5. Observe that the tree is constructed from left to right, and that the number of branches at each point corresponds to the number of ways the next event can occur. The twelve elements of $A \times B \times C$ are listed on the right of the diagram.

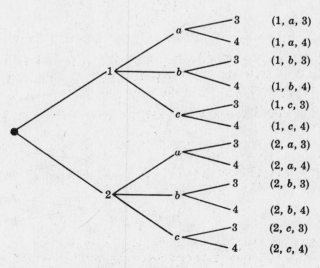

Fig. 11-5

11.24 Find the permutations of $\{a, b, c\}$.

Corollary 11.3 tells us that there are $3! = 3 \cdot 2 \cdot 1 = 6$ such permutations; a tree diagram can be used to display them. This is done in Fig. 11-6, where the six permutations are listed on the right of the diagram.

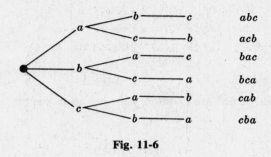

Fig. 11-6

11.25 Teams A and B play in a basketball tournament. The team that first wins three games wins the tournament. Find the possible courses of the tournament.

Construct the appropriate tree diagram, as shown in Fig. 11-7. There are twenty courses:

AAA, AABA, AABBA, AABBB, ABAA, ABABA, ABABB, ABBAA, ABBAB, ABBB
BAAA, BAABA, BAABB, BABAA, BABAB, BABB, BBAAA, BBAAB, BBAB, BBB

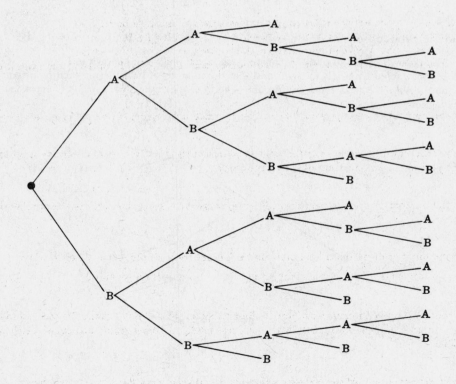

Fig. 11-7

Supplementary Problems

FACTORIALS, BINOMIAL COEFFICIENTS

11.26 Evaluate: (a) 9!, (b) 10!, (c) 11!

11.27 Evaluate: (a) 16!/14!, (b) 14!/11!, (c) 8!/10!, (d) 10!/13!

11.28 Write in terms of factorials: (a) $24 \cdot 23 \cdot 22 \cdot 21$, (b) $1/(10 \cdot 11 \cdot 12)$

11.29 Simplify: (a) $\dfrac{(n+1)!}{n!}$ (b) $\dfrac{n!}{(n-2)!}$ (c) $\dfrac{(n-1)!}{(n+2)!}$ (d) $\dfrac{(n-r+1)!}{(n-r-1)!}$

11.30 Evaluate: (a) $\binom{5}{2}$ (b) $\binom{7}{3}$ (c) $\binom{14}{2}$ (d) $\binom{6}{4}$ (e) $\binom{20}{17}$ (f) $\binom{18}{15}$

11.31 The eighth row of Pascal's triangle is as follows:

$$1 \quad 8 \quad 28 \quad 56 \quad 70 \quad 56 \quad 28 \quad 8 \quad 1$$

Compute the ninth and tenth rows of the triangle.

PERMUTATIONS, PARTITIONS

11.32 (a) How many automobile license plates can be made if each plate contains two different letters followed by three different digits?   (b) Solve the problem if the first digit cannot be 0.

11.33 There are six roads between A and B and four roads between B and C.
(a) In how many ways can one drive from A to C by way of B?
(b) In how many ways can one drive roundtrip from A to C by way of B?
(c) In how many ways can one drive roundtrip from A to C without using the same road more than once?

11.34 Find the number of ways in which six people can ride a toboggan if one of three must drive.

11.35 (a) Find the number of ways in which five persons can sit in a row.   (b) How many ways are there if two of the persons insist on sitting next to each other?

11.36 Find the number of ways in which a judge can award first, second and third places in a contest with ten contestants.

11.37 Find the number of permutations that can be formed from all the letters of each word:

(a) QUEUE   (b) COMMITTEE   (c) PROPOSITION   (d) BASEBALL

11.38 Consider all positive integers with three different digits.   (Note that 0 cannot be the first digit.)   (a) How many are greater than 700?   (b) How many are odd?   (c) How many are even?   (d) How many are divisible by 5?

11.39 (a) Find the number of permutations that can be formed from the letters of the word ELEVEN.
(b) How many of them begin and end with E?   (c) How many of them have the three E's together?
(d) How many begin with E and end with N?

11.40   In how many ways can nine toys be divided evenly among three children?

11.41   In how many ways can nine students be evenly divided into three teams?

11.42   In how many ways can ten students be divided into three teams, one containing four students and the others three?

11.43   In how many ways can a club with twelve members be partitioned into three committees containing 5, 4 and 3 members respectively?

11.44   (*a*) Assuming that a cell can be empty, in how many ways can a set with three elements be partitioned into (i) three ordered cells, (ii) three unordered cells?   (*b*) In how many ways can a set with four elements be partitioned into (i) three ordered cells, (ii) three unordered cells?

11.45   Denote by $f(n, k)$ the number of unordered partitions of n distinct elements into k nonempty cells, where $k \le n$.
(*a*)   Establish the relation

$$f(n, k) = f(n - 1, k - 1) + kf(n - 1, k)$$

and solve in a form similar to Pascal's triangle.
(*b*)   Referring to Problem 11.18, show that

$$\phi(n) = \sum_{k=1}^{n} f(n, k)$$

i.e. the sum of the nth row of the triangular array in (*a*) is $\phi(n)$.

COMBINATIONS

11.46   A class contains nine boys and three girls.   (*a*) In how many ways can the teacher choose a committee of four?   (*b*) How many committees will contain at least one girl?   (*c*) How many committees will contain exactly one girl?

11.47   A woman has eleven close friends.   (*a*) In how many ways can she invite five of them to dinner?   (*b*) In how many ways if two of the friends are married and will not attend separately?   (*c*) In how many ways if two of them are not on speaking terms and will not attend together?

11.48   There are ten points $A, B, \ldots$ in a plane, no three on the same line.   (*a*) How many lines are determined by the points?   (*b*) How many of these lines do not pass through A or B?   (*c*) How many triangles are determined by the points?   (*d*) How many of these triangles have the vertex A?   (*e*) How many of these triangles contain the side AB?

11.49   A student is to answer ten out of thirteen questions on an exam.   (*a*) How many choices has he?   (*b*) How many if he must answer the first two questions?   (*c*) How many if he must answer the first or second question but not both?   (*d*) How many if he must answer exactly three out of the first five questions?   (*e*) How many if he must answer at least three of the first five questions?

11.50   The English alphabet has twenty-six letters of which five are vowels.   Consider only the "words" consisting of five letters including three different consonants and two different vowels.   How many such words   (*a*) can be formed?   (*b*) contain the letter B?   (*c*) contain the letters B and C?   (*d*) begin with B and contain the letter C?   (*e*) begin with B and end with C?   (*f*) contain the letters A and B?   (*g*) begin with A and contain B?   (*h*) begin with B and contain A?   (*i*) begin with A and end with B?   (*j*) contain the letters A, B and C?

TREE DIAGRAMS

11.51   Find the product set $\{1, 2\} \times \{a, b, c\} \times \{3, 4\}$ by constructing the appropriate tree diagram.

11.52   Teams A and B play in baseball's world series, where the team that first wins four games wins the series. Assuming that A wins the first game and the team that wins the second game also wins the fourth game, how many ways can the series be played?

11.53   A man has time to play roulette five times. He wins or loses a dollar at each play. The man begins with two dollars and will stop playing before the five times if he loses all his money or wins three dollars (i.e. has five dollars). Find the number of ways the playing can occur.

11.54   Teams A and B play in a basketball tournament. The first team that wins two games in a row or a total of four games wins the tournament. Find the number of ways the tournament can occur.

Answers to Supplementary Problems

11.26   (*a*) 362 880, (*b*) 3 628 800, (*c*) 39 916 800

11.27   (*a*) 240, (*b*) 2184, (*c*) 1/90, (*d*) 1/1716

11.28   (*a*) 24!/20!, (*b*) 9!/12!

11.29   (*a*) $n + 1$, (*b*) $n(n - 1) = n^2 - n$, (*c*) $1/[n(n + 1)(n + 2)]$, (*d*) $(n - r)(n - r + 1)$

11.30   (*a*) 10, (*b*) 35, (*c*) 91, (*d*) 15, (*e*) 1140, (*f*) 816

11.31

		1		8		28		56		70		56		28		8		1	
	1		9		36		84		126		126		84		36		9		1	
1		10		45		120		210		252		210		120		45		10		1

11.32   (*a*) $26 \cdot 25 \cdot 10 \cdot 9 \cdot 8 = 468\,000$, (*b*) $26 \cdot 25 \cdot 9 \cdot 9 \cdot 8 = 421\,200$

11.33   (*a*) 24, (*b*) 576, (*c*) 360

11.34   360

11.35   (*a*) 120, (*b*) 48

11.36   720

11.37   (*a*) 30, (*b*) $\dfrac{9!}{2!\,2!\,2!} = 45\,360$, (*c*) $\dfrac{11!}{2!\,3!\,2!} = 1\,663\,200$, (*d*) $\dfrac{8!}{2!\,2!\,2!} = 5040$

11.38   (*a*) 216, (*b*) 320, (*c*) 328, (*d*) 136

11.39   (*a*) 120, (*b*) 24, (*c*) 24, (*d*) 12

11.40 $\dfrac{9!}{3!\,3!\,3!} = 1680$

11.41 $\dfrac{9!}{(3!)^3 3!} = 280$

11.42 $\dfrac{10!}{[(3!)^2 2!][(4!)^1 1!]} = 2100$

11.43 $\dfrac{12!}{5!\,4!\,3!} = 27\,720$

11.44 (a) (i) $3^3 = 27$ (Each element can be placed in any of the three cells.)

(ii) The numbers of elements in the three cells can be distributed as follows:

(a) [{3}, {0}, {0}], (b) [{2}, {1}, {0}], (c) [{1}, {1}, {1}]

Thus the number of partitions is $1 + 3 + 1 = 5$.

(b) (i) $3^4 = 81$.

(ii) The numbers of elements in the three cells can be distributed as follows:

(a) [{4}, {0}, {0}], (b) [{3}, {1}, {0}], (c) [{2}, {2}, {0}], (d) [{2}, {1}, {1}]

Thus the number of partitions is $1 + 4 + 3 + 6 = 14$.

11.45

```
                    1
                 1     1
              1     3     1
           1     7     6     1
        1    15    25    10    1
```
. . . .

11.46 (a) 495, (b) 369, (c) 252

11.47 (a) 462, (b) 210, (c) 378

11.48 (a) 45, (b) 28, (c) 120, (d) 36, (e) 8

11.49 (a) 286, (b) 165, (c) 110, (d) 80, (e) 276

11.50 (a) $\dbinom{21}{3}\dbinom{5}{2} \cdot 5! = 1\,596\,000$ (e) $19 \cdot \dbinom{5}{2} \cdot 3! = 1140$ (i) $4 \cdot \dbinom{20}{2} \cdot 3! = 4456$

(b) $\dbinom{20}{2}\dbinom{5}{2} \cdot 5! = 228\,000$ (f) $4 \cdot \dbinom{20}{2} \cdot 5! = 91\,200$ (j) $4 \cdot 19 \cdot 5! = 9120$

(c) $19 \cdot \dbinom{5}{2} \cdot 5! = 22\,800$ (g) $4 \cdot \dbinom{20}{2} \cdot 4! = 18\,240$

(d) $19 \cdot \dbinom{5}{2} \cdot 4! = 4560$ (h) $18\,240$ [same as (g)]

11.52 15

11.53 20

11.54 14

Chapter 12

Probability

12.1 INTRODUCTION

Probability is the study of random or nondeterministic experiments. If a die is tossed in the air, then it is certain that the die will come down, but it is not certain that, say, a 6 will appear. However, suppose we repeat this experiment of tossing a die; let s be the number of successes, i.e. the number of times a 6 appears, and let n be the number of tosses. Then it has been empirically observed that the ratio $f = s/n$, called the *relative frequency* of success, becomes stable in the long run, i.e. approaches a limit. This stability is the basis of probability theory.

In probability theory, we define a mathematical model of the above phenomenon by assigning "probabilities" (or: the limiting values of the relative frequencies) to all possible outcomes of an experiment. Furthermore, since the relative frequency of each outcome is nonnegative and the sum of the relative frequencies of all possible outcomes is unity, we require that our assigned "probabilities" also possess these two properties. The reliability of our mathematical model for a given experiment depends upon the closeness of the assigned probabilities to the actual limiting relative frequencies. This then gives rise to problems of testing and reliability which form the subject matter of statistics.

12.2 SAMPLE SPACES AND EVENTS

The set S of all possible outcomes of some given experiment is called the *sample space*. A particular outcome, i.e. an element in S, is called a *sample point* or *sample*. An *event A* is a set of outcomes or, in other words, a subset of the sample space S. In particular, the set $\{a\}$ consisting of a single sample $a \in S$ is an event, and is called an *elementary event*. Furthermore, the empty set $\varnothing$ and S itself are subsets of S and so are events; $\varnothing$ is sometimes called the *impossible event*.

Since an event is a set, we can combine events to form new events using the various set operations:

(1) $A \cup B$ is the event that occurs whenever A occurs or B occurs (or both).

(2) $A \cap B$ is the event that occurs whenever A and B both occur.

(3) A^c, the complement of A, also written $\bar{A}$, is the event that occurs whenever A does not occur.

Two events A and B are called *mutually exclusive* if they are disjoint, i.e. $A \cap B = \varnothing$. In other words, A and B are mutually exclusive if and only if they do not occur simultaneously.

EXAMPLE 12.1

(a) Experiment: Toss a die and observe the number that appears on top. Then the sample space consists of the six possible numbers:

$$S = \{1, 2, 3, 4, 5, 6\}$$

Let A be the event that an even number occurs, B that an odd number occurs and C that a prime number occurs:

$$A = \{2, 4, 6\}, \qquad B = \{1, 3, 5\}, \qquad C = \{2, 3, 5\}$$

Then:

$A \cup C = \{2, 3, 4, 5, 6\}$ is the event that an even or a prime number occurs;

$B \cap C = \{3, 5\}$ is the event that an odd prime number occurs;

$C^c = \{1, 4, 6\}$ is the event that a prime number does not occur.

Note that A and B are mutually exclusive: $A \cap B = \emptyset$; in other words, an even number and an odd number cannot occur simultaneously.

(b)   Experiment: Toss a coin 3 times and observe the sequence of heads (H) and tails (T) that appears.   The sample space S consists of eight elements:

$$S = \{HHH, HHT, HTH, HTT, THH, THT, TTH, TTT\}$$

Let A be the event that two or more heads appear consecutively, and B that all the tosses have the same outcome:

$$A = \{HHH, HHT, THH\} \qquad \text{and} \qquad B = \{HHH, TTT\}$$

Then $A \cap B = \{HHH\}$ is the elementary event in which only heads appear.   The event that 5 heads appear is the empty set $\emptyset$.

The sample spaces in Example 12.1 are finite.   There also exist infinite sample spaces; however, the theory concerning such spaces lies beyond the scope of this book.   Thus, unless otherwise stated, all our sample spaces S shall be finite.

12.3   FINITE PROBABILITY SPACES

Let S be a finite sample space: $S = \{a_1, a_2, \ldots, a_n\}$.   A finite *probability space* is obtained by assigning to each sample point $a_i \in S$ a real number p_i, called the *probability* of a_i, satisfying the following conditions:

(1)   Each p_i is nonnegative, $p_i \geq 0$.

(2)   The sum of the p_i is one, $p_1 + p_2 + \cdots + p_n = 1$.

The probability of any event A, written $P(A)$, is then defined to be the sum of the probabilities of the sample points in A.   For notational convenience we write $P(a_i)$ for $P(\{a_i\})$.

EXAMPLE 12.2   Three horses, A, B, and C, are in a race; A is twice as likely to win as B, and B is twice as likely to win as C.   We wish to find their respective probabilities of winning, which we denote as $P(A)$, $P(B)$, and $P(C)$.   We also wish to find the probability that B or C wins, i.e. $P(\{B, C\})$.

Let $P(C) = p$; since B is twice as likely to win as C, $P(B) = 2p$; and since A is twice as likely to win as B, $P(A) = 2P(B) = 2(2p) = 4p$.   Now the sum of the probabilities must be 1; hence

$$p + 2p + 4p = 1 \qquad \text{or} \qquad 7p = 1 \qquad \text{or} \qquad p = \tfrac{1}{7}$$

Accordingly,

$$P(A) = 4p = \tfrac{4}{7}, \qquad P(B) = 2p = \tfrac{2}{7}, \qquad P(C) = p = \tfrac{1}{7}$$

Also, $P(\{B, C\}) = P(B) + P(C) = \tfrac{2}{7} + \tfrac{1}{7} = \tfrac{3}{7}$.

Equiprobable Spaces

Frequently, the physical characteristics of an experiment suggest that the various outcomes be assigned equal probabilities.   Such a finite probability space S, where each sample point has the same probability, will be called an *equiprobable space*.   In particular, if S contains n points, then the probability of each point is $1/n$.   Furthermore, if an event A contains r points, then its probability is

$$r \cdot \frac{1}{n} = \frac{r}{n}$$

In other words,

$$P(A) = \frac{\text{number of elements in } A}{\text{number of elements in } S}$$

or

$$P(A) = \frac{\text{number of ways that the event } A \text{ can occur}}{\text{number of ways that the sample space } S \text{ can occur}}$$

We emphasize that the above formula for $P(A)$ holds only for equiprobable spaces.

The expression "at random" will be used only with respect to an equiprobable space; that is, the directive "Choose an element at random from a set S" shall mean that S, as the sample space of outcomes of the choice, is an equiprobable space.

EXAMPLE 12.3 Let a card be selected at random from an ordinary deck of 52 cards.   Let

$$A = \{\text{the card is a spade}\}$$

and $B = \{\text{the card is a face card, i.e. a jack, queen or king}\}$

We compute $P(A)$, $P(B)$ and $P(A \cap B)$.   Since we have an equiprobable space,

$$P(A) = \frac{\text{number of spades}}{\text{number of cards}} = \frac{13}{52} = \frac{1}{4} \qquad P(B) = \frac{\text{number of face cards}}{\text{number of cards}} = \frac{12}{52} = \frac{3}{13}$$

$$P(A \cap B) = \frac{\text{number of spade face cards}}{\text{number of cards}} = \frac{3}{52}$$

12.4 THEOREMS ON FINITE PROBABILITY SPACES

Finite probability spaces can also be defined by means of the following three axioms, which ensure that the probability of an event is the sum of the probabilities of the elementary events composing it.   That is, a finite probability space consists of a finite set S, together with a real-valued function $P(\ \ )$, defined on the class of all events (subsets) of S, that satisfies the following requirements:

[**P₁**] For every event A, $P(A) \geq 0$.

[**P₂**] $P(S) = 1$.

[**P₃**] If events A and B are mutually exclusive, then $P(A \cup B) = P(A) + P(B)$.

From these axioms one can prove

Theorem 12.1:   If $\emptyset$ is the empty set, and A and B are arbitrary events, then:
   (i) $P(\emptyset) = 0$;
   (ii) $P(A^c) = 1 - P(A)$;
   (iii) $P(A \setminus B) = P(A) - P(A \cap B)$, i.e. $P(A \cap B^c) = P(A) - P(A \cap B)$;
   (iv) $A \subset B$ implies $P(A) \leq P(B)$.
   (v) $P(A) \leq 1$

The *odds* that an event having probability p occurs is defined to be the ratio $p : q$, where q is the probability that it does not occur.   By (ii) of Theorem 12.1, the odds is therefore $p : (1 - p)$.   For example, if $P(A) = \frac{2}{3}$, then the odds that A occurs is

$$\frac{2}{3} : \frac{1}{3} = 2 : 1$$

which is read "2 to 1."

Observe that axiom [**P₃**] gives the probability of a union of events in the case that the events are mutually exclusive, i.e. disjoint.   We derive:

Theorem 12.2: For any events A and B, $P(A \cup B) = P(A) + P(B) - P(A \cap B)$.

Corollary 12.3: For any events A, B and C,

$$P(A \cup B \cup C) = P(A) + P(B) + P(C) - P(A \cap B)$$
$$- P(A \cap C) - P(B \cap C) + P(A \cap B \cap C)$$

with analogous formulas for four, five, six, ... events (compare Theorem 6.3 and Corollary 6.4).

12.5 CONDITIONAL PROBABILITY

Let E be an arbitrary event in a sample space S for which $P(E) > 0$. The probability that an event A occurs once E has occurred or, in other words, the *conditional probability of A given E*, written $P(A \mid E)$, is defined as follows:

$$P(A \mid E) = \frac{P(A \cap E)}{P(E)}$$

Figure 12-1 is a Venn diagram picturing a sample space S and events (sets) E and A. It could be said that $P(A \mid E)$ measures the probability of A relative to the reduced space E.

In an equiprobable space, the probability of an event is proportional to the number of sample points in the event, and so

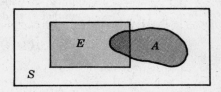

Fig. 12-1

$$P(A \mid E) = \frac{\text{number of elements in } A \cap E}{\text{number of elements in } E}$$

$$= \frac{\text{number of ways } A \text{ and } E \text{ can occur}}{\text{number of ways } E \text{ can occur}}$$

EXAMPLE 12.4 Let a pair of fair dice be tossed. If the sum is 6, find the probability that one of the dice is a 2. In other words, if

$$E = \{\text{sum is 6}\} = \{(1, 5), (2, 4), (3, 3), (4, 2), (5, 1)\}$$

and $A = \{\text{a 2 appears on at least one die}\}$

find $P(A \mid E)$.

Now E consists of five elements and two of them, $(2, 4)$ and $(4, 2)$, belong to A. Therefore, since the space is equiprobable,

$$P(A \mid E) = \frac{2}{5}$$

The definition of conditional probability may be rewritten as

$$P(A_1 \cap A_2) = P(A_1) P(A_2 \mid A_1)$$

where we have used the fact that $A_1 \cap A_2 = A_2 \cap A_1$. The above formula may be extended as follows:

Theorem 12.4 (Multiplication Theorem):

$$P(A_1 \cap A_2 \cap \cdots \cap A_n) = P(A_1) P(A_2 \mid A_1) P(A_3 \mid A_1 \cap A_2) P(A_4 \mid A_1 \cap A_2 \cap A_3)$$
$$\cdots P(A_n \mid A_1 \cap A_2 \cap \cdots \cap A_{n-1})$$

EXAMPLE 12.5 A lot contains 12 items of which 4 are defective. Three items are drawn at random from the lot, one after the other. Find the probability p that all three are nondefective.

The probability that the first item is nondefective is 8/12, since 8 of 12 items are nondefective. If the first item is nondefective, then the probability that the next item is nondefective is 7/11, since only 7 of the remaining 11 items are nondefective. If the first two items are nondefective, then the probability that the last item is non-defective is 6/10, since only 6 of the remaining 10 items are nondefective. Thus by the multiplication theorem,

$$p = \frac{8}{12} \cdot \frac{7}{11} \cdot \frac{6}{10} = \frac{14}{55}$$

12.6 INDEPENDENCE

An event B is said to be *independent* of an event A if the probability that B occurs is not influenced by whether A has or has not occurred. In other words, if the probability of B equals the conditional probability of B given A: $P(B) = P(B \mid A)$. Substituting $P(B)$ for $P(B \mid A)$ in the multiplication theorem $P(A \cap B) = P(A) P(B \mid A)$, we obtain

$$P(A \cap B) = P(A) P(B)$$

We formally use the above equation as our definition of independence.

Definition: Events A and B are independent if $P(A \cap B) = P(A) P(B)$; otherwise they are *dependent*.

EXAMPLE 12.6 Let a fair coin be tossed three times; we obtain the equiprobable space

$$S = \{HHH, HHT, HTH, HTT, THH, THT, TTH, TTT\}$$

Consider the events

$$A = \{\text{first toss is heads}\} \qquad B = \{\text{second toss is heads}\}$$

$$C = \{\text{exactly two heads are tossed in a row}\}$$

Clearly A and B are independent events; this fact is verified below. On the other hand, the relationship between A and C or B and C is not obvious. We claim that A and C are independent, but that B and C are dependent. We have

$$P(A) = P(\{HHH, HHT, HTH, HTT\}) = \frac{4}{8} = \frac{1}{2}$$

$$P(B) = P(\{HHH, HHT, THH, THT\}) = \frac{4}{8} = \frac{1}{2}$$

$$P(C) = P(\{HHT, THH\}) = \frac{2}{8} = \frac{1}{4}$$

Then

$$P(A \cap B) = P(\{HHH, HHT\}) = \frac{1}{4} \qquad P(A \cap C) = P(\{HHT\}) = \frac{1}{8}$$

$$P(B \cap C) = P(\{HHT, THH\}) = \frac{1}{4}$$

Accordingly,

$$P(A) P(B) = \frac{1}{2} \cdot \frac{1}{2} = \frac{1}{4} = P(A \cap B), \text{ and so } A \text{ and } B \text{ are independent;}$$

$$P(A) P(C) = \frac{1}{2} \cdot \frac{1}{4} = \frac{1}{8} = P(A \cap C), \text{ and so } A \text{ and } C \text{ are independent;}$$

$$P(B) P(C) = \frac{1}{2} \cdot \frac{1}{4} = \frac{1}{8} \neq P(B \cap C), \text{ and so } B \text{ and } C \text{ are dependent.}$$

Frequently, if the conditions of the experiment itself suggest that two events A and B "have no connection with each other," we *define* $P(A \cap B)$ as $P(A) P(B)$; that is, we postulate the independence of A and B. Again we emphasize that the independence or dependence of two given events is not determined by the events themselves, but by the probabilities which we assign to the events and to their intersection. See Problem 12.18.

12.7 REPEATED TRIALS

We have already encountered probability spaces associated with an experiment repeated a finite number of times, as in Example 12.6. This concept of repetition is formalized as follows:

Definition: Let S^* be a finite probability space. By n *independent* or *repeated trials*, we mean the probability space S consisting of all ordered n-tuples of elements of S^*, with the probability of an n-tuple defined to be the product of the probabilities of its components:

$$P((s_1, s_2, \ldots, s_n)) = P(s_1) P(s_2) \cdots P(s_n)$$

EXAMPLE 12.7 Suppose that whenever three horses a, b and c race together, their respective probabilities of winning are $\frac{1}{2}$, $\frac{1}{3}$ and $\frac{1}{6}$. In other words, $S^* = \{a, b, c\}$ with $P(a) = \frac{1}{2}$, $P(b) = \frac{1}{3}$ and $P(c) = \frac{1}{6}$. If the horses race twice, then the sample space of the two repeated trials is

$$S = \{aa, ab, ac, ba, bb, bc, ca, cb, cc\}$$

For notational convenience, we have written ac for the ordered pair (a, c). The probabilities of the sample points of S are:

$$P(aa) = P(a) P(a) = \frac{1}{2} \cdot \frac{1}{2} = \frac{1}{4} \qquad P(ba) = \frac{1}{6} \qquad P(ca) = \frac{1}{12}$$

$$P(ab) = P(a) P(b) = \frac{1}{2} \cdot \frac{1}{3} = \frac{1}{6} \qquad P(bb) = \frac{1}{9} \qquad P(cb) = \frac{1}{18}$$

$$P(ac) = P(a) P(c) = \frac{1}{2} \cdot \frac{1}{6} = \frac{1}{12} \qquad P(bc) = \frac{1}{18} \qquad P(cc) = \frac{1}{36}$$

Thus the probability of c winning the first race and a winning the second race is $P(ca) = 1/12$.

One frequently studies repeated trials with only two outcomes; we call one of the outcomes *success* and the other outcome *failure*. Let p be the probability of success, and so $q = 1 - p$ is the probability of failure. Frequently we are interested in the number of successes without regard to the order in which they occur. The following theorem, proved in Problem 12.23, applies.

Theorem 12.5: The probability of exactly k successes in n repeated trials is denoted and given by

$$b(k; n, p) = \binom{n}{k} p^k q^{n-k}$$

(See Section 11.3 for the definition of the binomial coefficient.)

EXAMPLE 12.8 A fair coin is tossed 6 times or, equivalently, six fair coins are tossed; call heads a success. Then $n = 6$ and $p = q = \frac{1}{2}$.

(a) The probability that exactly two heads occur (i.e. $k = 2$) is

$$b(2; 6, \tfrac{1}{2}) = \binom{6}{2}\left(\frac{1}{2}\right)^2\left(\frac{1}{2}\right)^4 = \frac{15}{64}$$

(b) The probability of getting at least four heads (i.e. $k = 4$, 5 or 6) is

$$b = (4; 6, \tfrac{1}{2}) + b(5; 6, \tfrac{1}{2}) + b(6; 6, \tfrac{1}{2}) = \binom{6}{4}\left(\frac{1}{2}\right)^4\left(\frac{1}{2}\right)^2 + \binom{6}{5}\left(\frac{1}{2}\right)^5\left(\frac{1}{2}\right)^1 + \binom{6}{6}\left(\frac{1}{2}\right)^6$$

$$= \frac{15}{64} + \frac{6}{64} + \frac{1}{64} = \frac{11}{32}$$

(c) The probability of no heads (i.e. $k = 0$) is

$$b(0; 6, \tfrac{1}{2}) = \left(\frac{1}{2}\right)^6 = \frac{1}{64}$$

and so the probability of at least one head is

$$1 - \frac{1}{64} = \frac{63}{64}$$

Suppose we regard n and p as constant. Then $b(k; n, p)$, a function of k, is called a *binomial distribution*, since for $k = 0, 1, \ldots, n$ it corresponds to the successive terms of the binomial expansion

$$(q + p)^n = q^n + \binom{n}{1}q^{n-1}p + \binom{n}{2}q^{n-2}p^2 + \cdots + p^n$$

$$= b(0; n, p) + b(1; n, p) + b(2; n, p) + \cdots + b(n; n, p)$$

The use of the word "distribution" will be explained in Chapter 13. The binomial distribution is also called the *Bernoulli distribution*, and independent trials with two outcomes are called *Bernoulli trials*.

Solved Problems

SAMPLE SPACES, FINITE PROBABILITY SPACES

12.1 Let A and B be events.   Find an expression and exhibit the Venn diagram for the event that:
(a) A but not B occurs; (b) either A or B, but not both, occurs.

(a) Since A but not B occurs, shade the area of A outside of B, as in Fig. 12-2(a).   Note that B^c, the complement of B, occurs, since B does not occur; hence A and B^c occur.   In other words, the event is $A \cap B^c$.

(b) Since A or B, but not both, occurs, shade the area of A and B except where they intersect, as in Fig. 12-2(b).   The event is equivalent to the occurrence of A but not B or B but not A.   Thus the given event is $(A \cap B^c) \cup (B \cap A^c)$.

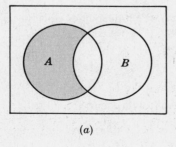

(a)

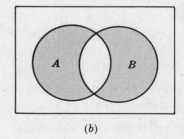

(b)

Fig. 12-2

12.2 Let a coin and a die be tossed; let the sample space S consist of the twelve elements:

$$S = \{H1, H2, H3, H4, H5, H6, T1, T2, T3, T4, T5, T6\}$$

(a) Express explicitly the following events:

$$A = \{\text{heads and an even number appears}\}$$

$$B = \{\text{a prime number appears}\} \quad C = \{\text{tails and an odd number appears}\}$$

(b) Express explicitly the event that: (i) A or B occurs, (ii) B and C occur, (iii) only B occurs.

(c) Which pair of the events A, B, and C are mutually exclusive?

(a) The elements of A are those elements of S consisting of an H and an even number:

$$A = \{H2, H4, H6\}$$

The elements of B are those points in S whose second component is a prime number:

$$B = \{H2, H3, H5, T2, T3, T5\}$$

The elements of C are those points in S consisting of a T and an odd number: $C = \{T1, T3, T5\}$.

(b) (i) $A \cup B = \{H2, H4, H6, H3, H5, T2, T3, T5\}$
  (ii) $B \cap C = \{T3, T5\}$
  (iii) $B \cap A^c \cap C^c = \{H3, H5, T2\}$

(c) A and C are mutually exclusive since $A \cap C = \emptyset$.

12.3 A sample space S consists of four elements: $S = \{a_1, a_2, a_3, a_4\}$. Under which of the following functions does S become a probability space?

(a) $P(a_1) = \frac{1}{2}$ $P(a_2) = \frac{1}{3}$ $P(a_3) = \frac{1}{4}$ $P(a_4) = \frac{1}{5}$

(b) $P(a_1) = \frac{1}{2}$ $P(a_2) = \frac{1}{4}$ $P(a_3) = -\frac{1}{4}$ $P(a_4) = \frac{1}{2}$

(c) $P(a_1) = \frac{1}{2}$ $P(a_2) = \frac{1}{4}$ $P(a_3) = \frac{1}{8}$ $P(a_4) = \frac{1}{8}$

(d) $P(a_1) = \frac{1}{2}$ $P(a_2) = \frac{1}{4}$ $P(a_3) = \frac{1}{4}$ $P(a_4) = 0$

(a) Since the sum of the values on the sample points is greater than one, the function does not define S as a probability space.
(b) Since $P(a_3)$ is negative, the function does not define S as a probability space.
(c) Since each value is nonnegative and the sum of the values is one, the function does define S as a probability space.
(d) The values are nonnegative and add up to one; hence the function does define S as a probability space.

12.4 A coin is weighted so that heads is twice as likely to appear as tails. Find $P(T)$ and $P(H)$.

Let $P(T) = p$; then $P(H) = 2p$. Now set the sum of the probabilities equal to one:

$$p + 2p = 1 \quad \text{or} \quad p = \frac{1}{3}$$

Thus $P(T) = 1/3$ and $P(H) = 2/3$.

12.5 Find the probability p of an event if the odds that it will occur are "3 to 2".

The odds that an event with probability p will occur is the ratio $p : (1-p)$. Hence

$$\frac{p}{1-p} = \frac{3}{2} \quad \text{or} \quad 2p = 3 - 3p \quad \text{or} \quad 5p = 3 \quad \text{or} \quad p = \frac{3}{5}$$

12.6   Determine the probability p of each event:
  (a)   an even number appears in the toss of a fair die;
  (b)   a king appears in drawing a single card from an ordinary deck of 52 cards;
  (c)   at least one tail appears in the toss of three fair coins;
  (d)   a white marble appears in drawing a single marble from an urn containing 4 white, 3 red and 5 blue marbles.

  (a)   The event can occur in three ways (a 2, 4 or 6) out of 6 equally likely cases; hence $p = 3/6 = 1/2$.
  (b)   There are 4 kings among the 52 cards; hence $p = 4/52 = 1/13$.
  (c)   If we consider the coins distinguished, then there are 8 equally likely cases: HHH, HHT, HTH, HTT, THH, THT, TTH, TTT.   Only the first case is not favorable to the given event; hence $p = 7/8$.
  (d)   There are $4 + 3 + 5 = 12$ marbles, of which 4 are white; hence $p = 4/12 = 1/3$.

12.7   Two cards are drawn at random from an ordinary deck of 52 cards.   Find the probability p that (a) both are spades, (b) one is a spade and one is a heart.

There are

$$\binom{52}{2} = 1326 \text{ ways}$$

to draw 2 cards from 52 cards.

(a). There are

$$\binom{13}{2} = 78 \text{ ways}$$

to draw 2 spades from 13 spades; hence

$$p = \frac{\text{number of ways 2 spades can be drawn}}{\text{number of ways 2 cards can be drawn}} = \frac{78}{1326} = \frac{1}{17}$$

(b)   Since there are 13 spades and 13 hearts, there are $13 \cdot 13 = 169$ ways to draw a spade and a heart; hence

$$p = \frac{169}{1326} = \frac{13}{102}$$

12.8   In a class of 10 men and 20 women, half the men and half the women have brown eyes.   Find the probability p that a person chosen at random is a man or has brown eyes.

Let $A = \{\text{person is a man}\}$ and $B = \{\text{person has brown eyes}\}$; we seek $P(A \cup B)$.   Now

$$P(A) = \frac{10}{30} = \frac{1}{3} \qquad P(B) = \frac{1}{2} \qquad P(A \cap B) = \frac{5}{30} = \frac{1}{6}$$

Thus, by Theorem 12.2,

$$p = P(A \cup B) = P(A) + P(B) - P(A \cap B) = \frac{1}{3} + \frac{1}{2} - \frac{1}{6} = \frac{2}{3}$$

CONDITIONAL PROBABILITY

12.9   Three fair coins, a penny, a nickel, and a dime, are tossed.   Find the probability p that they are all heads if (a) the penny is heads, (b) at least one of the coins is heads.

The sample space has eight elements: $S = \{\text{HHH, HHT, HTH, HTT, THH, THT, TTH, TTT}\}$.
  (a)   If the penny is heads, the reduced sample space is $A = \{\text{HHH, HHT, HTH, HTT}\}$.   Since the coins are all heads in 1 of 4 cases, $p = 1/4$.
  (b)   If one or more of the coins is heads, the reduced sample space is $B = \{\text{HHH, HHT, HTH, HTT, THH, THT, TTH}\}$.   Since the coins are all heads in 1 of 7 cases, $p = 1/7$.

12.10 A pair of fair dice is thrown. If the two numbers appearing are different, find the probability p that (a) the sum is six, (b) an ace appears, (c) the sum is 4 or less.

Of the 36 ways the pair of dice can be thrown, 6 will produce the same numbers: $(1, 1), (2, 2), \ldots,$ $(6, 6)$. Thus the reduced sample space will consist of $36 - 6 = 30$ elements.

(a) The sum six can appear in 4 ways: $(1, 5), (2, 4), (4, 2), (5, 1)$. (We cannot include $(3, 3)$ since the numbers are the same.) Hence $p = 4/30 = 2/15$.

(b) An ace can appear in 10 ways: $(1, 2), (1, 3), \ldots, (1, 6)$ and $(2, 1), (3, 1), \ldots, (6, 1)$. Therefore $p = 10/30 = 1/3$.

(c) The sum of 4 or less can occur in 4 ways: $(3, 1), (1, 3), (2, 1), (1, 2)$. Thus $p = 4/30 = 2/15$.

12.11 A class has 12 boys and 4 girls. If three students are selected at random from the class, what is the probability p that they are all boys?

The probability that the first student selected is a boy is 12/16 since there are 12 boys out of 16 students. If the first student is a boy, then the probability that the second is a boy is 11/15 since there are 11 boys left out of 15 students. Finally, if the first two students selected were boys, then the probability that the third student is a boy is 10/14 since there are 10 boys left out of 14 students. Thus, by the multiplication theorem, the probability that all three are boys is

$$p = \frac{12}{16} \cdot \frac{11}{15} \cdot \frac{10}{14} = \frac{11}{28}$$

Another Method

There are $C(16, 3) = 560$ ways to select 3 students out of the 16 students, and $C(12, 3) = 220$ ways to select 3 boys out of 12 boys; hence

$$p = \frac{220}{560} = \frac{11}{28}$$

Another Method

If the students are selected one after the other, then there are $16 \cdot 15 \cdot 14$ ways to select three students, and $12 \cdot 11 \cdot 10$ ways to select three boys; hence

$$p = \frac{12 \cdot 11 \cdot 10}{16 \cdot 15 \cdot 14} = \frac{11}{28}$$

12.12 Five cards are dealt one after the other from an ordinary deck of 52 cards. What is the probability p that they are all spades?

The probability that the first card is a spade is 13/52, the second is a spade is 12/51, the third is a spade is 11/50, the fourth is a spade is 10/49, and the last is a spade is 9/48. (We assumed in each case that the previous cards were spades.) Thus

$$p = \frac{13}{52} \cdot \frac{12}{51} \cdot \frac{11}{50} \cdot \frac{10}{49} \cdot \frac{9}{48} = \frac{33}{66\,640}$$

12.13 In a certain college, 25% of the students failed mathematics, 15% of the students failed chemistry, and 10% of the students failed both mathematics and chemistry. A student is selected at random.

(a) If he failed chemistry, what is the probability that he failed mathematics?

(b) If he failed mathematics, what is the probability that he failed chemistry?

(c) What is the probability that he failed mathematics or chemistry?

Let $M = \{$students who failed mathematics$\}$ and $C = \{$students who failed chemistry$\}$; then

$$P(M) = 0.25 \qquad P(C) = 0.15 \qquad P(M \cap C) = 0.10$$

(a)   The probability that a student failed mathematics, given that he failed chemistry, is

$$P(M \mid C) = \frac{P(M \cap C)}{P(C)} = \frac{0.10}{0.15} = \frac{2}{3}$$

(b)   The probability that a student failed chemistry, given that he failed mathematics is

$$P(C \mid M) = \frac{P(C \cap M)}{P(M)} = \frac{0.10}{0.25} = \frac{2}{5}$$

(c) $$P(M \cup C) = P(M) + P(C) - P(M \cap C) = 0.25 + 0.15 - 0.10 = 0.30 = \frac{3}{10}$$

12.14 Find $P(B \mid A)$ if (a) A is a subset of B, (b) A and B are mutually exclusive.   Assume that $P(A) > 0$.

(a)   If A is a subset of B, then whenever A occurs B must occur; hence $P(B \mid A) = 1$.   Alternately, $A \cap B = A$; hence

$$P(B \mid A) = \frac{P(A \cap B)}{P(A)} = \frac{P(A)}{P(A)} = 1$$

(b)   If A and B are mutually exclusive, i.e. disjoint, then whenever A occurs B does not occur; hence $P(B \mid A) = 0$.   Alternately, $A \cap B = \emptyset$; hence

$$P(B \mid A) = \frac{P(A \cap B)}{P(A)} = \frac{P(\emptyset)}{P(A)} = \frac{0}{P(A)} = 0$$

INDEPENDENCE

12.15 The probability that a man will live 10 more years is 1/4, and the probability that his wife will live 10 more years is 1/3.   Find the probability that (a) both will be alive in 10 years, (b) at least one will be alive in 10 years, (c) neither will be alive in 10 years, (d) only the wife will be alive in 10 years.

   We shall assume that $A \equiv$ event that the man is alive in 10 years, and $B \equiv$ event that his wife is alive in 10 years, are independent events; this may or may not correspond to actual vital statistics.

(a) $$P(A \cap B) = P(A) P(B) = \frac{1}{4} \cdot \frac{1}{3} = \frac{1}{12}$$

(b) $$P(A \cup B) = P(A) + P(B) - P(A \cap B) = \frac{1}{4} + \frac{1}{3} - \frac{1}{12} = \frac{1}{2}$$

(c)   We seek $P(A^c \cap B^c)$.   Now

$$P(A^c) = 1 - P(A) = 1 - \frac{1}{4} = \frac{3}{4} \qquad \text{and} \qquad P(B^c) = 1 - P(B) = 1 - \frac{1}{3} = \frac{2}{3}$$

Furthermore, A^c and B^c are independent (see Problem 12.17).   Thus,

$$P(A^c \cap B^c) = P(A^c) P(B^c) = \frac{3}{4} \cdot \frac{2}{3} = \frac{1}{2}$$

Alternately, since $(A \cup B)^c = A^c \cap B^c$,

$$P(A^c \cap B^c) = P((A \cup B)^c) = 1 - P(A \cup B) = 1 - \frac{1}{2} = \frac{1}{2}$$

(d)   We seek $P(A^c \cap B)$.   Since

$$P(A^c) = \frac{3}{4} \qquad \text{and} \qquad P(B) = \frac{1}{3}$$

and A^c and B are independent,

$$P(A^c \cap B) = P(A^c) P(B) = \frac{3}{4} \cdot \frac{1}{3} = \frac{1}{4}$$

12.16 Box A contains 8 items of which 3 are defective, and box B contains 5 items of which 2 are defective. An item is drawn at random from each box. (*a*) What is the probability p that both items are nondefective? (*b*) What is the probability p that one item is defective and one not? (*c*) If one item is defective and one is not, what is the probability p that the defective item came from box A?

(*a*)   The probability of choosing a nondefective item from A is 5/8 and from B is 3/5. Since the events are independent,

$$p = \frac{5}{8} \cdot \frac{3}{5} = \frac{3}{8}$$

(*b*)   **Method 1.**   The probability of choosing two defective items is

$$\frac{3}{8} \cdot \frac{2}{5} = \frac{3}{20}$$

From (*a*), the probability of choosing two nondefective items is 3/8. Hence

$$p = 1 - \frac{3}{8} - \frac{3}{20} = \frac{19}{40}$$

Method 2.   The probability p_1 of choosing a defective item from A and a nondefective item from B is

$$\frac{3}{8} \cdot \frac{3}{5} = \frac{9}{40}$$

The probability p_2 of choosing a nondefective item from A and a defective item from B is

$$\frac{5}{8} \cdot \frac{2}{5} = \frac{1}{4}$$

Hence

$$p = p_1 + p_2 = \frac{9}{40} + \frac{1}{4} = \frac{19}{40}$$

(*c*)   Consider the events $X = \{$defective item from $A\}$ and $Y = \{$one item is defective and one nondefective$\}$. We seek $P(X \mid Y)$. By (*b*), $P(X \cap Y) = p_1 = 9/40$ and $P(Y) = 19/40$. Hence

$$p = P(X \mid Y) = \frac{P(X \cap Y)}{P(Y)} = \frac{9/40}{19/40} = \frac{9}{19}$$

12.17 Prove: If A and B are independent events, then A^c and B^c are independent events.

Let $P(A) = x$ and $P(B) = y$. Then $P(A^c) = 1 - x$ and $P(B^c) = 1 - y$. Since A and B are independent, $P(A \cap B) = P(A) P(B) = xy$. Furthermore,

$$P(A \cup B) = P(A) + P(B) - P(A \cap B) = x + y - xy$$

By DeMorgan's law, $(A \cup B)^c = A^c \cap B^c$; hence

$$P(A^c \cap B^c) = P((A \cup B)^c) = 1 - P(A \cup B) = 1 - x - y + xy$$

On the other hand,

$$P(A^c) P(B^c) = (1 - x)(1 - y) = 1 - x - y + xy$$

Thus $P(A^c \cap B^c) = P(A^c) P(B^c)$, and so A^c and B^c are independent.

In similar fashion, we can show that A and B^c, as well as A^c and B, are independent.

REPEATED TRIALS

12.18 A family contains three children. Let $A \equiv$ event that the family has children of both sexes, and let $B \equiv$ event that the family has at most one boy. Are A and B independent events?

We consider the three births as three independent trials, with probability p of a girl and probability $q = 1 - p$ of a boy on each trial. The points of the sample space,

$$S = \{bbb, bbg, bgb, bgg, gbb, gbg, ggb, ggg\}$$

thus have probabilities $P(bbb) = q^3$, $P(bbg) = pq^2$, etc. We have

$$A = \{bbg, bgb, bgg, gbb, gbg, ggb\}$$
$$B = \{bgg, gbg, ggb, ggg\}$$
$$A \cap B = \{bgg, gbg, ggb\}$$

and so

$$P(A) = pq^2 + pq^2 + p^2q + pq^2 + p^2q + p^2q = 3pq(p+q) = 3pq$$
$$P(B) = p^2q + p^2q + p^2q + p^3 = (3q+p)p^2 = (2q+1)p^2$$
$$P(A \cap B) = p^2q + p^2q + p^2q = 3p^2q$$

Thus, A and B are independent events if and only if

$$3p^2q = 3pq \cdot (2q+1)p^2$$

which reduces to

$$1 = 2pq + p \qquad \text{or} \qquad q = 2pq \qquad \text{or} \qquad p = \frac{1}{2}$$

We conclude that A and B are independent if female and male births are equally likely; otherwise the events are dependent.

12.19 Calculate (a) $b(2; 5, \frac{1}{3})$, (b) $b(3; 6, \frac{1}{2})$, (c) $b(3; 4, \frac{1}{4})$.

(a)
$$b(2; 5, \tfrac{1}{3}) = \binom{5}{2}\left(\tfrac{1}{3}\right)^2\left(\tfrac{2}{3}\right)^3 = \frac{5 \cdot 4}{1 \cdot 2}\left(\tfrac{1}{3}\right)^2\left(\tfrac{2}{3}\right)^3 = \frac{80}{243}$$

(b)
$$b(3; 6, \tfrac{1}{2}) = \binom{6}{3}\left(\tfrac{1}{2}\right)^3\left(\tfrac{1}{2}\right)^3 = \frac{6 \cdot 5 \cdot 4}{1 \cdot 2 \cdot 3}\left(\tfrac{1}{2}\right)^6 = \frac{5}{16}$$

(c)
$$b(3; 4, \tfrac{1}{4}) = \binom{4}{3}\left(\tfrac{1}{4}\right)^3\left(\tfrac{3}{4}\right)^1 = \binom{4}{1}\left(\tfrac{1}{4}\right)^3\left(\tfrac{3}{4}\right) = 4\left(\tfrac{1}{4}\right)^3\left(\tfrac{3}{4}\right) = \frac{3}{64}$$

12.20 A fair coin is tossed three times. Find the probability p that there will appear (a) three heads, (b) two heads, (c) one head, (d) no heads.

Method 1.

We obtain the following equiprobable space of eight elements:

$$S = \{HHH, HHT, HTH, HTT, THH, THT, TTH, TTT\}$$

(a) Three heads (HHH) occurs only once among the eight sample points; hence $p = 1/8$.
(b) Two heads occurs 3 times (HHT, HTH, and THH); hence $p = 3/8$.
(c) One head occurs 3 times (HTT, THT and TTH); hence $p = 3/8$.
(d) No heads, i.e. three tails (TTT), occurs only once; hence $p = 1/8$.

Method 2.

Use Theorem 12.5, with $n = 3$ and $p = q = 1/2$.

290   PROBABILITY   [CHAP. 12

(a)
$$p = b(3; 3, \tfrac{1}{2}) = \binom{3}{3}\left(\tfrac{1}{2}\right)^3 = 1 \cdot \tfrac{1}{8} = \tfrac{1}{8}$$

(b)
$$p = b(2; 3, \tfrac{1}{2}) = \binom{3}{2}\left(\tfrac{1}{2}\right)^3 = 3 \cdot \tfrac{1}{8} = \tfrac{3}{8}$$

(c)
$$p = b(1; 3, \tfrac{1}{2}) = \binom{3}{1}\left(\tfrac{1}{2}\right)^3 = 3 \cdot \tfrac{1}{8} = \tfrac{3}{8}$$

(d)
$$p = b(0; 3, \tfrac{1}{2}) = \binom{3}{0}\left(\tfrac{1}{2}\right)^3 = 1 \cdot \tfrac{1}{8} = \tfrac{1}{8}$$

12.21 Suppose that 20% of the items produced by a factory are defective. If 4 items are chosen at random, what is the probability p that (a) 2 are defective, (b) 3 are defective, (c) at least one is defective?

Use Theorem 12.5, with $n = 4$, $p = 0.2$ and $q = 1 - p = 0.8$.

(a)
$$p = b(2; 4, 0.2) = \binom{4}{2}(0.2)^2(0.8)^2 = 0.1536$$

(b)
$$p = b(3; 4, 0.2) = \binom{4}{3}(0.2)^3(0.8)^1 = 0.0256$$

(c)
$$p = 1 - b(0; 4, 0.2) = 1 - (0.8)^4 = 0.5904$$

12.22 Team A has probability 2/3 of winning whenever it plays. If A plays 4 games, find the probability that A wins more than half of the games.

Here $n = 4$, $p = 2/3$, and $q = 1 - p = 1/3$. A wins more than half the games if it wins 3 or 4 games. Hence the required probability is

$$b(3; 4, \tfrac{2}{3}) + b(4; 4, \tfrac{2}{3}) = \binom{4}{3}\left(\tfrac{2}{3}\right)^3\left(\tfrac{1}{3}\right)^1 + \binom{4}{4}\left(\tfrac{2}{3}\right)^4 = \frac{32}{81} + \frac{16}{81} = \frac{16}{27}$$

12.23 Prove Theorem 12.5.

The sample space of the n repeated trials consists of all ordered n-tuples whose components are either s (success) or f (failure). The event A of k successes consists of all ordered n-tuples of which k components are s and the other $n - k$ components are f. The number of n-tuples in the event A is equal to the number of ways that k letters s can be distributed among the n components of an n-tuple; hence A consists of $C(n, k)$ sample points. Since the probability of each point in A is $p^k q^{n-k}$, we have

$$P(A) = \binom{n}{k}p^k q^{n-k}$$

Supplementary Problems

PROBABILITY SPACES

12.24 Let A and B be events. Find an expression and exhibit the Venn diagram for the event that (a) A occurs or B does not occur, (b) neither A nor B occurs.

12.25 Let a penny, a dime and a die be tossed.
  (a) Describe a suitable sample space S.
  (b) Express explicitly the following events: $A = \{$two heads and an even number appear$\}$, $B = \{$a 2 appears$\}$, $C = \{$exactly one head and an odd number appear$\}$.
  (c) Express explicitly the event that (1) A and B occur, (2) only B occurs, (3) B and C occur.

12.26 Which functions, defined on $S = \{a_1, a_2, a_3\}$, make S a probability space?

(a) $P(a_1) = \frac{1}{4}$, $P(a_2) = \frac{1}{3}$, $P(a_3) = \frac{1}{2}$ (c) $P(a_1) = \frac{1}{6}$, $P(a_2) = \frac{1}{3}$, $P(a_3) = \frac{1}{2}$

(b) $P(a_1) = \frac{2}{3}$, $P(a_2) = -\frac{1}{3}$, $P(a_3) = \frac{2}{3}$ (d) $P(a_1) = 0$, $P(a_2) = \frac{1}{3}$, $P(a_3) = \frac{2}{3}$

12.27 Let $P(\ )$ be a probability function on $S = \{a_1, a_2, a_3\}$. Find $P(a_1)$ if (a) $P(a_2) = 1/3$ and $P(a_3) = 1/4$, (b) $P(a_1) = 2P(a_2)$ and $P(a_3) = 1/4$.

12.28 A coin is weighted so that heads is three times as likely to appear as tails. Find $P(H)$ and $P(T)$.

12.29 Three students A, B and C are in a swimming race. A and B have the same probability of winning and each is twice as likely to win as C. Find the probability that (a) B wins, (b) C wins, (c) B or C wins.

12.30 Let A and B be events with $P(A \cup B) = 7/8$, $P(A \cap B) = 1/4$, and $P(A^c) = 5/8$. Find $P(A)$, $P(B)$ and $P(A \cap B^c)$.

12.31 A class contains 5 freshmen, 4 sophomores, 8 juniors and 3 seniors. A student is chosen at random to represent the class. Find the probability that the student is (a) a sophomore, (b) a senior, (c) a junior or a senior.

12.32 One card is selected at random from 50 cards numbered 1 to 50. Find the probability that the number on the card is (a) greater than 10, (b) divisible by 5, (c) ends in the digit 2.

12.33 Three bolts and three nuts are put in a box. If two parts are chosen at random, find the probability that one is a bolt and one a nut.

12.34 Ten students, A, B, ..., are in a class. If a committee of 3 is chosen at random from the class, find the probability that (a) A belongs to the committee, (b) B belongs to the committee, (c) A and B belong to the committee, (d) A or B belongs to the committee.

12.35 A pair of fair dice is tossed. Find the probability that the maximum of the two numbers is greater than 4.

12.36 A die is tossed 50 times. The following table gives the six numbers and their frequency of occurrence:

Number	1	2	3	4	5	6
Frequency	7	9	8	7	9	10

Find the relative frequency of the event (a) a 4 appears, (b) an odd number appears, (c) a number greater than 4 appears.

CONDITIONAL PROBABILITY, INDEPENDENCE

12.37 A fair die is tossed. Consider the events

$$A = \{2, 4, 6\} \qquad B = \{1, 2\} \qquad C = \{1, 2, 3, 4\}$$

(a) Find $P(A$ and $B)$, $P(A$ or $C)$. (b) Find $P(A \mid B)$, $P(B \mid A)$. (c) Find $P(A \mid C)$, $P(C \mid A)$. (d) Find $P(B \mid C)$, $P(C \mid B)$. (e) Are A and B independent? A and C? B and C?

12.38 A pair of dice is tossed. If the numbers appearing are different, find the probability that the sum is even.

12.39   Two different digits are selected at random from the digits 1 through 9.   If

$$E = \{\text{sum is odd}\} \qquad F = \{\text{the digit 2 is selected}\}$$

find (a) $P(E \mid F)$, (b) $P(F \mid E)$.   (c) Are E and F independent?

12.40   Let A and B be events with $P(A) = 1/3$, $P(B) = 1/4$, and $P(A \cup B) = 1/2$.   (a) Find $P(A \mid B)$ and $P(B \mid A)$.   (b) Are A and B independent?

12.41   In a certain college, 25% of the male students and 10% of the female students are studying mathematics.   Women constitute 60% of the student body.   A student is chosen at random.   (a) Find the probability that the student is studying mathematics.   (b) If the student is studying mathematics, find the probability that the student is female.

12.42   A box contains 5 transistors of which 2 are defective.   The transistors are tested one after the other until the two defectives are identified.   (a) Find the probability that the process stopped on (i) the second test, (ii) the third test.   (*Hint*: If the first three items prove nondefective, ...)   (b) If the process stopped on the third test, what is the probability that the first transistor tested was defective?

12.43   Let A and B be events with $P(A) = 1/4$, $P(A \cup B) = 1/3$, and $P(B) = p$.   Find p if (a) A and B are mutually exclusive, (b) A and B are independent, (c) A is a subset of B.

12.44   Let three fair coins be tossed.   Let $A = \{\text{all heads or all tails}\}$, $B = \{\text{at least two heads}\}$ and $C = \{\text{at most two heads}\}$.   Of the pairs (A, B), (A, C) and (B, C), which are independent and which are dependent?

12.45   Let A and B be independent events with $P(A) = 0.3$ and $P(B) = 0.4$.   (a) Find $P(A \cap B)$ and $P(A \cup B)$.   (b) Find $P(A \mid B)$ and $P(B \mid A)$.

INDEPENDENT TRIALS

12.46   Show that if the family of Problem 12.18 contains two children, A and B must be dependent.

12.47   Find (a) $b(1; 5, \frac{1}{3})$, (b) $b(2; 7, \frac{1}{2})$, (c) $b(2; 4, \frac{1}{4})$.

12.48   A card is drawn and replaced three times from an ordinary deck of 52 cards.   Find the probability that (a) two hearts are drawn, (b) three hearts are drawn, (c) at least one heart is drawn.

12.49   A baseball player's batting average is .300.   He comes to bat 4 times.   Find the probability that he will get (a) two hits, (b) at least one hit.

12.50   A team wins (W) with probability 0.5, loses (L) with probability 0.3 and ties (T) with probability 0.2.   The team plays twice.   (a) Determine the sample space S and the probabilities of the elementary events.   (b) Find the probability that the team wins at least once.

12.51   The probability that A hits a target is 1/3.   (a) If he fires 3 times, find the probability that he hits the target at least once.   (b) How many times must he fire so that the probability of hitting the target at least once is greater than 90%?

Answers to Supplementary Problems

12.24   (a) $A \cup B^c$, (b) $(A \cup B)^c$ or $A^c \cap B^c$

12.25   (a) S = {HH1, HH2, HH3, HH4, HH5, HH6, HT1, HT2, HT3, HT4, HT5, HT6,
      TH1, TH2, TH3, TH4, TH5, TH6, TT1, TT2, TT3, TT4, TT5, TT6}

   (b) A = {HH2, HH4, HH6}, B = {HH2, HT2, TH2, TT2}, C = {HT1, TH1, HT3, TH3, HT5, TH5}

   (c)   (1) $A \cap B$ = {HH2}
      (2) $B \cap (A \cup C)^c$ = {HT2, TH2, TT2}
      (3) $B \cap C$ = $\emptyset$

12.26   (c) and (d)

12.27   (a) $\dfrac{5}{12}$, (b) $\dfrac{1}{2}$

12.28 $P(\text{H}) = \dfrac{3}{4}$, $P(\text{T}) = \dfrac{1}{4}$

12.29   (a) 2/5, (b) 1/5, (c) 3/5

12.30 $P(A) = \dfrac{3}{8}$, $P(B) = \dfrac{3}{4}$, $P(A \cap B^c) = \dfrac{1}{8}$

12.31   (a) $\dfrac{1}{5}$, (b) $\dfrac{3}{20}$, (c) $\dfrac{11}{20}$

12.32   (a) $\dfrac{4}{5}$, (b) $\dfrac{1}{5}$, (c) $\dfrac{1}{10}$

12.33   3/5

12.34   (a) $\dfrac{3}{10}$, (b) $\dfrac{3}{10}$, (c) $\dfrac{1}{15}$, (d) $\dfrac{8}{15}$

12.35   5/9

12.36   (a) $\dfrac{7}{50}$, (b) $\dfrac{24}{50}$, (c) $\dfrac{19}{50}$

12.37   (a)   1/6, 5/6
   (b)   1/2, 1/3
   (c)   1/2, 2/3
   (d)   1/2, 1
   (e)   yes, yes, no

12.38 $\dfrac{12}{30} = \dfrac{2}{5}$

12.39   (a) $\dfrac{5}{8}$, (b) $\dfrac{1}{4}$, (c) no

12.40   (a) $\dfrac{1}{3}, \dfrac{1}{4}$; (b) yes

12.41   (a) $\dfrac{4}{25}$, (b) $\dfrac{3}{8}$

12.42   (a) (i) $\dfrac{1}{10}$, (ii) $\dfrac{3}{10}$; (b) $\dfrac{1}{3}$

12.43   (a) $\dfrac{1}{12}$, (b) $\dfrac{1}{9}$, (c) $\dfrac{1}{3}$

12.44   Only A and B are independent.

12.45   (a) 0.12, 0.58; (b) 0.3, 0.4

12.47   (a) $\dfrac{80}{243}$, (b) $\dfrac{21}{128}$, (c) $\dfrac{27}{128}$

12.48   (a) $\dfrac{9}{64}$, (b) $\dfrac{1}{64}$, (c) $1 - \dfrac{27}{64} = \dfrac{37}{64}$

12.49   (a) 0.2646, (b) $1 - (0.7)^4 = 0.7599$

12.50   (a) {WW, WL, WT, LW, LL, LT, TW, TL, TT}
   (b) 0.75

12.51   (a) $1 - \dfrac{8}{27} = \dfrac{19}{27}$, (b) 6

Chapter 13

Statistics; Random Variables

13.1 INTRODUCTION

Statistics means, on the one hand, lists of numerical values. For example, the salaries of the employees of a company, or the number of children per family in a city. Statistics as a science, on the other hand, is that branch of mathematics which organizes, analyzes, and interprets such raw data. Statistical methods are applicable to any area of human endeavor where numerical data are collected for some type of decision-making process.

This chapter will first cover some elementary topics of descriptive statistics. Then we discuss the important notion of a *random variable*, related both to statistical concepts and to the subject of probability, studied in Chapter 12.

13.2 FREQUENCY TABLES, HISTOGRAMS

One of the first things one usually does with a large list of numerical data is to form some type of *frequency table*, which shows the number of times an individual item occurs or the number of items that fall within a given interval. These *frequency distributions* can then be pictured as *histograms*. We illustrate the technique with two examples.

EXAMPLE 13.1 An apartment house has 45 apartments, with the following numbers of tenants:

$$2, \ 1, \ 3, \ 5, \ 2, \ 2, \ 2, \ 1, \ 4, \ 2, \ 6, \ 2, \ 4, \ 3, \ 1$$
$$2, \ 4, \ 3, \ 1, \ 4, \ 4, \ 2, \ 4, \ 4, \ 2, \ 2, \ 3, \ 1, \ 4, \ 2$$
$$3, \ 1, \ 5, \ 2, \ 4, \ 1, \ 3, \ 2, \ 4, \ 4, \ 2, \ 5, \ 1, \ 3, \ 4$$

Observe that the only numbers which appear in the list are 1, 2, 3, 4, 5, and 6. The frequency distribution of these numbers appears in column 3 of Fig. 13-1. Column 2 is the tally count. The last column gives the *cumulative frequency*, which is obtained by adding the frequencies row by row beginning with the top row. This column gives the number of tenant numbers not exceeding the given number. For instance, there are 29 apartments with 3 or fewer tenants.

Number of People	Tally	Frequency	Cumulative Frequency
1	//// ///	8	8
2	### ### ////	14	22
3	### //	7	29
4	### ### //	12	41
5	///	3	44
6	/	1	45
SUM		45	

Fig. 13-1

As an alternative to Fig. 13-1, we can represent the frequency distribution by its histogram, Fig. 13-2. A histogram is simply a bar graph where the height of the bar gives the number of times the given number appears in the list.   Similarly, the cumulative frequency distribution could be presented as a histogram; the heights of the bars would be 8, 22, 29, . . . , 45.

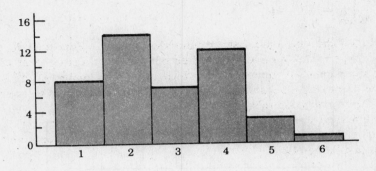

Fig. 13-2

EXAMPLE 13.2 Suppose the 6:00 P.M. temperatures (in degrees Fahrenheit) for a 35-day period are as follows:

$$72,\ 78,\ 86,\ 93,\ 106,\ 107,\ 98,\ 82,\ 81,\ 77,\ 87,\ 82$$
$$91,\ 95,\ 92,\ 83,\ 76,\ 78,\ 73,\ 81,\ 86,\ 92,\ 93,\ 84$$
$$107,\ 99,\ 94,\ 86,\ 81,\ 77,\ 73,\ 76,\ 80,\ 88,\ 91$$

Rather than find the frequency of each individual data item, it is more useful to construct a frequency table which counts the number of times the observed temperature falls in a given class, i.e. an interval with certain limits.   This is done in Fig. 13-3.

Class Boundaries, °F	Class Value, °F	Tally	Frequency	Cumulative Frequency
70–75	72.5	///	3	3
75–80	77.5	### /	6	9
80–85	82.5	### ///	8	17
85–90	87.5	###	5	22
90–95	92.5	### //	7	29
95–100	97.5	///	3	32
100–105	102.5		0	32
105–110	107.5	///	3	35
		SUM	35	

Fig. 13-3

The numbers 70, 75, 80, ..., are called the *class boundaries* or *class limits*. If a data item falls on a class boundary, it is usually assigned to the higher class; e.g. the number 95° was placed in the 95–100 class. Sometimes a frequency table also lists each *class value*, i.e. the midpoint of the class interval, which serves as an approximation to the values in the interval. The histogram corresponding to Fig. 13-3 appears in Fig. 13-4.

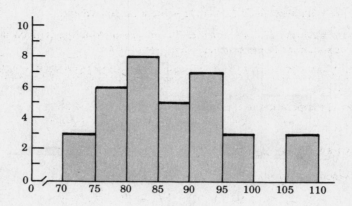

Fig. 13-4

13.3 MEAN

Suppose we are given a list of numerical values, say the eight numbers

$$7, \quad 11, \quad 11, \quad 8, \quad 12, \quad 7, \quad 6, \quad 6$$

The *arithmetic average*, or *arithmetic mean*, or, simply, the *mean*, is defined to be the sum of the values divided by the number of values; that is,

$$\text{mean} = \frac{7 + 11 + 11 + 8 + 12 + 7 + 6 + 6}{8} = \frac{68}{8} = 8.5$$

Generally speaking, if $x_1, x_2, \ldots, x_n$ is a list of n numerical values, then the mean of the numbers, denoted by $\bar{x}$, is defined as

$$\bar{x} = \frac{x_1 + x_2 + \cdots + x_n}{n} = \frac{\sum x_i}{n} \tag{13.1}$$

(For the summation symbol, see Section 9.5.)

Now suppose that our data are organized into a frequency table; let there be t *distinct* numerical values, $x_1, x_2, \ldots, x_t$, occurring with respective frequencies $f_1, f_2, \ldots, f_t$. Then the product $f_1 x_1$ gives the sum of the x_1's, $f_2 x_2$ gives the sum of the x_2's, and so on. Note also that

$$f_1 + f_2 + \cdots + f_t = n$$

the total number of data items. Accordingly, formula (13.1) for the mean $\bar{x}$ can be rewritten as

$$\bar{x} = \frac{f_1 x_1 + f_2 x_2 + \cdots + f_t x_t}{f_1 + f_2 + \cdots + f_t} = \frac{\sum f_i x_i}{\sum f_i} \tag{13.2}$$

Conversely, (13.2) reduces to (13.1) in the special case $t = n$ and all $f_i = 1$.

For data organized into classes, one normally applies (13.2) with f_i interpreted as the frequency of the ith class value, x_i.

EXAMPLE 13.3

(a) Consider the data of Example 13.1, of which the frequency distribution is given in Fig. 13-1. The mean is

$$\bar{x} = \frac{8(1) + 14(2) + 7(3) + 12(4) + 3(5) + 1(6)}{45} = \frac{126}{45} = 2.8$$

In other words, there is an average of 2.8 people living in an apartment.

(b) Consider the data of Example 13.2, of which the frequency distribution is given in Fig. 13-3. Using the class values as approximations to the original values, we obtain

$$\bar{x} = \frac{3(72.5) + 6(77.5) + 8(82.5) + 5(87.5) + 7(92.5) + 0(102.5) + 3(107.5)}{35} = \frac{3042}{35} = 86.9$$

i.e. the mean 6:00 P.M. temperature is approximately 86.9 °F.

13.4 VARIANCE, STANDARD DEVIATION

Consider the following two lists of numerical values:

List A: 12, 10, 9, 9, 10

List B: 7, 10, 14, 11, 8

For each list, the mean is $\bar{x} = 10$. Observe that the values in List A are clustered more closely about the mean than the values in List B. This section will discuss an important way of measuring such dispersions of data.

Let $\bar{x}$ be the mean of the n values $x_1, x_2, \ldots, x_n$. The difference $x_i - \bar{x}$ is called the *deviation* of the data value x_i from the mean $\bar{x}$; it is positive or negative according as x_i is greater or less than $\bar{x}$. The average of the squares of the deviations is called the *variance* of the data, and the square root of the variance is called the *standard deviation*. That is,

$$\text{variance} = \frac{(x_1 - \bar{x})^2 + (x_2 - \bar{x})^2 + \cdots + (x_n - \bar{x})^2}{n} = \frac{\sum (x_i - \bar{x})^2}{n} \qquad (13.3)$$

$$\text{standard deviation} = \sqrt{\text{variance}} \qquad (13.4)$$

Since each squared deviation is nonnegative, so is the variance. Moreover, the variance is zero if and only if the data values are all equal (and therefore are all equal to the mean).

A formula equivalent to (13.3) is

$$\text{variance} = \frac{x_1^2 + x_2^2 + \cdots + x_n^2}{n} - \bar{x}^2 = \frac{\sum x_i^2}{n} - \bar{x}^2 \qquad (13.5)$$

Sometimes we let s or s_x denote the standard deviation of the x_i, in which case s^2 or s_x^2 will denote the variance of the x_i.

EXAMPLE 13.4

(a) Consider List A above, whose mean is $\bar{x} = 10$. The deviations of the five data values are:

$$12 - 10 = 2 \qquad 10 - 10 = 0 \qquad 9 - 10 = -1 \qquad 9 - 10 = -1 \qquad 10 - 10 = 0$$

The squares of the deviations are then

$$2^2 = 4 \qquad 0^2 = 0 \qquad (-1)^2 = 1 \qquad (-1)^2 = 1 \qquad 0^2 = 0$$

Accordingly,

$$\text{variance} = \frac{4 + 0 + 1 + 1 + 0}{5} = \frac{7}{5} = 1.4$$

$$\text{standard deviation} = \sqrt{1.4} \approx 1.2$$

(b)   Consider List B above, whose mean is $\bar{x} = 10$.   By (13.3),

$$\text{variance} = \frac{(7-10)^2 + (10-10)^2 + (14-10)^2 + (11-10)^2 + (8-10)^2}{5} = \frac{9 + 0 + 16 + 1 + 4}{5} = \frac{30}{5} = 6$$

Alternatively, by (13.5),

$$\text{variance} = \frac{7^2 + 10^2 + 14^2 + 11^2 + 8^2}{5} - 10^2 = \frac{49 + 100 + 196 + 121 + 64}{5} - 100 = 106 - 100 = 6$$

Hence $$\text{standard deviation} = \sqrt{6} \approx 2.4$$

Note that List B, which exhibits much more scatter than List A, has a much larger variance (and standard deviation).

For data organized into a frequency distribution—say, t distinct values $x_1, x_2, \ldots, x_t$ with respective frequencies $f_1, f_2, \ldots, f_t$—the product $f_1(x_1 - \bar{x})^2$ gives the sum of the squares of the deviations of the x_1's from $\bar{x}$, etc.   Hence we can rewrite (13.3) and (13.5) as

$$\text{variance} = \frac{f_1(x_1 - \bar{x})^2 + f_2(x_2 - \bar{x})^2 + \cdots + f_t(x_t - \bar{x})^2}{f_1 + f_2 + \cdots + f_t} = \frac{\sum f_i(x_i - \bar{x})^2}{\sum f_i} \qquad (13.6)$$

$$\text{variance} = \frac{f_1 x_1^2 + f_2 x_2^2 + \cdots + f_t x_t^2}{f_1 + f_2 + \cdots + f_t} - \bar{x}^2 = \frac{\sum f_i x_i^2}{\sum f_i} - \bar{x}^2 \qquad (13.7)$$

For computations, (13.7), which involves only one subtraction, is usually preferred to (13.6).

Again, if the data are organized into classes, one uses the class values as approximations to the original values.

EXAMPLE 13.5   For the data of Example 13.1, Fig. 13-1 is expanded into Fig. 13-5, from which we obtain:

$$\bar{x} = \frac{\sum f_i x_i}{\sum f_i} = \frac{126}{45} = 2.8$$

$$\text{variance} = \frac{\sum f_i x_i^2}{\sum f_i} - \bar{x}^2 = \frac{430}{45} - (2.8)^2 = 9.56 - 7.84 = 1.72$$

$$\text{standard deviation} = \sqrt{\text{variance}} = \sqrt{1.72} = 1.31$$

Number of People, x_i	Frequency, f_i	$f_i x_i$	$x_i \cdot f_i x_i$ $= f_i x_i^2$	Cumulative Frequency
1	8	8	8	8
2	14	28	56	22
3	7	21	63	29
4	12	48	192	41
5	3	15	75	44
6	1	6	36	45
SUMS	45	126	430	

Fig. 13-5

EXAMPLE 13.6 Three hundred incoming students take a mathematics exam consisting of 75 multiple-choice questions. If the distribution of the scores on the exam is

Test Scores	5–15	15–25	25–35	35–45	45–55	55–65	65–75
Number of Students	2	0	8	36	110	78	66

find the mean $\bar{x}$, the variance s^2, and the standard deviation s of the distribution.
We complete the first four columns in Fig. 13-6. This gives

$$\bar{x} = \frac{16\,500}{300} = 55$$

Using this value for the mean, we complete the remaining columns in Fig. 13-6 and obtain

$$s^2 = \frac{36\,700}{300} = 122.3$$
$$s = \sqrt{122.3} = 11.1$$

Observe that here we used (13.6) for the variance.

Class Limits	Class Value, x_i	Frequency, f_i	$f_i x_i$	$x_i - \bar{x}$	$(x_i - \bar{x})^2$	$f_i(x_i - \bar{x})^2$	Cumulative Frequency
5–15	10	2	20	−45	2025	4 050	2
15–25	20	0	0	−35	1525	0	2
25–35	30	8	240	−25	625	5 000	10
35–45	40	36	1 440	−15	225	8 100	46
45–55	50	110	5 500	−5	25	2 750	156
55–65	60	78	4 680	5	25	1 950	234
65–75	70	66	4 620	15	225	14 850	300
SUMS		300	16 500			36 700	

Fig. 13-6

Figure 13-7 is the flowchart of a computer program that calculates the mean MEAN, the variance VAR [on the basis of (13.7)], and the standard deviation SD, where the input data are the values X_1, $X_2, \ldots, X_T$ with respective frequencies $F_1, F_2, \ldots, F_T$. Observe that we require the sums

$$N = \sum F_K \qquad SUM = \sum F_K X_K \qquad SUMSQ = \sum F_K X_K^2$$

each of which is initialized at zero.

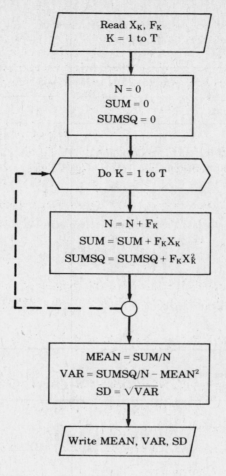

Fig. 13-7

13.5 MEDIAN, MODE

Besides the mean, there are two other measurements which serve to define what might be called the "center" of a set of numerical data.

Median

Consider a collection of n data values which are sorted in increasing order. The *median* of the data is the "middle value." That is, if n is odd, then the median is the $[(n + 1)/2]$th term; but if n is even, then the median is the average of the $(n/2)$th and $[(n/2) + 1]$th terms. For example, consider the following two lists of sorted numbers:

$$\text{List A:} \quad 11, \ 11, \ 16, \ 17, \ 25$$
$$\text{List B:} \quad 1, \ 4, \ 8, \ 8, \ 10, \ 16, \ 16, \ 19$$

List A has five terms; its median is 16, the middle or third term. List B has eight terms; its median is 9, the average of the fourth term, 8, and the fifth term, 10. For any collection of data values (sorted or not), we note that just as many of the numbers are less than or equal to the median as are greater than or equal to the median. The cumulative frequency distribution can be used to find the median of an arbitrary set of data.

Mode

The *mode* of a collection of data values is that value or class value which occurs most often. Some collections have more than one mode; they are said to be *multimodal*. For example, the unique mode of List A above is 11, since that number occurs twice whereas all the other numbers occur only once. On the other hand, List B is bimodal; it has two modes, 8 and 16.

EXAMPLE 13.7

(a) Consider the data in Fig. 13-1. There are $n = 45$ values. The cumulative frequency column tells us that the median is 3, the 23rd value. The mode is 2, since 2 has the highest frequency.

(b) Consider the data in Fig. 13-3. The median is 87.5, the approximate 18th value; the mode is 82.5, since its class has the highest frequency.

(c) Consider the data in Fig. 13-6. Note that $n = 300$. The median and mode for this distribution is the class value 50.

13.6 RANDOM VARIABLES

We frequently wish to assign a specific number to each element of a finite sample space S, particularly when the elements are not themselves numbers. Such an assignment, X, is called a *random variable* (on S). That is,

Definition: A random variable X on a finite sample space S is a function from S into the real numbers **R**.

EXAMPLE 13.8 Consider the sample space of outcomes of the toss of two coins:

$$S = \{HH, HT, TH, TT\}$$

(a) Let X_1 denote the random variable on S defined by

$$X_1(HH) = 1, \quad X_1(HT) = 2, \quad X_1(TH) = 3, \quad X_1(TT) = 4$$

i.e. the image of X_1 is the set of real numbers

$$S' = \{1, 2, 3, 4\}$$

We might, in any further consideration of the experiment, find it convenient to take S' as the sample space, instead of S.

(b) Suppose we were interested only in the number of heads that occurred. Then we could define a random variable X_2 on S by

$$X_2(HH) = 2, \quad X_2(HT) = 1, \quad X_2(TH) = 1, \quad X_2(TT) = 0$$

and could consider $S'' = \{0, 1, 2\}$, the image of X_2, as the new sample space of outcomes.

Now let S be a finite probability space (Section 12.3), and let X be a random variable on S whose values are the real numbers $x_1, x_2, x_3, \ldots, x_t$. (Since S is finite, X can have only a finite number of values; we call X a *discrete* random variable.) Then the existing assignment of probabilities to the sample points of S induces an assignment of probabilities to the points x_i of the image of X, as follows:

$$p_i = P(x_i) = \text{sum of probabilities of points in } S \text{ whose image is } x_i \qquad (13.8)$$

In other words, a random variable maps a finite probability space into a finite probability space of real numbers, where probabilities are assigned to the points of the new space according to the rule (13.8).

The function assigning p_i to x_i, i.e. the set of ordered pairs $(x_1, p_1), \ldots, (x_t, p_t)$, is usually given by a table

x_1	x_2	$\cdots$	x_t
p_1	p_2	$\cdots$	p_t

and is called the *distribution* of the random variable X. We also say that "X takes on the value x_i with probability p_i," or, in terms of the new sample space, that "x_i occurs with probability p_i."

In the case that S is an equiprobable space, we can easily obtain the distribution of a random variable on S from the following result.

Theorem 13.1: Let S be a finite equiprobable space and let X be a random variable on S, with values $x_1, x_2, \ldots, x_t$. Then

$$p_i = P(x_i) = \frac{\text{number of points in } S \text{ whose image is } x_i}{\text{number of points in } S}$$

$(i = 1, 2, \ldots, t)$ gives the distribution of X.

EXAMPLE 13.9

(a) A pair of fair dice is tossed. We obtain the equiprobable space S consisting of the 36 ordered pairs of integers from 1 to 6: $S = \{(1, 1), (1, 2), \ldots, (6, 6)\}$. Let X be the random variable which assigns to each element of S the sum of the two integers. Then X takes on the values

$$2, \ 3, \ 4, \ 5, \ 6, \ 7, \ 8, \ 9, \ 10, \ 11, \ 12$$

We use Theorem 13.1 to obtain the distribution of X. There is only one point, $(1, 1)$, whose image is 2; hence $P(2) = 1/36$. There are two points in S, $(1, 2)$ and $(2, 1)$, having image 3; hence $P(3) = 2/36$. There are three points in S, $(1, 3)$, $(2, 2)$, and $(3, 1)$, having image 4; hence $P(4) = 3/36$. And so on. The distribution of X consists of its values and their respective probabilities:

x_i	2	3	4	5	6	7	8	9	10	11	12
p_i	1/36	2/36	3/36	4/36	5/36	6/36	5/36	4/36	3/36	2/36	1/36

(b) Three items are selected at random from a box containing 12 items of which 3 are defective. The sample space S consists of the distinct, equally likely samples of size 3. Let X be the random variable which counts the number of defective items in a sample; the values of X are 0, 1, 2, and 3.

Now the number of sample points in S corresponding to x_i defectives equals the number of ways of choosing x_i defectives from among 3 defectives and choosing $3 - x_i$ nondefectives from 9 nondefectives:

$$\binom{3}{x_i}\binom{9}{3 - x_i}$$

The total number of sample points in S is

$$\binom{12}{3}$$

Hence, by Theorem 13.1, the probability of the value x_i of X is

$$p_i = \binom{3}{x_i}\binom{9}{3 - x_i} \bigg/ \binom{12}{3} \qquad (x_i = 0, 1, 2, 3)$$

This is the distribution of X, in functional form. (It is called a *hypergeometric distribution*.) In tabular form, we have:

x_i	0	1	2	3
p_i	84/220	108/220	27/220	1/220

Functions of Random Variables

If X is a (discrete) random variable on a finite probability space S, then so too is $Y = f(X)$, where $f(\ )$ is any real-valued function. The distribution of Y is obtained from the distribution of X by a rule analogous to (13.8); namely,

$$P(y_k) = \text{sum of probabilities of all } x_i \text{ such that } y_k = f(x_i) \qquad (13.9)$$

EXAMPLE 13.10 Let us calculate the distribution of $Y = (X - 2)^2$, where X is the random variable of Example 13.9(b).

The values $x_1 = 0$, $x_2 = 1$, $x_3 = 2$, and $x_4 = 3$ of X have as their respective images $(0 - 2)^2 = 4$, $(1 - 2)^2 = 1$, $(2 - 2)^2 = 0$, and $(3 - 2)^2 = 1$. Thus the values of Y are $y_1 = 0$, $y_2 = 1$, and $y_3 = 4$; and, by (13.9),

$$P(y_1) = P(x_3) = \frac{27}{220}$$

$$P(y_2) = P(x_2) + P(x_4) = \frac{108}{220} + \frac{1}{220} = \frac{109}{220}$$

$$P(y_3) = P(x_1) = \frac{84}{220}$$

In tabular form, the distribution of Y is:

y_k	0	1	4
p_k	27/220	109/220	84/220

13.7 EXPECTATION AND VARIANCE OF A RANDOM VARIABLE

Let X be a discrete random variable on a sample space S; X takes on the values $x_1, x_2, \ldots, x_t$ with respective probabilities $p_1, p_2, \ldots, p_t$. Suppose now that the experiment which generates S is repeated n times, and that the numbers $x_1, x_2, \ldots, x_t$ occur with respective frequencies $f_1, f_2, \ldots, f_t$ ($\Sigma f_i = n$). If n is large, one expects that

$$\frac{f_1}{n} \approx p_1 \qquad \frac{f_2}{n} \approx p_2 \qquad \cdots \qquad \frac{f_t}{n} \approx p_t$$

so that (13.2) becomes

$$\bar{x} = \frac{f_1 x_1 + f_2 x_2 + \cdots + f_t x_t}{n} = \frac{f_1}{n} x_1 + \frac{f_2}{n} x_2 + \cdots + \frac{f_t}{n} x_t$$

$$\approx x_1 p_1 + x_2 p_2 + \cdots + x_t p_t$$

The last expression depends only on the distribution of the random variable X; it is denoted by μ (or μ_X) or $E(X)$, and is called the *mean* or *expectation* or *expected value* of X.

Similarly, for large n we have from (13.6):

$$\text{variance} = \frac{f_1(x_1 - \bar{x})^2 + f_2(x_2 - \bar{x})^2 + \cdots + f_t(x_t - \bar{x})^2}{n}$$

$$= \frac{f_1}{n}(x_1 - \bar{x})^2 + \frac{f_2}{n}(x_2 - \bar{x})^2 + \cdots + \frac{f_t}{n}(x_t - \bar{x})^2$$

$$\approx (x_1 - \mu)^2 p_1 + (x_2 - \mu)^2 p_2 + \cdots + (x_t - \mu)^2 p_t$$

Again, the last expression depends only on the distribution of X; it is denoted by σ^2 (or σ_X^2) or Var(X) and is called the *variance* of X.

We formally restate the above results as a

Definition: Let a discrete random variable X take on the values $x_1, x_2, \ldots, x_t$ with respective probabilities $p_1, p_2, \ldots, p_t$. Then

$$E(X) \equiv x_1 p_1 + x_2 p_2 + \cdots + x_t p_t = \sum x_i p_i \tag{13.10}$$

(also denoted μ or μ_X) is called the *mean* or *expectation* or *expected value* of X. Moreover,

$$\text{Var}(X) \equiv (x_1 - \mu)^2 p_1 + (x_2 - \mu)^2 p_2 + \cdots + (x_t - \mu)^2 p_t = \sum (x_i - \mu)^2 p_i \tag{13.11}$$

(also denoted σ^2 or σ_X^2) is called the *variance* of X. The quantity

$$\sigma \equiv \sqrt{\text{Var}(X)} \tag{13.12}$$

(also denoted σ_X) is called the *standard deviation* of X.

An important property of the expectation is its *linearity*; that is, if X_1 and X_2 are random variables on a finite probability space S, and c_1 and c_2 are constants, then $c_1 X_1 + c_2 X_2$ is a random variable on S and

$$E(c_1 X_1 + c_2 X_2) = c_1 E(X_1) + c_2 E(X_2) \tag{13.13}$$

(see Problem 13.8). Furthermore, given any function of a random variable, $Y = f(X)$, it is possible to calculate the expectation of Y directly from the distribution of X, without first calculating the distribution of Y. In fact, (13.9) implies

Theorem 13.2: If $Y = f(X)$, where the discrete random variable X takes on the values $x_1, x_2, \ldots, x_t$ with probabilities $p_1, p_2, \ldots, p_t$, then

$$E(Y) = f(x_1) p_1 + f(x_2) p_2 + \cdots + f(x_t) p_t = \sum f(x_i) p_i$$

EXAMPLE 13.11 From (13.11) and Theorem 13.2, we see that the variance of X may be written as

$$\text{Var}(X) = E[(X - \mu)^2]$$

Then, by linearity,

$$\text{Var}(X) = E[(X - \mu)^2] = E(X^2 - 2\mu X + \mu^2) = E(X^2) - 2\mu E(X) + E(\mu^2)$$
$$= E(X^2) - 2\mu \cdot \mu + \mu^2 = E(X^2) - \mu^2$$

and so, again using Theorem 13.2,

$$\text{Var}(X) = x_1^2 p_1 + x_2^2 p_2 + \cdots + x_t^2 p_t - \mu^2 = \left(\sum x_i^2 p_i\right) - \mu^2 \tag{13.14}$$

Equation (13.14) is the analog of (13.7); it may be used as a simpler alternative to the definition (13.11).

EXAMPLE 13.12 Consider the random variable X of Example 13.9(b), which gives the number of defective items in a sample of size 3. We have

$$\mu = E(X) = 0\left(\frac{84}{220}\right) + 1\left(\frac{108}{220}\right) + 2\left(\frac{27}{220}\right) + 3\left(\frac{1}{220}\right) = \frac{3}{5}$$

That is, 3/5 is the expected number of defective items in a sample of size 3. We also have, from (13.14),

$$\text{Var}(X) = 0^2\left(\frac{84}{220}\right) + 1^2\left(\frac{108}{220}\right) + 2^2\left(\frac{27}{220}\right) + 3^2\left(\frac{1}{220}\right) - \left(\frac{3}{5}\right)^2 \approx 0.663$$

and, from this, $\sigma \approx \sqrt{0.663} \approx 0.81$.

EXAMPLE 13.13 A player tosses a fair die. If a prime number appears, he wins that number of dollars; but if a nonprime number appears, he loses that number of dollars. Is this game *fair*?

Denote the player's gain by the random variable X, defined on the equiprobable space

$$S = \{1, 2, 3, 4, 5, 6\}$$

of outcomes of the toss. The distribution of X is as follows:

x_i	2	3	5	-1	-4	-6
p_i	1/6	1/6	1/6	1/6	1/6	1/6

The negative values -1, -4 and -6 correspond to the fact that the player loses if a nonprime comes up. Then

$$E(X) = 2(1/6) + 3(1/6) + 5(1/6) + (-1)(1/6) + (-4)(1/6) + (-6)(1/6) = -1/6$$

For a fair game, $E(X) = 0$; this game is *unfavorable* to the player, since he can expect to lose one-sixth dollar each time he plays. The quantity $E(X)$ is often called the *value of the game* to the player.

Binomially-Distributed Random Variables

Consider the random variable X_n, defined on the sample space of outcomes of n repeated trials of a success-or-failure experiment, that gives the number of successes. As was shown in Section 12.7, the distribution of X_n is the *binomial distribution*:

$$P(k) = b(k; n, p) \qquad (k = 0, 1, 2, \ldots, n)$$

i.e.

k	0	1	2	$\cdots$	n
$P(k)$	q^n	$\binom{n}{1}pq^{n-1}$	$\binom{n}{2}p^2q^{n-2}$	$\cdots$	p^n

where $p = 1 - q$ is the probability of a success in a single trial.

The mean and variance of X_n may now be calculated from (*13.10*) and (*13.14*); we state the results as

Theorem 13.3:
$$E(X_n) = np$$
$$\mathrm{Var}(X_n) = npq$$

EXAMPLE 13.14 If a fair die is tossed 180 times, the expected number of sixes is

$$\mu = np = 180 \left(\frac{1}{6}\right) = 30$$

Also, the standard deviation of the number of sixes is

$$\sigma = \sqrt{npq} = \sqrt{180 \left(\frac{1}{6}\right)\left(\frac{5}{6}\right)} = 5$$

Solved Problems

STATISTICS

13.1 Find the (*a*) mean, (*b*) variance, (*c*) standard deviation, (*d*) median, and (*e*) mode, of the six numbers 4, 6, 6, 7, 9, 10.

(*a*) The mean or arithmetic average is equal to the sum of the six numbers divided by six:

$$\text{mean} = \frac{4+6+6+7+9+10}{6} = \frac{42}{6} = 7$$

(*b*) The variance is the average of the squares of the deviations from the mean:

$$\text{variance} = \frac{(4-7)^2 + (6-7)^2 + (6-7)^2 + (7-7)^2 + (9-7)^2 - (10-7)^2}{6}$$

$$= \frac{9+1+1+0+4+9}{6} = \frac{24}{6} = 4$$

(*c*) The standard deviation is the square root of the variance:

$$\text{standard deviation} = \sqrt{4} = 2$$

(*d*) There are two middle numbers, the third and fourth numbers, which are 6 and 7. The median is their average:

$$\text{median} = \frac{6+7}{2} = 6.5$$

(*e*) The mode is 6, since 6 occurs most often.

13.2 The class results on an exam with 20 questions follows:

Number of Correct Answers, x	20	19	18	17	16	15	14	13	12	10	9
Number of Students	4	6	2	7	1	2	7	2	1	2	1

Find the (*a*) mean $\bar{x}$, (*b*) variance s^2, (*c*) standard deviation s, (*d*) median, (*e*) mode.

Complete a table as in Fig. 13-8. Then:

(*a*)
$$\bar{x} = \frac{560}{35} = 16$$

(*b*)
$$s^2 = \frac{9278}{35} - \bar{x}^2 = 265.1 - 256 = 9.1$$

(*c*)
$$s = \sqrt{9.1} = 3.0$$

(*d*) The median is 17, since this is the score of the 18th student (18th from the top or 18th from the bottom).

(*e*) There are two modes, 14 and 17.

13.3 The yearly rainfall, measured to the nearest tenth of a centimeter, for a 30-year period is as follows:

```
42.3   35.7   47.6   31.2   28.3   37.0   41.3   32.4   41.3   29.3
34.3   35.2   43.0   36.3   35.7   41.5   43.2   30.7   38.4   46.5
43.2   31.7   36.8   43.6   45.2   32.8   30.7   36.2   34.7   35.3
```

Test Score, x_i	Frequency, f_i	Cumulative Frequency	$f_i x_i$	$f_i x_i^2$
20	4	4	80	1600
19	6	10	114	2166
18	2	12	36	648
17	7	19	119	2023
16	1	20	16	256
15	2	22	30	450
14	7	29	98	1372
13	2	31	26	338
12	1	32	12	144
10	2	34	20	200
9	1	35	9	81
SUMS	35		560	9278

Fig. 13-8

Class Limits, cm	Class Value, x_i	Tally	Frequency f_i	$f_i x_i$	$f_i x_i^2$
28–30	29	//	2	58	1 682
30–32	31	////	4	124	3 844
32–34	33	//	2	66	2 178
34–36	35	### /	6	210	7 350
36–38	37	////	4	148	5 476
38–40	39	/	1	39	1 521
40–42	41	///	3	123	5 043
42–44	43	###	5	215	9 245
44–46	45	/	1	45	2 025
46–48	47	//	2	94	4 418
		SUMS	30	1122	42 782

Fig. 13-9

(*a*) Classify the data and construct a frequency distribution.　(*b*) Find the mean, variance, and standard deviation of the classed data.

(*a*)　First find the *range* (maximum value minus minimum value) of the data:

$$\text{range} = 47.6 - 28.3 = 19.3 \text{ cm}$$

The choice of classes is arbitrary, although one normally covers the range with 7 to 12 classes. We choose 10 classes with class limits 28–30, 30–32, 32–34, . . . , 46–48. We can then complete a table as in Fig. 13-9.

(*b*)　From Fig. 13-9,

$$\text{mean} = \frac{1122}{30} = 37.4 \text{ cm}$$

$$\text{variance} = \frac{42782}{30} - (37.4)^2 = 27.3 \text{ cm}^2$$

$$\text{standard deviation} = \sqrt{27.3} = 5.2 \text{ cm}$$

Notice that the standard deviation, like the mean, carries the same units as the original data.

RANDOM VARIABLES, EXPECTATION

13.4　A pair of fair dice is thrown. Let X be the maximum of the two numbers which come up. (*a*) Find the distribution of X. (*b*) Find the expectation $E(X)$, variance $\text{Var}(X)$, and standard deviation σ_X.

(*a*)　The sample space S is the equiprobable space consisting of the 36 ordered pairs of integers from 1 to 6; that is,

$$S = \{(1, 1), (1, 2), (1, 3), \ldots, (6, 6)\}$$

Since X assigns to each element of S the larger of the two integers, the values of X are the integers from 1 to 6. Now there is only one point of S, $(1, 1)$, giving a maximum of 1; hence (Theorem 13.1) $P(1) = 1/36$. Each of three points in S, $(1, 2)$, $(2, 2)$ and $(2, 1)$, gives a maximum of 2; hence $P(2) = 3/36$. Each of five points in S, $(1, 3)$, $(2, 3)$, $(3, 3)$, $(3, 2)$ and $(3, 1)$, gives 3; hence $P(3) = 5/36$. Similarly, $P(4) = 7/36$, $P(5) = 9/36$ and $P(6) = 11/36$.

The distribution of X consists of its values with their respective probabilities; it is given in the following table:

x_i	1	2	3	4	5	6
p_i	1/36	3/36	5/36	7/36	9/36	11/36

(*b*)　We find the expectation (mean) of X by multiplying each x_i by its probability p_i and then summing:

$$\mu = E(X) = 1\left(\frac{1}{36}\right) + 2\left(\frac{3}{36}\right) + 3\left(\frac{5}{36}\right) + 4\left(\frac{7}{36}\right) + 5\left(\frac{9}{36}\right) + 6\left(\frac{11}{36}\right) = \frac{161}{36} \approx 4.5$$

We find $E(X^2)$ by multiplying x_i^2 by p_i and taking the sum:

$$E(X^2) = 1\left(\frac{1}{36}\right) + 4\left(\frac{3}{36}\right) + 9\left(\frac{5}{36}\right) + 16\left(\frac{7}{36}\right) + 25\left(\frac{9}{36}\right) + 36\left(\frac{11}{36}\right) = \frac{791}{36} = 22.0$$

Then

$$\text{Var}(X) = E(X^2) - \mu^2 = 22.0 - (4.5)^2 = 1.7 \qquad \text{and} \qquad \sigma_X = \sqrt{1.7} = 1.3$$

13.5 Let X be a random variable with the following distribution:

x_i	1	3	4	5
p_i	0.4	0.1	0.2	0.3

(a) Find the mean μ_X, variance σ_X^2, and standard deviation σ_X.
(b) Find the distribution of the random variable $Y = X^2 + 2$.
(c) Find the distribution of the random variable $Z = \max \{X, 4\}$ (i.e. Z is the larger of X and 4). Compute μ_Z from this distribution and show that the same result is given by Theorem 13.2.

(a)
$$\mu_X = \sum x_i p_i = 1(0.4) + 3(0.1) + 4(0.2) + 5(0.3) = 3$$

$$E(X^2) = \sum x_i^2 p_i = 1(0.4) + 9(0.1) + 16(0.2) + 25(0.3) = 12$$

$$\sigma_X^2 = E(X^2) - \mu_X^2 = 12 - 9 = 3$$
$$\sigma_X = \sqrt{3} = 1.7$$

(b) For this function, distinct values of X give distinct values of Y:

$$y_1 = x_1^2 + 2 = 1^2 + 2 = 3 \qquad y_3 = x_3^2 + 2 = 4^2 + 2 = 18$$
$$y_2 = x_2^2 + 2 = 3^2 + 2 = 11 \qquad y_4 = x_4^2 + 2 = 5^2 + 2 = 27$$

Hence each y_i is assigned the probability of the corresponding x_i, and the distribution of Y is

y_i	3	11	18	27
p_i	0.4	0.1	0.2	0.3

(c) Here $x_1 = 1$, $x_2 = 3$, and $x_3 = 4$ each give the value $z_1 = 4$, while $x_4 = 5$ gives the value $z_2 = 5$. Hence we make the assignments

$$P(z_1) = P(x_1) + P(x_2) + P(x_3) = 0.4 + 0.1 + 0.2 = 0.7$$
$$P(z_2) = P(x_4) = 0.3$$

and the distribution of Z is

z_k	4	5
p_k	0.7	0.3

From this distribution,

$$\mu_Z = 4(0.7) + 5(0.3) \quad (= 4.3)$$

On the other hand, in terms of the X-distribution,

$$\mu_Z = [\max \{1, 4\}](0.4) + [\max \{3, 4\}](0.1) + [\max \{4, 4\}](0.2) + [\max \{5, 4\}](0.3)$$
$$= 4(0.4) + 4(0.1) + 4(0.2) + 5(0.3)$$
$$= 4(0.7) + 5(0.3)$$

as before.

13.6 Five cards are numbered 1 to 5. Two cards are drawn at random. Let X denote the sum of the numbers drawn. Find (a) the distribution of X; (b) the mean, variance, and standard deviation of X.

(a) There are $C(5, 2) = 10$ ways of drawing two cards at random. The ten equiprobable sample points, with their corresponding X-values, are shown below:

$$\{1, 2\} \to 3 \quad \{1, 3\} \to 4 \quad \{1, 4\} \to 5 \quad \{1, 5\} \to 6 \quad \{2, 3\} \to 5$$
$$\{2, 4\} \to 6 \quad \{2, 5\} \to 7 \quad \{3, 4\} \to 7 \quad \{3, 5\} \to 8 \quad \{4, 5\} \to 9$$

Observe that the values of X are the seven numbers 3, 4, 5, 6, 7, 8, and 9; of these 3, 4, 8, and 9 are each assumed at one sample point, while 5, 6, and 7 are each assumed at two sample points. Hence the distribution of X is:

x_i	3	4	5	6	7	8	9
p_i	0.1	0.1	0.2	0.2	0.2	0.1	0.1

(b)
$$\mu = E(X) = \sum x_i p_i = 3(0.1) + 4(0.1) + 5(0.2) + 6(0.2) + 7(0.2) + 8(0.1) + 9(0.1) = 6$$

$$E(X^2) = \sum x_i^2 p_i = 9(0.1) + 16(0.1) + 25(0.2) + 36(0.2) + 49(0.2) + 64(0.1) + 81(0.1) = 39$$

$$\text{Var}(X) = E(X^2) - \mu^2 = 39 - 6^2 = 3$$
$$\sigma = \sqrt{\text{Var}(X)} = \sqrt{3} = 1.7$$

13.7 A fair die is tossed. Let X denote twice the number appearing, and let Y be 1 or 3 according as an odd or an even number appears. Find the distribution (a) of X, (b) of Y.

The sample space is $S = \{1, 2, 3, 4, 5, 6\}$, with each sample point having probability 1/6.
(a) The images of the sample points are:

$$X(1) = 2 \quad X(2) = 4 \quad X(3) = 6 \quad X(4) = 8 \quad X(5) = 10 \quad X(6) = 12$$

As these are distinct, the distribution of X is

x_i	2	4	6	8	10	12
$P(x_i)$	1/6	1/6	1/6	1/6	1/6	1/6

(b) The images of the sample points are:

$$Y(1) = 1 \quad Y(2) = 3 \quad Y(3) = 1 \quad Y(4) = 3 \quad Y(5) = 1 \quad Y(6) = 3$$

The two Y-values, 1 and 3, are each assumed at three sample points. Hence we have the distribution

y_i	1	3
$P(y_i)$	3/6	3/6

13.8 Let X and Y be random variables defined on the same finite probability space S. Then $X + Y$ and XY, defined by

$$(X + Y)(s) = X(s) + Y(s) \qquad \text{and} \qquad (XY)(s) = X(s)\,Y(s)$$

are also random variables on S. In particular, let X and Y be the random variables of Problem 13.7. (a) Find the distribution of $X + Y$. (b) Find the distribution of XY. (c) Verify that $E(X + Y) = E(X) + E(Y)$. (d) Is is true that $E(XY) = E(X)\,E(Y)$?

The sample space is still $S = \{1, 2, 3, 4, 5, 6\}$, and each sample point still has probability 1/6.

(a) Using $(X + Y)(s) = X(s) + Y(s)$ and the values of X and Y from Problem 13.7, we obtain:

$$(X + Y)(1) = 2 + 1 = 3 \qquad (X + Y)(3) = 6 + 1 = 7 \qquad (X + Y)(5) = 10 + 1 = 11$$
$$(X + Y)(2) = 4 + 3 = 7 \qquad (X + Y)(4) = 8 + 3 = 11 \qquad (X + Y)(6) = 12 + 3 = 15$$

The image set is $\{3, 7, 11, 5\}$. The values 3 and 15 are each assumed at only one sample point and hence have probability 1/6; the values 7 and 11 are each assumed at two sample points and hence have probability 2/6. Thus the distribution of $Z = X + Y$ is:

z_i	3	7	11	15
$P(z_i)$	1/6	2/6	2/6	1/6

(b) Using $(XY)(s) = X(s)\,Y(s)$, we obtain:

$$(XY)(1) = 2 \cdot 1 = 2 \qquad (XY)(3) = 6 \cdot 1 = 6 \qquad (XY)(5) = 10 \cdot 1 = 10$$
$$(XY)(2) = 4 \cdot 3 = 12 \qquad (XY)(4) = 8 \cdot 3 = 24 \qquad (XY)(6) = 12 \cdot 3 = 36$$

Each value of XY is assumed at just one sample point; hence the distribution of $W = XY$ is:

w_i	2	6	10	12	24	36
$P(w_i)$	1/6	1/6	1/6	1/6	1/6	1/6

(c) Using the distributions found in Problem 13.7,

$$E(X) = \sum x_i P(x_i) = \frac{2}{6} + \frac{4}{6} + \frac{6}{6} + \frac{8}{6} + \frac{10}{6} + \frac{12}{6} = 7$$

$$E(Y) = \sum y_i P(y_i) = \frac{3}{6} + \frac{9}{6} = 2$$

and using the distribution found in (a),

$$E(X + Y) = E(Z) = \sum z_i P(z_i) = \frac{3}{6} + \frac{14}{6} + \frac{22}{6} + \frac{15}{6} = 9 = 7 + 2$$

(d) Using the distribution found in (b),

$$E(XY) = E(W) = \sum w_i P(w_i) = \frac{2}{6} + \frac{6}{6} + \frac{10}{6} + \frac{12}{6} + \frac{24}{6} + \frac{36}{6} = 15$$

and $15 \ne 7 \cdot 2$. (For *independent* random variables, $E(XY) = E(X)\,E(Y)$; we must, however, omit any discussion of this topic.)

13.9 A fair coin is tossed until a head or five tails occurs. Find the expected number of tosses of the coin.

As in any probability problem, one first identifies the possible outcomes of the underlying experiment, i.e. the sample points. Here there are six:

$$\text{H} \qquad \text{TH} \qquad \text{TTH} \qquad \text{TTTH} \qquad \text{TTTTH} \qquad \text{TTTTT}$$

with respective probabilities (independent trials)

$$\frac{1}{2} \qquad \left(\frac{1}{2}\right)^2 = \frac{1}{4} \qquad \left(\frac{1}{2}\right)^3 = \frac{1}{8} \qquad \left(\frac{1}{2}\right)^4 = \frac{1}{16} \qquad \left(\frac{1}{2}\right)^5 = \frac{1}{32} \qquad \left(\frac{1}{2}\right)^5 = \frac{1}{32}$$

The random variable of interest, X, is the number of components in a sample point. Thus,

$$X(H) = 1 \qquad X(TTH) = 3 \qquad X(TTTTH) = 5$$
$$X(TH) = 2 \qquad X(TTTH) = 4 \qquad X(TTTTT) = 5$$

and these X-values are assigned the probabilities

$$P(1) = P(H) = \frac{1}{2} \qquad P(3) = P(TTH) = \frac{1}{8} \qquad P(5) = P(TTTTH) + P(TTTTT)$$
$$P(2) = P(TH) = \frac{1}{4} \qquad P(4) = P(TTTH) = \frac{1}{16} \qquad \qquad = \frac{1}{32} + \frac{1}{32} = \frac{1}{16}$$

Accordingly,

$$E(X) = 1\left(\frac{1}{2}\right) + 2\left(\frac{1}{4}\right) + 3\left(\frac{1}{8}\right) + 4\left(\frac{1}{16}\right) + 5\left(\frac{1}{16}\right) = 1.9$$

This expectation is quite close to 2, which (it can be shown) is the expected number of tosses *until the first head*.

13.10 A box contains 8 light bulbs of which 3 are defective. A bulb is selected from the box and tested, until a nondefective bulb is chosen. Find the expected number of bulbs chosen.

Writing D and N for defective and nondefective, respectively, we have as the sample points

$$\text{N} \qquad\qquad \text{DN} \qquad\qquad \text{DDN} \qquad\qquad \text{DDDN}$$

with corresponding probabilities given by Theorem 12.4 as

$$\frac{5}{8} \qquad \frac{3}{8} \cdot \frac{5}{7} = \frac{15}{56} \qquad \frac{3}{8} \cdot \frac{2}{7} \cdot \frac{5}{6} = \frac{5}{56} \qquad \frac{3}{8} \cdot \frac{2}{7} \cdot \frac{1}{6} \cdot \frac{5}{5} = \frac{1}{56}$$

The number of bulbs chosen, X, has the values

$$X(N) = 1 \qquad X(DN) = 2 \qquad X(DDN) = 3 \qquad X(DDDN) = 4$$

with the above respective probabilities. Hence

$$E(X) = 1\left(\frac{5}{8}\right) + 2\left(\frac{15}{56}\right) + 3\left(\frac{5}{56}\right) + 4\left(\frac{1}{56}\right) = \frac{3}{2}$$

13.11 A box contains 10 items of which 2 are defective. If four items are selected from the box, what is the expected number of defective items in the sample of size 4?

As shown in Example 13.9(b), the random variable X that counts the number of defective items in a sample has a hypergeometric distribution, which in this case is given by

$$p_i = \binom{2}{x_i}\binom{8}{4-x_i} \Big/ \binom{10}{4} \qquad (x_i = 0, 1, 2)$$

or

x_i	0	1	2
p_i	70/210	112/210	28/210

Therefore,

$$E(X) = \sum x_i p_i = 0\left(\frac{70}{210}\right) + 1\left(\frac{112}{210}\right) + 2\left(\frac{28}{210}\right) = \frac{4}{5}$$

Note that $E(X) = 4(2/10) = np$, where n is the size of the sample and p is the proportion of defectives in the box. Compare this result, which holds in general for the hypergeometric distribution, with Theorem 13.3 for the binomial distribution.

13.12 A player tosses two fair coins. He wins \$2 if 2 heads occur, and \$1 if 1 head occurs. On the other hand, he loses \$3 if no heads occur. Determine the expected value of the game and if it is favorable to the player.

The sample space is $S = \{HH, HT, TH, TT\}$ and each sample point has probability 1/4. For the player's gain, we have

$$X(HH) = \$2 \qquad X(HT) = X(TH) = \$1 \qquad X(TT) = -\$3$$

and so the distribution of X is

x_i, \$	2	1	-3
p_i	1/4	2/4	1/4

and

$$E(X) = 2\left(\frac{1}{4}\right) + 1\left(\frac{2}{4}\right) - 3\left(\frac{1}{4}\right) = \$0.25$$

Since $E(X) > 0$, the game is favorable to the player.

13.13 A player tosses two fair coins. He wins \$3 if 2 heads occur, and \$1 if 1 head occurs. If the game is to be fair, how much should he lose if no heads occur?

The player's gain, X, has the distribution (see Problem 13.12)

x_i, \$	3	1	x_3
p_i	1/4	2/4	1/4

and

$$E(X) = 3\left(\frac{1}{4}\right) + 1\left(\frac{2}{4}\right) + x_3\left(\frac{1}{4}\right) = \frac{5 + x_3}{4}$$

For a fair game, $E(X) = 0$, or $x_3 = -\$5$. The player should lose \$5 if no heads occur.

13.14 Prove the first half of Theorem 13.3: $E(X_n) = np$.

On the sample space of n repeated trials define the random variables $Y_1, Y_2, \ldots, Y_n$ as the number of successes (0 or 1) on the first, second, $\ldots$, nth trial, respectively. Then each Y_j has the distribution

y	0	1
$P(y)$	q	p

and the expected value $E(Y_j) = 0(q) + 1(p) = p$. Now the total number of successes is just

$$X_n = Y_1 + Y_2 + \cdots + Y_n$$

so that, by linearity of the mean,

$$E(X_n) = E(Y_1 + Y_2 + \cdots + Y_n) = E(Y_1) + E(Y_2) + \cdots + E(Y_n)$$
$$= p + p + \cdots + p = np$$

13.15 Assuming that the variances of the Y_j in Problem 13.14 can be added like their means, prove the second half of Theorem 13.3:

$$\text{Var}(X_n) = npq$$

From the common distribution of the Y_j found in Problem 13.14,

$$E(Y_j^2) = 0^2(q) + 1^2(p) = p$$

and so

$$\text{Var}(Y_j) = E(Y_j^2) - [E(Y_j)]^2 = p - p^2 = p(1-p) = pq$$

Consequently,

$$\text{Var}(X_n) = \text{Var}(Y_1 + Y_2 + \cdots + Y_n) = \text{Var}(Y_1) + \text{Var}(Y_2) + \cdots + \text{Var}(Y_n)$$
$$= pq + pq + \cdots + pq = npq$$

13.16 Four fair coins are tossed. Let X_4 denote the number of heads occurring. Calculate the expectation of X_4 directly, and compare with Theorem 13.3.

X_4 is binomially distributed, with $n = 4$ and $p = 1/2$. We have:

$$P(0) = b(0; 4, \tfrac{1}{2}) = 1/16 \qquad P(3) = b(3; 4, \tfrac{1}{2}) = 4/16$$
$$P(1) = b(1; 4, \tfrac{1}{2}) = 4/16 \qquad P(4) = b(4; 4, \tfrac{1}{2}) = 1/16$$
$$P(2) = b(2; 4, \tfrac{1}{2}) = 6/16$$

Thus the distribution is

x_i	0	1	2	3	4
p_i	1/16	4/16	6/16	4/16	1/16

and the expectation is

$$E(X_4) = 0\left(\frac{1}{16}\right) + 1\left(\frac{4}{16}\right) + 2\left(\frac{6}{16}\right) + 3\left(\frac{4}{16}\right) + 4\left(\frac{1}{16}\right) = 2$$

This agrees with Theorem 13.3, which states that $E(X_4) = np = 4(1/2) = 2$.

13.17 A fair die is tossed 300 times. Find the expected value, μ, and the standard deviation, σ, of the number of sixes.

The number of sixes is binomially distributed, with $n = 300$ and $p = 1/16$. By Theorem 13.3,

$$\mu = np = 300\left(\frac{1}{6}\right) = 50 \qquad \sigma = \sqrt{npq} = \sqrt{300\left(\frac{1}{6}\right)\left(\frac{5}{6}\right)} = 6.45$$

13.18 A family has 8 children. (a) Determine the expected number of girls, if male and female children are equally probable. (b) Find the probability P that the expected number of girls does occur.

(a) The number of girls is binomially distributed, with $n = 8$ and $p = 0.5$. By Theorem 13.3, the expected number of girls is

$$\mu = np = 8(0.5) = 4$$

(b)

$$P = b(4; 8, 0.5) = \binom{8}{4}(0.5)^4(0.5)^4 = 0.27$$

13.19 There is a fixed probability that an item produced by Factory A is defective. A shipment of 10 000 items from Factory A is sent to its warehouse. Show that the standard deviation of the shipment (i.e. of the number of defectives or of the number of nondefectives) cannot exceed 50 items.

We have to do with a binomial distribution, with $n = 10\,000$ and with p (denoting either the probability of success or the probability of failure) unknown. The variance of the shipment is then

$$\sigma^2 = npq = np(1-p) = n\left[\frac{1}{4} - \left(p - \frac{1}{2}\right)^2\right]$$

The right-hand side is as large as possible when $p = 1/2$. Hence,

$$\sigma^2 \le \frac{n}{4} = 2500 \qquad \text{and} \qquad \sigma \le \sqrt{2500} = 50$$

Supplementary Problems

STATISTICS

13.20 The prices of a pound of coffee in seven stores are: $2.58, $3.18, $2.84, $2.75, $2.67, $2.95, $2.62. Find (a) the mean price, (b) the median price.

13.21 During a given month, ten salesmen in an automobile dealership sold 10, 14, 7, 15, 9, 14, 6, 14, 10, and 11 cars, respectively. Find (a) the mean, median, and mode of the data; (b) the variance and standard deviation of the data.

13.22 During a 30-day period, the daily number of station wagons rented by an automobile rental agency was as follows:

$$7, \ 10, \ 6, \ 7, \ 9, \ 4, \ 7, \ 9, \ 9, \ 8, \ 5, \ 5, \ 7, \ 8, \ 4$$
$$6, \ 9, \ 7, \ 12, \ 7, \ 9, \ 10, \ 4, \ 7, \ 5, \ 9, \ 8, \ 9, \ 5, \ 7$$

(a) Give the frequency distribution of the data. (b) Find the mean, variance, and standard deviation.

13.23 The amounts of 45 personal loans from a loan company follow:

$$\$700, \ \$450, \ \$725, \ \$1125, \ \$675, \ \$1650, \ \$750, \ \$400, \ \$1050$$
$$\$500, \ \$750, \ \$850, \ \$1250, \ \$725, \ \$475, \ \$925, \ \$1050, \ \$925$$
$$\$850, \ \$625, \ \$900, \ \$1750, \ \$700, \ \$825, \ \$550, \ \$925, \ \$850$$
$$\$475, \ \$750, \ \$550, \ \$725, \ \$575, \ \$575, \ \$1450, \ \$700, \ \$450$$
$$\$700, \ \$1650, \ \$925, \ \$500, \ \$675, \ \$1300, \ \$1125, \ \$775, \ \$850$$

Group the data into $200-classes, beginning with $400. (a) Find the frequency distribution of the classed data. (b) Find the mean, variance, and standard deviation of the data, using the class values.

13.24 The weekly wages of a group of unskilled workers follow:

Weekly Wages, $	140–160	160–180	180–200	200–220	220–240	240–260	260–280
Number of Workers	18	24	32	20	8	6	2

Find the mean and standard deviation of the data.

13.25 The following distribution gives the numbers of hours of overtime during one month for the employees of a company:

Overtime, h	0	1	2	3	4	5	6	7	8	9	10
Employees	10	2	4	2	6	4	2	4	6	2	8

Find the mean, variance, and standard deviation of the data.

RANDOM VARIABLES, EXPECTATION

13.26 Find the mean μ, variance σ^2 and standard deviation σ of each distribution:

(a)

x_i	2	3	8
p_i	1/4	1/2	1/4

(b)

x_i	-2	-1	7
p_i	1/3	1/2	1/6

(c)

x_i	-1	0	1	2	3
p_i	0.3	0.1	0.1	0.3	0.2

13.27 A pair of fair dice is thrown. Let X be the random variable which denotes the minimum of the two numbers which appear. Find the distribution, mean, variance and standard deviation of X.

13.28 A fair coin is tossed four times. Let Y denote the longest string of heads occurring. Find the distribution, mean, variance and standard deviation of Y.

13.29 Two cards are selected at random from a box which contains five cards numbered 1, 1, 2, 2 and 3. Let X denote the sum and Y the maximum of the two numbers drawn. Find the distribution, mean, variance, and standard deviation of (a) X, (b) Y, (c) $X + Y$, (d) XY. (See Problem 13.8 for the definitions of $X + Y$ and XY.)

13.30 A fair coin is tossed until a head or four tails occur. Find the expected number of tosses of the coin.

13.31 A box contains 8 items of which 2 are defective. A person selects 3 items from the box. Find the expected number of defective items drawn.

13.32 A box contains 10 transistors of which 2 are defective. A transistor is selected from the box and tested, until a nondefective one is chosen. Find the expected number of transistors chosen.

13.33 The probability of team A winning any game is 0.5. A plays team B in a tournament. The first team to win 2 games in a row or a total of three games wins the tournament. Find the expected number of games in the tournament.

13.34 A player tosses three fair coins. He wins $5 if 3 heads occur, $3 if 2 heads occur, and $1 if only 1 head occurs. On the other hand, he loses $15 if 3 tails occur. Find the value of the game to the player.

13.35 A player tosses three fair coins. He wins $8 if 3 heads occur, $3 if 2 heads occur, and $1 if only 1 head occurs. If the game is to be fair, how much should he lose if no heads occur?

13.36 Evaluate (a) $b(3; 6, 0.4)$, (b) $b(2; 5, 0.7)$, (c) $b(4; 6, 0.3)$.

13.37 A geology quiz consists of 10 multiple-choice questions, there being four choices for each question. An unprepared student finds he must guess the answer to every question. What is the probability that the student will pass the quiz, if 70% is the passing grade?

13.38 Team A has probability $p = 0.4$ of winning whenever it plays. Let X denote the number of times A wins in four games. (a) Find the distribution of X. (b) Find the mean, variance, and standard deviation of X.

13.39 A card is drawn and replaced in an ordinary deck of 52 cards. Find the number of times a card must be drawn so that (a) there is at least an even chance of drawing a heart, (b) the probability of drawing a heart is greater than 3/4.

13.40 Let X be a binomially distributed random variable with $E(X) = 2$ and $\text{Var}(X) = 4/3$. Tabulate the distribution of X.

Answers to Supplementary Problems

13.20 (a) \$2.80, (b) \$2.75

13.21 (a) mean = 11, median = 10.5, mode = 10 and 14; (b) variance = 9.6, standard deviation = 3.1

13.22 (a)

Wagons	4	5	6	7	8	9	10	11	12
Days	3	4	2	8	3	7	2	0	1

(b) mean = 7.3, variance = 3.88, standard deviation = 1.97

13.23 (a)

Amount ÷ \$200	2–3	3–4	4–5	5–6	6–7	7–8	8–9
Number of Loans	11	14	10	4	2	1	3

(b) mean = \$842, variance = 109 925 \$^{2}, standard deviation = \$332

13.24 mean = \$190, standard deviation = \$31

13.25 mean = 2.46 h, variance = 12.41 h^2, standard deviation = 3.52 h

13.26 (a) $\mu = 4$, $\sigma^2 = 5.5$, $\sigma = 2.3$; (b) $\mu = 0$, $\sigma^2 = 10$, $\sigma = 3.2$; (c) $\mu = 1$, $\sigma^2 = 2.4$, $\sigma = 1.5$

13.27

x_i	1	2	3	4	5	6
p_i	11/36	9/36	7/36	5/36	3/36	1/36

, $E(X) = 2.5$, $\text{Var}(X) = 2.1$, $\sigma_X = 1.4$

13.28

y_j	0	1	2	3	4
p_j	1/16	7/16	5/16	2/16	1/16

, $E(Y) = 1.7$, $\text{Var}(Y) = 0.9$, $\sigma_Y = 0.95$

13.29 (a)

x_i	2	3	4	5
$P(x_i)$	0.1	0.4	0.3	0.2

, $E(X) = 3.6$, $\text{Var}(X) = 0.84$, $\sigma_X = 0.9$

(b)

y_j	1	2	3
$P(y_j)$	0.1	0.5	0.4

, $E(Y) = 2.3$, $\text{Var}(Y) = 0.41$, $\sigma_Y = 0.64$

(c)

z_k	3	5	6	7	8
$P(z_k)$	0.1	0.4	0.1	0.2	0.2

, $E(X+Y) = E(X) + E(Y) = 5.9$, $\text{Var}(X+Y) = 2.3$, $\sigma_{X+Y} = 1.5$

(d)

w_k	2	6	8	12	15
$P(w_k)$	0.1	0.4	0.1	0.2	0.2

, $E(XY) = 8.8$, $\text{Var}(XY) = 17.6$, $\sigma_{XY} = 4.2$

13.30 15/8

13.31 3/4

13.32 11/9

13.33 23/8

13.34 $0.25

13.35 $20

13.36 (a) 0.276, (b) 0.132, (c) 0.060

13.37 $b(7; 10, \frac{1}{4}) + b(8; 10, \frac{1}{4}) + b(9; 10, \frac{1}{4}) + b(10; 10, \frac{1}{4}) = 0.0035$

13.38 (a)

x_i	0	1	2	3	4
p_i	0.13	0.35	0.35	0.15	0.03

(b) $n = 4$, $p = 0.4$, so that $\mu = np = 1.6$, $\sigma^2 = npq = 0.96$, and $\sigma = 0.98$.

13.39 (a) three, (b) five

13.40

x_i	0	1	2	3	4	5	6
p_i	64/729	192/729	240/729	160/729	60/729	12/729	1/729

Chapter 14

Graphs, Directed Graphs, Machines

14.1 INTRODUCTION

The term "graph" in mathematics has different meanings. We have already introduced the "graph" of a function and of a relation. In the present chapter we shall use the word "graph" in another sense—one alluded to in Section 6.13, where we spoke of the directed graph of a relation.

Graphs, especially tree graphs, and directed graphs appear in many different places in the computer and information sciences. Flowcharts, for example, discussed in Chapter 5, are directed graphs. Other examples are treated in this chapter. We close the chapter with the definition of a *finite state machine*, of which the computer is an example.

14.2 GRAPHS AND MULTIGRAPHS

A *graph* G consists of two things:
(i)   A set V whose elements are called *vertices*, *points* or *nodes*.
(ii) A set E of unordered pairs of distinct vertices, called *edges*.
We denote a graph by $G(V, E)$ when we want to emphasize the two parts of G.

Vertices u and v are said to be *adjacent* if there is an edge $\{u, v\}$.

We picture graphs by diagrams in the plane in a natural way. That is, each vertex v in V is represented by a dot (or small circle) and each edge $e = \{v_1, v_2\}$ is represented by a curve which connects its *endpoints* v_1 and v_2.

EXAMPLE 14.1

(a) Figure 14-1(a) represents the graph G with four vertices, A, B, C, and D, and five edges, $e_1 = \{A, B\}$, $e_2 = \{B, C\}$, $e_3 = \{C, D\}$, $e_4 = \{A, C\}$, $e_5 = \{B, D\}$. We usually denote a graph by drawing its diagram rather than explicitly listing its vertices and edges.

(b) Figure 14-1(b) is not a graph but a *multigraph*. The reason is that e_4 and e_5 are *multiple edges*, i.e. edges connecting the same endpoints, and e_6 is a *loop*, i.e. an edge whose endpoints are the same vertex. The definition of a graph permits neither multiple edges nor loops. In other words, we may define a graph as *a multigraph without multiple edges or loops*. A multigraph is said to be *finite* if it has a finite number of vertices and a finite number of edges. Unless otherwise specified, the multigraphs considered in this book shall be finite. Note that a graph with a finite number of vertices must automatically have a finite number of edges and so must be finite.

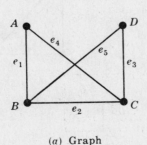

(a) Graph   (b) Multigraph

Fig. 14-1

Let $G(V, E)$ be a graph. Let V' be a subset of V and let E' be a subset of E whose endpoints belong to V'. Then $G(V', E')$ is a graph and is called a *subgraph* of $G(V, E)$. If E' contains all the edges of E whose endpoints lie in V', then $G(V', E')$ is called the subgraph *generated by* V'.

14.3 DEGREE OF A VERTEX

If v is an endpoint of an edge e, then we say that e is *incident* on v. The *degree* of a vertex v, written $\deg(v)$, is equal to the number of edges which are incident on v. (A vertex of degree zero, i.e. one that does not belong to any edge, is called an *isolated* vertex.) Since each edge is counted twice in summing the degrees of the vertices of a graph, we have the following simple but important result.

Theorem 14.1: The sum of the degrees of the vertices of a graph is equal to twice the number of edges.

EXAMPLE 14.2 In Fig. 14-1(a),

$$\deg(A) = 2 \qquad \deg(B) = 3 \qquad \deg(C) = 3 \qquad \deg(D) = 2$$

The sum of the degrees equals ten, which, as expected, is twice the number of edges. A vertex is said to be *even* or *odd* according as its degree is an even or an odd number. Thus A and D are even vertices, whereas B and C are odd vertices.

Theorem 14.1 also holds for multigraphs if a loop is counted twice towards the degree of its endpoint. Thus, in Fig. 14-1(b), we have $\deg(D) = 4$, since the edge e_6 is counted twice; hence D is an even vertex.

14.4 CONNECTIVITY

A *walk* in a multigraph consists of an alternating sequence of vertices and edges of the form

$$v_0, e_1, v_1, e_2, v_2, \ldots, e_{n-1}, v_{n-1}, e_n, v_n$$

where each edge e_i is incident on v_{i-1} and v_i. The number n of edges is called the *length* of the walk. When there is no ambiguity we denote a walk by its sequence of edges $(e_1, e_2, \ldots, e_n)$ or by its sequence of vertices $(v_0, v_1, \ldots, v_n)$. The walk is said to be *closed* if $v_0 = v_n$. Otherwise, we say that the walk is from v_0 to v_n, or *between v_0 and v_n*, or *connects v_0 to v_n*.

A *trail* is a walk in which all edges are distinct. A *path* is a walk in which all vertices are distinct; hence a path must be a trail. A *cycle* is a closed walk such that all vertices are distinct except $v_0 = v_n$. A cycle of length k is called a *k-cycle*. In a graph, any cycle must have length (number of edges or number of vertices) three or more.

EXAMPLE 14.3 Consider the graph in Fig. 14-2. Then the sequence

$$(P_4, P_1, P_2, P_5, P_1, P_2, P_3, P_6)$$

is a walk from P_4 to P_6. It is not a trail since the edge $\{P_1, P_2\}$ is used twice. The sequence

$$(P_4, P_1, P_5, P_2, P_6)$$

is not a walk since there is no edge $\{P_2, P_6\}$. The sequence

$$(P_4, P_1, P_5, P_2, P_3, P_5, P_6)$$

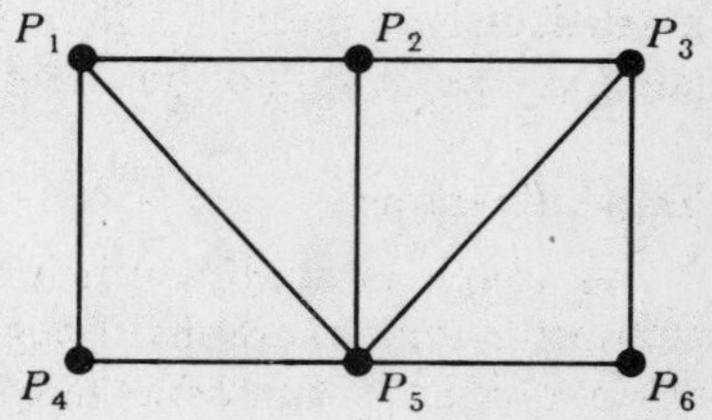

Fig. 14-2

is a trail since no edge is used twice; but it is not a path since the vertex P_5 is used twice. The sequence

$$(P_4, P_1, P_5, P_3, P_6)$$

is a path from P_4 to P_6. The shortest path (with respect to length) from P_4 to P_6 is (P_4, P_5, P_6), which has length 2.

By eliminating unnecessary edges, it is not difficult to show that any walk from a vertex u to a vertex v can be replaced by a path from u to v.   We state this result formally.

Theorem 14.2:   There is a walk from a vertex u to a vertex v if and only if there is a path from u to v.

A graph is said to be *connected* if there is a path between any two of its vertices.   The graph in Fig. 14-2 is connected, but the graph in Fig. 14-3(a) is not connected, because, for example, there is no path between D and E.   A connected subgraph of a graph G is called a connected *component* of G if it is not contained in any larger connected subgraph.   It is intuitively clear that any graph can be partitioned into its connected components.   For example, the graph of Fig. 14-3(a) has three connected components.   Any connected graph has itself as its single connected component.

The *distance* between vertices u and v of a connected graph G, written $d(u, v)$, is the length of the shortest path between u and v.   The *diameter* of a connected graph G is the maximum distance between any two of its vertices.   In Fig. 14-3(b), we have $d(A, F) = 2$ and the diameter of the graph is 3.   (Although the edges $\{A, D\}$ and $\{B, C\}$ are pictured crossing in Fig. 14-3(b), they do not meet at a vertex.)

Let v be a vertex of a graph G.   By $G - v$ we mean the graph obtained from G by deleting v and all edges incident on v.   A vertex v in a connected graph G is called a *cut point* if $G - v$ is disconnected.   The vertex D in Fig. 14-3(b) is a cut point.

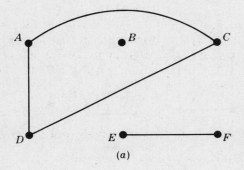

(a)

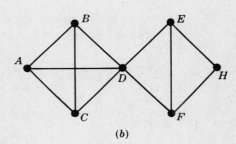

(b)

Fig. 14-3

14.5   SPECIAL TYPES OF GRAPHS

There are many different types of graphs; we mention four of them here.

Complete Graphs

A graph is *complete* if each vertex is connected to every other vertex.   The complete graph with n vertices is denoted by K_n.   Figure 14-4 shows the graphs $K_1, K_2, \ldots, K_6$.   The graph K_1, an isolated vertex, is called the *trivial graph*.

Regular Graphs

A graph or multigraph is *regular of degree k* or *k-regular* if every vertex has degree k.   The connected regular graphs of degrees 0, 1 or 2 are easily described.   The connected 0-regular graph is the trivial graph.   The connected 1-regular graph is the graph with two vertices and one edge connecting them.   The connected 2-regular graph with $n > 2$ vertices is the graph which consists of a single n-cycle.   (The 2-cycle is a multigraph.)   See Fig. 14-5.

The 3-regular graphs must, by Theorem 14.1, have an even number of vertices.   Figure 14-6 gives two examples of connected 3-regular graphs with six vertices.   In general, regular graphs can be quite complicated.   For example, there are nineteen 3-regular graphs with ten vertices.   We note that the complete graph with n vertices, K_n, is regular of degree $n - 1$.

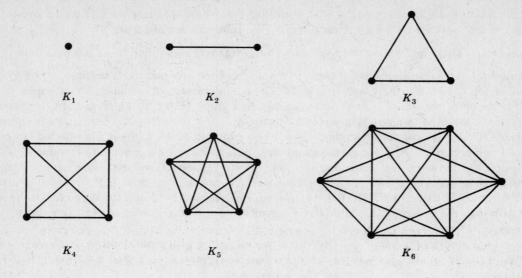

K_1 K_2 K_3

K_4 K_5 K_6

Fig. 14-4

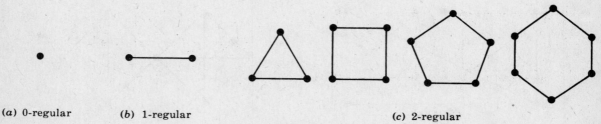

(a) 0-regular   (b) 1-regular   (c) 2-regular

Fig. 14-5

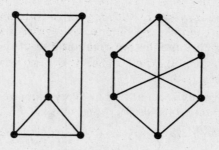

Fig. 14-6

Bipartite Graphs

A graph G is said to be *bipartite* if its vertex set V can be partitioned into two subsets M and N such that each edge of G connects a vertex of M to a vertex of N. By a complete bipartite graph, we mean that each vertex of M is connected to each vertex of N; this graph is denoted by $K_{m,n}$ where m is the number of vertices in M and n is the number of vertices in N, and, for standardization, we assume $m \leq n$. Figure 14-7 shows the graphs $K_{2,3}$, $K_{3,3}$ and $K_{2,4}$. Clearly, $K_{m,n}$ has mn edges.

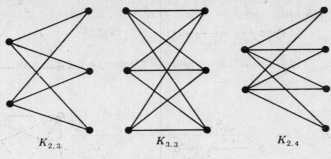

Fig. 14-7

Planar Graphs

A graph or multigraph which can be drawn in the plane so that its edges do not cross is said to be *planar*. Although the complete graph with four vertices, K_4, is usually pictured with crossing edges, as in Fig. 14-8(a), it can also be drawn with noncrossing edges, as in Fig. 14-8(b). Thus K_4 is a planar graph.

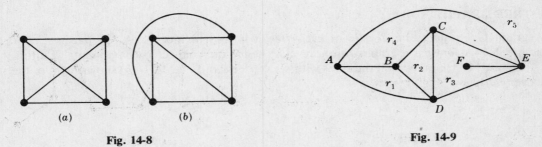

(a)	(b)

Fig. 14-8   **Fig. 14-9**

A particular planar representation of a finite planar multigraph is called a *map*. We say that the map is *connected* if the underlying multigraph is connected. A given map divides the plane into various regions. For example, the map in Fig. 14-9 divides the plane into five regions. Observe that four of the regions are bounded, but the fifth region, outside the diagram, is unbounded. Thus there is no loss in generality in counting the number of regions if we assume that our map is contained in some large rectangle rather than in the entire plane.

Euler gave a formula which connects the number V of vertices, the number E of edges and the number R of regions of any connected map.

Theorem 14.3 (Euler): $V - E + R = 2$

EXAMPLE 14.4   In Fig. 14-9 we have $V = 6$, $E = 9$, $R = 5$; and, in accordance with Euler's formula,

$$V - E + R = 6 - 9 + 5 = 2$$

Euler's formula holds for disconnected maps as well, provided the constant 2 is replaced by $\nu + 1$, where ν is the number of connected components of the map.

14.6   LABELED GRAPHS

A graph G is called a *labeled graph* if its edges and/or vertices are assigned data of one kind or another. In particular, if each edge e of G is assigned a nonnegative number $\ell(e)$ then $\ell(e)$ is called the *weight* or *length* of e. Figure 14-10 shows a labeled graph where the weight of each edge is indicated in the obvious way. It is often important to find a path of least weight between two given vertices of a labeled graph. A minimum path between P and Q in Fig. 14-10 is given by the sequence

$$(P, A_1, A_2, A_5, A_3, A_6, Q)$$

The path has weight 14. (Find another minimum path between the same two vertices.)

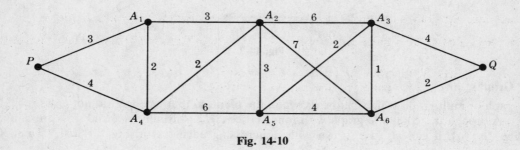

Fig. 14-10

14.7 TREE GRAPHS

A graph G is said to be *acyclic* or *cycle-free* if it contains no cycles. A *tree* is a connected acyclic graph. A *forest* is a graph with no cycles; hence the connected components of a forest are trees. Figure 14-11(*a*) shows all the trees with six vertices, and Fig. 14-11(*b*) shows eight of the trees with seven vertices.

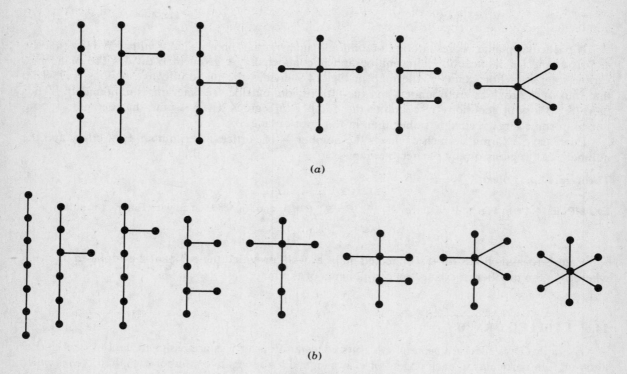

(*a*)

(*b*)

Fig. 14-11

There are a number of equivalent ways of defining a tree, as shown by the following theorem.

Theorem 14.4:   Let G be a graph with more than one vertex.   Then the following are equivalent:
   (i) G is a tree.
   (ii)   Each pair of vertices is connected by exactly one path.
   (iii) G is connected, but if any edge is deleted then the resulting graph is not connected.
   (iv) G is cycle-free, but if any edge is added to the graph then the resulting graph has exactly one cycle.

In the case that our graphs are finite, then we have additional ways of defining a tree.

Theorem 14.5:   Let G be a finite graph with $n > 1$ vertices.   Then the following are equivalent:
   (i) G is a tree.
   (ii) G is cycle-free and has $n - 1$ edges.
   (iii) G is connected and has $n - 1$ edges.

In particular, Theorem 14.5 tells us that a finite tree has one more vertex than edge.   (This holds also for the *trivial tree*, $n = 1$.)

A few general properties of trees are:

(1) A tree is a bipartite graph.

(2) A tree is a planar graph.

(3) A finite nontrivial tree has at least two *endpoints* (vertices of degree 1).

(4) In a nontrivial tree, each vertex is either an endpoint or a cut point.

Spanning Trees

A subgraph T of a graph G is called a *spanning tree* of G if T is a tree and T includes all the vertices of G.   Figure 14-12 shows a graph G and spanning trees T_1, T_2 and T_3 of G.   If G is a graph whose edges have weights, then a *minimal spanning tree* of G is a spanning tree of G such that the sum of the weights of its edges is minimal among all spanning trees of G.

We give two algorithms to find a minimal spanning tree of a finite connected labeled graph G with m vertices.   First, order the edges of G according to decreasing weight.   Proceeding sequentially, delete each edge which does not disconnect the graph until $m - 1$ edges remain.   These edges will then form a minimal spanning tree of G.   This algorithm depends upon knowing whether or not a graph is connected, which, in general, is not easily programmable.

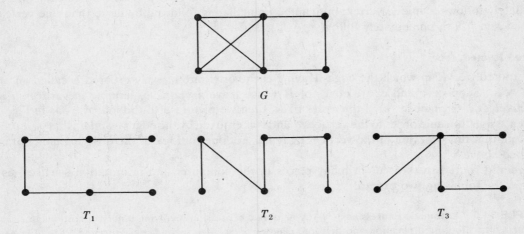

Fig. 14-12

For the second algorithm, the edges are ordered according to increasing weight.   Then, begin-ning with only the vertices of G, we add one edge after another where each edge has minimal weight and does not form any cycle.   After adding $m-1$ edges we obtain a minimal spanning tree.

We emphasize that since some edges can be of the same weight, we can obtain different minimal spanning trees.   Figure 14-13 gives a labeled connected graph G and a minimal spanning tree M.

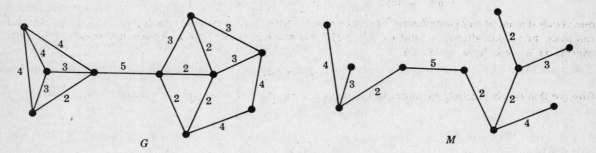

Fig. 14-13

14.8   ROOTED TREES

A *rooted tree* R consists of a tree graph together with a designated vertex r called the *root* of the tree.   The length of the unique path from the root r to any other vertex v is called the *level* or *depth* or *generation* of v.   The endpoints of R (excepting r, if it is an endpoint) are called the *leaves* of the rooted tree.   Figure 14-14 shows a rooted tree where the root r is pictured at the top of the tree.   The tree has five leaves, d, f, h, i and j.   The level of a is 1, the level of f is 2, and the level of j is 3.   We emphasize that any tree may be made into a rooted tree by simply picking any one of the vertices as the root.

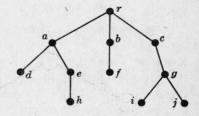

Fig. 14-14

The fact that there is a unique path from the root to any vertex of R induces an orientation of the edges of R.   A continuously directed path from a vertex to a leaf is called a *branch* of R.   We will say that a vertex u *precedes* a vertex v, or that v *follows* u, if the path from the root r to v includes u.   In particular, we say that v *immediately follows* u if v follows u and is adjacent to u.   In Fig. 14-14, the vertex j follows c, but immediately follows g.   Observe that every vertex other than the root r immediately follows a unique vertex, but can be immediately followed by more than one vertex, e.g. vertices i and j both immediately follow g.

Ordered Rooted Trees

A rooted tree R in which the edges leaving each vertex are linearly ordered is called an *ordered rooted tree*.   Suppose e and e' are edges of R which leave a vertex v, running to vertices a and b, respectively.   If e precedes e' in the order of R, then we picture e to the left of e', as in Fig. 14-15. We then assign the same order to the vertices a and b as enjoyed by the edges e and e'; that is, a precedes b.   (Note that this ordering of the vertices of R has nothing to do with the ordering along branches, described above.)

Ordered rooted trees occur in many places in the computer and information sciences, as illus-trated in the following two examples.

EXAMPLE 14.5 (Arithmetic Expressions).   Any algebraic expression involving binary operations, for example, addition, subtraction, multiplication and division, can be represented by an ordered rooted tree.   For example, the arithmetic expression

$$(a - b)/((c \times d) + e)$$

can be represented by the ordered rooted tree in Fig. 14-16(*a*). Observe that the variables in the expression, *a*, *b*, *c*, *d* and *e*, appear as leaves, and the operations appear as the other vertices. The tree must be ordered; *a* − *b* and *b* − *a* yield the same tree but not the same ordered tree.

The Polish mathematician Lukasiewicz observed that by placing the binary operational symbol *before* its arguments, e.g.

$$+ a\,b \quad \text{instead of} \quad a + b \quad \text{and} \quad /\,c\,d \quad \text{instead of} \quad c/d$$

one avoids the use of any parentheses. This notation is called *Polish notation* in *prefix form*. (Analogously, one can place the symbol after its arguments, called *Polish notation* in *postfix form*.) Rewriting the above arithmetic expression in prefix form, we obtain

$$/ - a\,b + \times c\,d\,e$$

Observe that this is precisely the order of the vertices when the tree is scanned as indicated in Fig. 14-16(*b*).

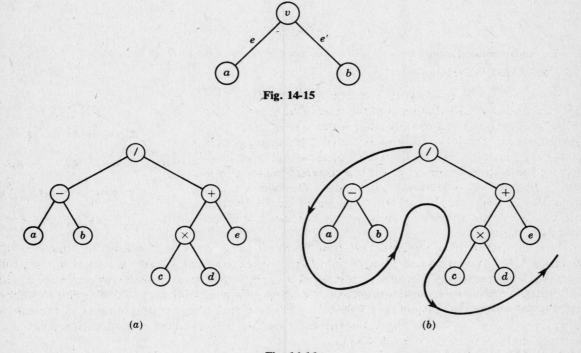

Fig. 14-15

(*a*)   (*b*)

Fig. 14-16

EXAMPLE 14.6 (Record Structure).   Data are frequently organized into a hierarchy of fields, records, and files, as follows.   A *record* is a collection of related data items, also called *fields*, which are treated as a unit; and a *file* is a collection of similar records.   For example, an employee personnel record may contain the data items:

Social Security Number, Name, Address, Age, Salary, Dependents

The company's employee file would contain the list of employee records.

Although a file is normally a linear list of records, the data items of a record usually form an ordered rooted tree.   The reason is that some of the data items may be *group items*, i.e. items consisting of two or more subitems, rather than *elementary items*, i.e. items which are not subdivided.   For example, the above employee personnel record may form the ordered rooted tree shown in Fig. 14-17.   Observe that Name is a group item, with subitems Last, First, and MI (Middle Initial).   Also, Address is a group item, with subitems Street address and Area address, where Area is itself a group item having subitems City, State and Zip code number.   There are eleven elementary items, the leaves of the tree.

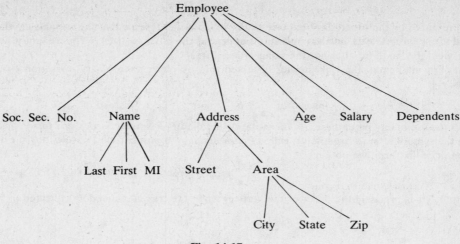

Fig. 14-17

The above ordered rooted tree can also be described in terms of the levels of the vertices:

```
00  Employee
        01  Social Security Number
        01  Name
                02  Last
                02  First
                02  Middle Initial
        01  Address
                02  Street
                02  Area
                        03  City
                        03  State
                        03  Zip
        01  Age
        01  Salary
        01  Dependents
```

This listing is equivalent to scanning the tree in the manner of Fig. 14-16(*b*); by means of it, we can read the entire record in linear fashion into the computer.

14.9   DIRECTED GRAPHS

We have just seen, in Section 14.8, that rooted trees have a natural direction defined along each edge; they may thus be considered directed graphs.   Generally speaking, a *directed graph*, also called a *digraph*, is a multigraph with a direction assigned to each edge.   We call the directed edges *arcs*, and write $a = \langle u, v \rangle$ for any one of the arcs joining the *initial point u* to the *terminal point v*.

EXAMPLE 14.7   Figure 14-18 represents a digraph that has four vertices and seven arcs.   Note the directed loop, $a_7 = \langle B, B \rangle$.   Arcs like a_2 and a_3, which have the same initial point and the same terminal point, are said to be *parallel*.

Let *D* be a directed graph.   An arc $a = \langle u, v \rangle$ is said to *begin* at its initial point *u* and *end* at its terminal point *v*.   The *outdegree* and *indegree* of a vertex *v* are equal respectively to the number of arcs beginning and ending at *v*.   Since each arc begins and ends at a vertex, we see that the sum of the outdegrees of the vertices equals the sum of the indegrees of the vertices, which equals the number of arcs.   A vertex with zero indegree is called a *source*, and a vertex with zero outdegree is called a *sink*.   In Fig. 14-18, the vertex C is a sink, but the digraph has no sources.   Also, the outdegree of D is 2 and its indegree is 1.

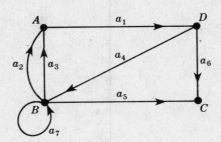

Fig. 14-18

If the arcs and/or vertices of a directed graph are labeled with some type of data, then we have a *labeled directed graph*.   Such graphs are frequently used to picture dynamic situations.   For example, flowcharts are directed graphs in which the vertices (boxes) are labeled, and the arcs from a decision box are labeled.   Another example follows.

EXAMPLE 14.8   Three boys, A, B, and C, are throwing a ball amongst themselves such that A always throws the ball to B, but B and C are just as likely to throw the ball to A as they are to each other.   Figure 14-19 illustrates this dynamic situation, where arcs are labeled with the respective probabilities, i.e. A throws the ball to B with probability 1, B throws the ball to A and C each with probability $\frac{1}{2}$, and C throws the ball to A and B each with probability $\frac{1}{2}$.

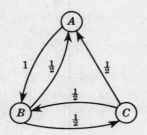

Fig. 14-19

Digraphs and Relations

Consider a digraph D which possesses no parallel arcs; let V be the set of its vertices and A be the set of its arcs.   Since the arcs represent distinct ordered pairs of vertices, A is simply a subset of $V \times V$ and hence A is a relation on V.   Conversely, if R is a relation on a set V, then V may be taken as the vertex set and R as the arc set of a digraph, $D(V, R)$, that has no parallel arcs.   Thus the concepts of relations on a set and digraphs without parallel arcs are one and the same.   In fact, in Chapter 6 we have already introduced the directed graph corresponding to a relation on a set.

Digraphs and Matrices

Let D be a directed graph with vertices $v_1, v_2, \ldots, v_m$.   The *matrix* of D is the $m \times m$ matrix $M_D = (m_{ij})$, where

$$m_{ij} = \text{the number of arcs beginning at } v_i \text{ and ending at } v_j$$

If D has no parallel arcs, then the entries of M_D will be only zeros and ones; otherwise the entries will be nonnegative integers. Conversely, every $m \times m$ matrix M with nonnegative integer entries uniquely defines a digraph with m vertices. Figure 14-20 shows a digraph D and the corresponding matrix M.

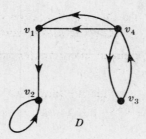

Fig. 14-20

14.10   CONNECTED DIGRAPHS

The concepts of walk, trail, path, and cycle carry over from undirected graphs except that the direction of the walk, etc., must agree with the directions of the arcs. Specifically, a (directed) *walk* W in a digraph D is an alternating sequence of vertices and arcs,

$$W = (v_0, a_1, v_1, a_2, v_2, \ldots, a_n, v_n)$$

such that each arc a_i begins at v_{i-1} and ends at v_i. A *semiwalk* is the same as a walk except that the arc a_i may begin at either v_{i-1} or v_i and end at the other vertex; in other words, a semiwalk is the same as a walk on the undirected multigraph D. *Semitrails* and *semipaths* have analogous definitions.

There are three types of connectivity in a digraph D. We say that D is *weakly connected* or *weak* if there is a semipath between any two vertices u and v of D. We say that D is *unilaterally connected* or *unilateral* if, for any vertices u and v of D, there is either a path from u to v or a path from v to u. We say that D is *strongly connected* or *strong* if, for any vertices u and v of D, there is a path from u to v and one from v to u. Observe that strongly connected implies unilaterally connected, and that unilaterally connected implies weakly connected. We say that D is *strictly unilateral* if it is unilateral but not strong, and we say that D is *strictly weak* if it is weak but not unilateral. For example, Fig. 14-21 shows a graph (a) which is strictly weak, a graph (b) which is strictly unilateral, and a graph (c) which is strong.

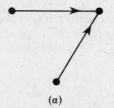   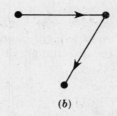   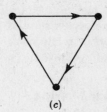

(a)   (b)   (c)

Fig. 14-21

In terms of *spanning walks* (walks containing all the vertices of the digraph), connectivity can be characterized as follows:

Theorem 14.6:　Let D be a finite directed graph.　Then
　　(a)　D is weak if and only if D has a spanning semiwalk.
　　(b)　D is unilateral if and only if D has a spanning walk.
　　(c)　D is strong if and only if D has a closed spanning walk.

The matrix M of a digraph D is useful in counting walks in D.　In fact, we have:

Theorem 14.7:　Let M be the matrix of a digraph D.　Then the (i, j)-entry of the matrix M^n gives the number of walks of length n from the vertex v_i to the vertex v_j.

Digraphs with sources and sinks appear in many applications (e.g. flow diagrams).　A sufficient condition for such vertices to exist follows:

Theorem 14.8:　If a finite digraph D contains no (directed) cycles, then D contains at least one source and at least one sink.

14.11　FINITE STATE MACHINES

We may view a digital computer as a machine which is in a certain "internal state" at any given moment.　The computer "reads" an input symbol, and then "prints" an output symbol and changes its "state".　The output symbol depends solely upon the input symbol and the internal state of the machine, and the internal state of the machine depends solely upon the preceding state of the machine and the preceding input symbol.　The number of states, input symbols and output symbols are assumed to be finite.　These ideas are formalized in the following definition.

A *finite state machine* (or *complete sequential machine*) M consists of five things:

(1)　A finite set A of *input symbols*.

(2)　A finite set S of *internal states*.

(3)　A finite set Z of *output symbols*.

(4)　A *next-state function f* from $S \times A$ into S.

(5)　An *output function g* from $S \times A$ into Z.

This machine M is denoted by $M = \langle A, S, Z, f, g \rangle$ when we want to designate its five parts.　Sometimes we are also given an *initial state* q_0 in S, and then the machine M is designated by the sextuple $M = \langle A, S, Z, q_0, f, g \rangle$.

EXAMPLE 14.9　The following define a finite state machine with two input symbols, three internal states, and three output symbols:

(1)　$A = \{a, b\}$
(2)　$S = \{q_0, q_1, q_2\}$
(3)　$Z = \{x, y, z\}$
(4)　Next-state function $f : S \times A \to S$ defined by

$$f(q_0, a) = q_1 \quad f(q_1, a) = q_2 \quad f(q_2, a) = q_0$$
$$f(q_0, b) = q_2 \quad f(q_1, b) = q_1 \quad f(q_2, b) = q_1$$

(5)　Output function $g : S \times A \to Z$ defined by

$$g(q_0, a) = x \quad g(q_1, a) = x \quad g(q_2, a) = z$$
$$g(q_0, b) = y \quad g(q_1, b) = z \quad g(q_2, b) = y$$

It is traditional to use the letter q for the states of a machine and to use the symbol q_0 for the initial state.

There are two ways of representing a finite state machine in compact form. The *state diagram* D of a finite state machine $M = \langle A, S, Z, f, g \rangle$ is a labeled directed graph D. The vertices of D are the states S of M, and if

$$f(q_i, a_j) = q_k \qquad \text{and} \qquad g(q_i, a_j) = z_r$$

then there is an arc from q_i to q_k which is labeled with the pair a_j, z_r. We usually put the input symbol a_j near the base of the arrow representing the arc (near q_i) and the output symbol z_r near the center of the arrow. In case an initial state q_0 is given, then we label the vertex q_0 by drawing an extra arrow into q_0. For example, Fig. 14-22(a) is the state diagram of the machine in Example 14.9, with q_0 being the initial state.

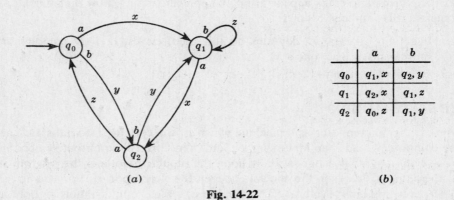

	a	b
q_0	q_1, x	q_2, y
q_1	q_2, x	q_1, z
q_2	q_0, z	q_1, y

(a)   (b)

Fig. 14-22

Alternatively, the machine can be represented by its *state table*, which for each combination of state and input lists the next state and the output. Figure 14-22(b) is the state table for the machine of Example 14.9.

14.12 STRINGS. INPUT AND OUTPUT TAPES

The previous section did not show the dynamic quality of a machine. Suppose a finite state machine M is given a string of input symbols:

$$U = a_1 a_2 \ldots a_n$$

We visualize these symbols on an "input tape"; the machine M "reads" these input symbols one by one and, simultaneously, changes through a string of states

$$V = s_0 s_1 s_2 \ldots s_n$$

where s_0 is the initial state, while "printing" a string of output symbols

$$W = z_1 z_2 \ldots z_n$$

on an "output tape". Formally, the initial state s_0 and the input string U determine the strings V and W by

$$s_i = f(s_{i-1}, a_i) \qquad \text{and} \qquad z_i = g(s_{i-1}, a_i)$$

where $i = 1, 2, \ldots, n$. (Observe that the word "string" is used for a finite sequence instead of "*n*-tuple" or "list.")

EXAMPLE 14.10 Suppose q_0 is the initial state of the machine in Example 14.9, and suppose the machine is given the input string *abaab*. We calculate the string of states and the string of output symbols from the state diagram by beginning at the vertex q_0 and following the arrows which are labeled with the input symbols:

$$q_0 \xrightarrow{a,\,x} q_1 \xrightarrow{b,\,z} q_1 \xrightarrow{a,\,x} q_2 \xrightarrow{a,\,z} q_0 \xrightarrow{b,\,y} q_2$$

This yields the following strings of states and output symbols:

$$q_0 q_1 q_1 q_2 q_0 q_2 \quad \text{and} \quad xzxzy$$

We now want to describe a machine which can do binary addition. By adding 0s at the beginning of our numbers, we can ensure that all numbers have the same number of digits. If the machine is given the input

$$1101011$$
$$+\,0111011$$

then we want the output to be the binary sum

$$10100110$$

Specifically, the input is the string of pairs of digits to be added:

$$11,\ 11,\ 00,\ 11,\ 01,\ 11,\ 10,\ b$$

where b denotes blank spaces, and the output should be the string

$$0,\ 1,\ 1,\ 0,\ 0,\ 1,\ 0,\ 1$$

We also want the machine to enter a state called "stop" when the machine finishes the addition. The input symbols are

$$A = \{00,\ 01,\ 10,\ 11,\ b\}$$

and the output symbols are

$$Z = \{0,\ 1,\ b\}$$

The machine that we "construct" will have three states:

$$S = \{\text{carry } (c),\ \text{no carry } (n),\ \text{stop } (s)\}$$

Here n is the initial state. The machine is diagramed in Fig. 14-23.

In order to show the limitations of our machines, we state the following theorem.

Theorem 14.9: There is no finite state machine which can do binary multiplication.

However, if we limit the size of the numbers to be multiplied, then such machines do exist. Computers are important examples of finite state machines which multiply restricted numbers.

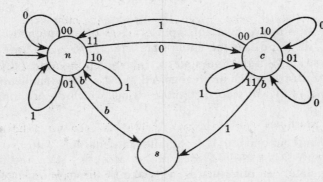

Fig. 14-23

14.13 FINITE AUTOMATA

A finite automaton is similar to a finite state machine except that an automaton has "accepting" and "rejecting" states rather than an output. Specifically, a *finite automaton M* consists of five things:

(1) A finite set A of *input symbols.*

(2) A finite set S of *internal states.*

(3) A subset T of S (whose elements are called *accepting states*).

(4) An *initial state* q_0 in S.

(5) A *next-state function f* from $S \times A$ into S.

The automaton M is denoted by $M = \langle A, S, T, q_0, f \rangle$ when we want to designate its five parts.

EXAMPLE 14.11 The following define a finite automaton with two input symbols and three states:

(1) $A = \{a, b\}$, input symbols.
(2) $S = \{q_0, q_1, q_2\}$, states.
(3) $T = \{q_0, q_1\}$, accepting states.
(4) q_0, the initial state.
(5) Next-state function $f : S \times A \to S$ defined by the table at right.

	a	b
q_0	q_0	q_1
q_1	q_0	q_2
q_2	q_2	q_2

We can concisely describe a finite automaton M by its state diagram as was done with finite state machines, except that here we use double circles for accepting states, and each edge is labeled only by the input symbol. Specifically, the *state diagram D* of M is a labeled directed graph whose vertices are the states of S; accepting states are labeled by having a double circle; and if $f(q_j, a_i) = q_k$, then there is an arc from q_j to q_k which is labeled with a_i. Also, the initial state q_0 is denoted by having an arrow entering the vertex q_0. For example, the state diagram of the automaton M of Example 14.11 is given in Fig. 14-24.

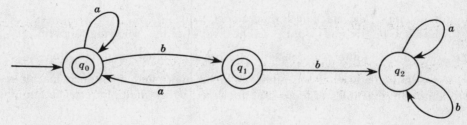

Fig. 14-24

Given a finite string $W = a_1 a_2 a_3 \ldots a_n$ of input symbols of an automaton M, we obtain a sequence of states $s_0 s_1 s_2 \ldots s_n$, where s_0 is the initial state and $s_i = f(s_{i-1}, a_i)$ for $i > 0$. We say that M *recognizes* or *accepts* the string W if the final state, s_n, is an accepting state, i.e. if $s_n \in T$. We will let $L(M)$ denote the set of all strings which are recognized by M. For example, one can show that the automaton M of Example 14.11 will recognize those strings which do not have two successive b's.

Automata as Finite State Machines

We may also view a finite automaton M as a finite state machine with two output symbols, say, YES and NO, where the output is YES if M goes into an accepting state and the output is NO if M goes into a nonaccepting state. In other words, we make M into a finite state machine by defining an output function g from $S \times A$ into $Z = \{$YES, NO$\}$ as follows:

$$g(q_i, a_j) = \begin{cases} \text{YES} & \text{if } f(q_i, a_j) \text{ is accepting (belongs to } T) \\ \text{NO} & \text{if } f(q_i, a_j) \text{ is nonaccepting} \end{cases}$$

Conversely, a finite state machine with two output symbols may be viewed as a finite automaton in an analogous way.

Solved Problems

GRAPHS, CONNECTIVITY

14.1 Draw the diagram of the graph G with: (a) vertices A, B, C, D, and edges $\{A, B\}$, $\{A, C\}$, $\{B, C\}$, $\{B, D\}$, $\{C, D\}$; (b) vertices a, b, c, d, e, and edges $\{a, b\}$, $\{a, c\}$, $\{b, c\}$, $\{d, e\}$. Which of the graphs, if any, are connected?

Draw a dot for each vertex v, and for each edge $\{x, y\}$ draw a curve between vertex x and vertex y, as shown in Fig. 14-25. Graph (a) is connected. However, graph (b) is not connected, since, for example, there is no path from vertex a to vertex d.

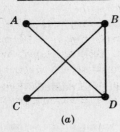
(a)

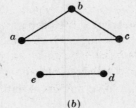
(b)

Fig. 14-25

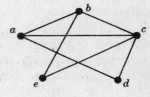

Fig. 14-26

14.2 Consider the graph in Fig. 14-26. Find the degree of each vertex and verify Theorem 14.1 for this graph.

The degree of a vertex is the number of edges to which it belongs; e.g. deg $(a) = 3$, since a belongs to the three edges $\{a, b\}$, $\{a, c\}$, $\{a, d\}$. Similarly, deg $(b) = 3$, deg $(c) = 4$, deg $(d) = 2$, deg $(e) = 2$.

The sum of the degrees of the vertices is $3 + 3 + 4 + 2 + 2 = 14$, which does equal twice the number of edges (2×7).

14.3 Consider the graph in Fig. 14-27. Find (a) all paths from the vertex a to the vertex f, (b) all trails from a to f, (c) the distance between a and f, (d) the diameter of the graph.

(a) A path from a to f is a walk such that no vertex and hence no edge is repeated. There are seven such paths:

(a, b, c, f) (a, d, e, f)

(a, b, c, e, f) (a, d, e, b, c, f)

(a, b, e, f) (a, d, e, c, f)

(a, b, e, c, f)

(b) A trail from a to f is a walk such that no edge is repeated. There are nine such trails, the seven paths from (a), together with

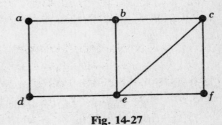

Fig. 14-27

$$(a, d, e, b, c, e, f) \quad \text{and} \quad (a, d, e, c, b, e, f)$$

(c) The distance from a to f is 3 since there is a path, e.g. (a, b, c, f), from a to f of length 3 and no shorter path from a to f.

(d) The distance between any two vertices is not greater than 3, and the distance between a and f is 3; hence the diameter of the graph is 3.

14.4 Which of the multigraphs in Fig. 14-28 are (a) connected, (b) loop-free, (c) graphs?

(a) Only (i) and (iii) are connected.

(b) Only (iv) has a loop, i.e. an edge with the same endpoints.

(c) Only (i) and (ii) are graphs. The multigraph (iii) has multiple edges and (iv) has multiple edges and a loop.

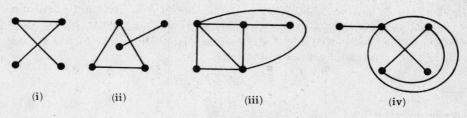

(i)   (ii)   (iii)   (iv)

Fig. 14-28

14.5 Consider the graph G in Fig. 14-27 (Problem 14.3). Find the subgraphs obtained when each vertex is deleted. Does the graph have any cut points?

When we delete a vertex from a graph, we also have to delete all edges incident on the vertex. The six graphs obtained by deleting each of the vertices in Fig. 14-27 are shown in Fig. 14-29. All six graphs are connected; hence no vertex is a cut point.

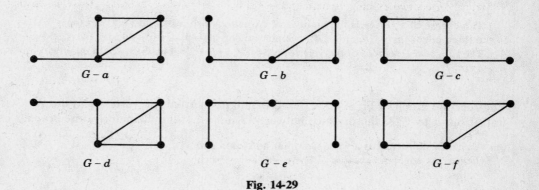

$G-a$ $G-b$ $G-c$

$G-d$ $G-e$ $G-f$

Fig. 14-29

14.6 Draw the graph $K_{2,5}$.

$K_{2,5}$ consists of seven vertices partitioned into a set M of two vertices, say u_1 and u_2, and a set N of five vertices, say v_1, v_2, v_3, v_4 and v_5, and all possible edges from a vertex u_i to a vertex v_j. See Fig. 14-30.

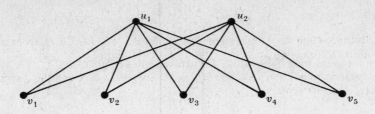

Fig. 14-30

14.7 Which connected graphs are both regular and bipartite?

The bipartite graph $K_{m,m}$ is connected and is regular of degree m, since each vertex is connected to m other vertices. If m disjoint edges are deleted from $K_{m,m}$, the resulting graph G_{m-1} is necessarily bipartite and regular of degree $m-1$, but it is not necessarily connected. For example, the subgraph of $K_{4,4}$ shown in Fig. 14-31(a) is 3-regular and connected, while the subgraph of $K_{2,2}$ in Fig. 14-31(b) is 1-regular and disconnected. In any event, the connected components of G_{m-1} have all the desired properties. In any event, the connected components of G_{m-1} have all the desired properties.

We can continue the process by deleting m disjoint edges from G_{m-1}, thereby obtaining regular, bipartite, connected components of degree $m-2$. And so on.

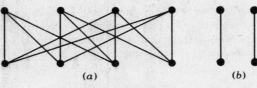

(a)　　　　　　　(b)

Fig. 14-31

14.8 Prove Theorem 14.3.

If our connected map M consists of a single vertex P, as in Fig. 14-32(a), then $V = 1$, $E = 0$, and there is one region, i.e. $R = 1$. Thus in this case $V - E + R = 2$. Otherwise M can be built up from a single vertex by the following two constructions:

(1) Add a new vertex Q_2 and connect it to an existing vertex Q_1 by an edge which does not cross any existing edge, as in Fig. 14-32(b).

(2) Connect two existing vertices Q_1 and Q_2 by an edge which does not cross any existing edge, as in Fig. 14-32(c).

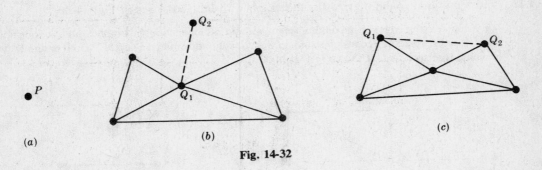

(a)　　　　　　　(b)　　　　　　　(c)

Fig. 14-32

The first operation does not change the value of $V - E + R$, since both V and E are increased by 1 and the number R of regions is not changed. The second operation also does not change the value of $V - E + R$, since V does not change, E is increased by 1, and (it can be shown) the number R of regions is also increased by 1. Accordingly, M must have the same value of $V - E + R$ as the map consisting of a single vertex; that is, $V - E + R = 2$, and the theorem is proved.

TREE GRAPHS

14.9 Draw all trees with five or fewer vertices.

There are eight such trees, which are exhibited in Fig. 14-33.

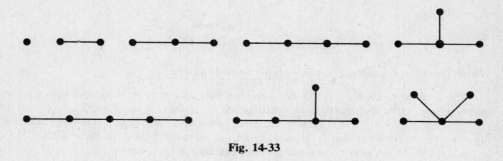

Fig. 14-33

14.10 Find all spanning trees of the graph G shown in Fig. 14-34.

There are eight such spanning trees, as shown in Fig. 14-35. Each spanning tree must have $4 - 1 = 3$ edges since G has four vertices. Thus each tree can be obtained by deleting two of the five edges of G. This can be done in ten ways, except that two of the ways lead to disconnected graphs. Hence the above eight spanning trees are all the spanning trees of G.

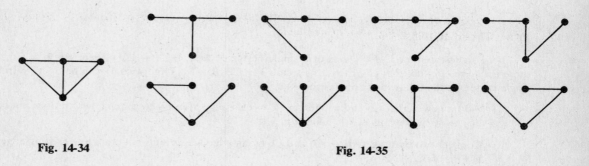

Fig. 14-34   **Fig. 14-35**

14.11 Find a minimal spanning tree for the graph with labeled edges in Fig. 14-36.

Keep deleting edges of maximum weight without disconnecting the graph. Or else, begin with the nine vertices and keep adding edges of minimum weight without forming any cycle. Both methods give a minimal spanning tree such as that shown in Fig. 14-37.

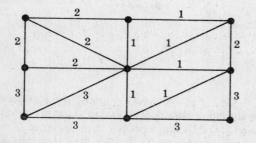

Fig. 14-36

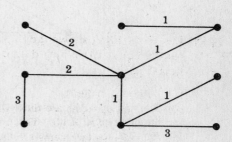

Fig. 14-37

14.12 Consider the tree shown in Fig. 14-38. (*a*) Which vertices, if any, are cut points? (*b*) Find all vertices at level 3 if the vertex picked as a root is (i) *u*, (ii) *w*.

  (*a*)   Each vertex of degree greater than 1 is a cut point in a tree; hence *c*, *r*, *u*, *w*, and *y*.

  (*b*)   Find all paths of length 3 from the root to obtain vertices at level (or depth) 3. Thus: (i) *a*, *b*, and *z*; (ii) *c*, *d*, *s*, and *t*.

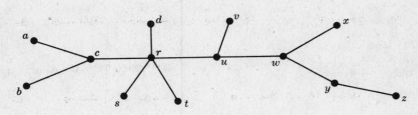

Fig. 14-38

14.13 Consider the algebraic expression $(2x + y)(5a - b)^3$. (*a*) Draw the corresponding ordered rooted tree. (*b*) Find the *scope* of the exponentiation operation. (The scope of a vertex *v* in a rooted tree is the subtree, with root *v*, generated by *v* and the vertices which follow *v* in the tree.) (*c*) Rewrite the expression in prefix Polish notation.

  (*a*)   Use an arrow (↑) for exponentiation and an asterisk (∗) for multiplication to obtain the tree shown in Fig. 14-39.

  (*b*)   The scope of ↑ is the tree circled in the Fig. 14-39. It corresponds to the expression $(5a - b)^3$.

  (*c*)   Scan the tree as in Fig. 14-16(*b*) to obtain

$$* + * 2\,x\,y \uparrow - * 5\,a\,b\,3$$

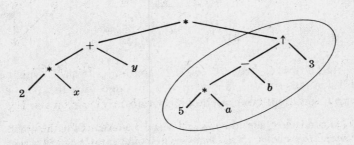

Fig. 14-39

14.14 Figure 14-40(*a*) shows the entries in a student record. (*a*) Draw the corresponding ordered rooted tree. (*b*) Which data items in the record are group items? (*c*) Which are elementary items?

  (*a*)   See Fig. 14-40(*b*).

  (*b*)   The group items are those data items which consist of two or more subitems. These are Name, Birthday, and SAT. These are the vertices which are neither leaves nor the root.

  (*c*)   The elementary items are the data items which are not further partitioned into subitems. These are Number, Last, First, MI, Sex, Day, Month, Year, Math, Verbal, Admittance. These are the leaves of the tree.

```
00  Student
     01  Number
     01  Name
          02  Last
          02  First
          02  MI
     01  Sex
     01  Birthday
          02  Day
          02  Month
          02  Year
     01  SAT score
          02  Math
          02  Verbal
     01  Admittance (date)
```

(a)

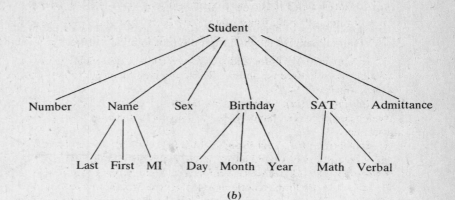

(b)

Fig. 14-40

14.15 Suppose there are two distinct paths, say P_1 and P_2, from a vertex u to a vertex v in a graph G. Prove that G contains a cycle.

Let w be a vertex on P_1 and P_2 such that the next vertices on P_1 and P_2 are distinct. Let w' be the first vertex following w which lies on both P_1 and P_2. (See Fig. 14-41.) Then the subpaths of P_1 and P_2 between w and w' have no vertices in common except w and w'; hence these two subpaths form a cycle.

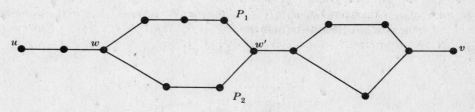

Fig. 14-41

14.16 For a connected graph G, prove: (a) If G contains a cycle C which contains an edge e, then $G - e$ is still connected. (b) If $e = \{u, v\}$ is an edge such that $G - e$ is disconnected, then u and v belong to different connected components of $G - e$.

(a) Since G is connected, any two vertices u and v are joined by a path P. We may always suppose that P does not include e; for otherwise we could construct such a path from subpaths of P and a subpath of $C - e$ (see Fig. 14-42). Hence removal of e cannot disconnect G.

(b) Assume, on the contrary, that u and v belong to the same connected component of $G - e$. Then there exists a path P in $G - e$ joining u and v. Together, P and e form a cycle C of G. But then, by (a) above, $G - e$ is connected. This contradiction yields the proof.

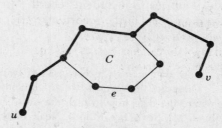

Fig. 14-42

14.17 Prove Theorem 14.4:   Let G be a graph with more than one vertex.   Then the following are equivalent: (i) G is a tree.   (ii) Each pair of vertices is connected by exactly one path.   (iii) G is connected; but if any edge is deleted then the resulting graph is disconnected.   (iv) G is cycle-free, but if any edge is added to the graph then the resulting graph has exactly one cycle.

(*i*) *implies* (*ii*).   Let u and v be two vertices in G.   Since G is a tree, G is connected and there is at least one path between u and v.   There can only be one path between u and v, otherwise G will contain a cycle (Problem 14.15).

(*ii*) *implies* (*iii*).   Suppose we delete an edge $e = \{u, v\}$ from the connected graph G.   Since e is the only path from u to v, $G - e$ is disconnected.

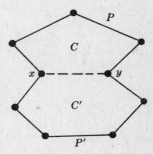

Fig. 14-43

(*iii*) *implies* (*iv*).   By Problem 14.16(*a*), G is cycle-free.   Now let x and y be vertices of G and let H be the graph obtained by adjoining the edge $e = \{x, y\}$ to G.   Since G is connected, there is a path P from x to y in G; hence $C = Pe$ forms a cycle in H.   Suppose H contains another cycle C'.   Since G is cycle-free, C' must contain the edge e, say $C' = P'e$.   Then P and P' are two paths in G from x to y.   (See Fig. 14-43.)   By Problem 14.15, G contains a cycle, which contradicts the fact that G is cycle-free.   Hence H contains only one cycle.

(*iv*) *implies* (*i*).   Since adding any edge $e = \{x, y\}$ to G produces a cycle, the vertices x and y must already be connected in G.   Hence G is connected and, by hypothesis, G is cycle-free; that is, G is a tree.

DIRECTED GRAPHS

14.18 Consider the directed graph D pictured in Fig. 14-44.   (*a*) Describe D formally.   (*b*) Find the number of paths from X to Z.   (*c*) Find the number of paths from Y to Z.   (*d*) Are there any sources or sinks?   (*e*) Find the matrix M_D of the digraph D.   (*f*) Is D weakly connected?   unilaterally connected?   strongly connected?

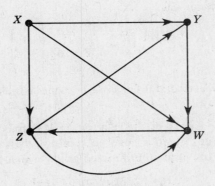

Fig. 14-44

(*a*)   There are four vertices: X, Y, Z, W, and there are seven arcs: $\langle X, Y \rangle, \langle X, W \rangle, \langle X, Z \rangle, \langle Y, W \rangle$, $\langle Z, Y \rangle, \langle Z, W \rangle, \langle W, Z \rangle$.

(*b*)   There are three paths from X to Z: $(X, Z), (X, W, Z)$ and (X, Y, W, Z).

(*c*)   There is only one path from Y to Z: (Y, W, Z).

(*d*) X is a source since it is not the terminal point of any arc, i.e. its indegree is zero.   There are no sinks since every vertex has nonzero outdegree, i.e. each vertex is the initial point of some arc.

(*e*)
$$M_D = \begin{pmatrix} 0 & 1 & 1 & 1 \\ 0 & 0 & 0 & 1 \\ 0 & 1 & 0 & 1 \\ 0 & 0 & 1 & 0 \end{pmatrix}$$

(Here the rows and columns of M_D are labeled by X, Y, Z, W respectively.)   The entry m_{ij} denotes the number of arcs from the ith vertex to the jth vertex.

(f)   The digraph is not strongly connected since X is a source and hence there is no path from any other vertex, say Y, to X.   However, D is unilaterally connected since the path (X, Y, W, Z) passes through all the vertices, and so there is a subpath connecting any pair of vertices.

14.19   Consider the directed graph D pictured in Fig. 14-45.   (a) Are there any sources or sinks?   (b) Find the matrix M of D.   (c) Is D unilaterally connected?   strongly connected?

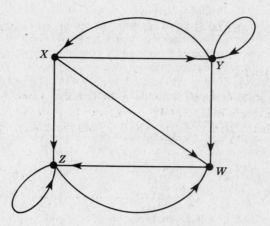

Fig. 14-45

(a)   There are no sources or sinks.

(b)
$$M = \begin{pmatrix} 0 & 1 & 1 & 1 \\ 1 & 1 & 0 & 1 \\ 0 & 0 & 1 & 1 \\ 0 & 0 & 1 & 0 \end{pmatrix}$$

(c) D is unilaterally connected; it is not strongly connected, since there is no path from Z to X.

14.20   Let $V = \{2, 3, 4, 5, 6\}$.   Let R be the relation on V defined by: xRy if x is less than y and x is relatively prime to y (i.e. x and y have no common divisor other than 1).   (a) Write R as a set of ordered pairs.   (b) Draw the directed graph which corresponds to R.

(a) $R = \{(2, 3), (2, 5), (3, 4), (3, 5), (4, 5), (5, 6)\}$

(b)   We draw an arc from x to y if (x, y) belongs to R, as shown in Fig. 14-46.

14.21   Draw the digraph D corresponding to the following matrix M (which has nonnegative integer entries):

$$M = \begin{pmatrix} 0 & 2 & 0 & 1 \\ 0 & 0 & 1 & 1 \\ 2 & 1 & 1 & 0 \\ 0 & 0 & 1 & 1 \end{pmatrix}$$

Since M is a 4×4 matrix, D has four vertices, say, v_1, v_2, v_3, v_4.   For each entry m_{ij}, draw m_{ij} arcs from the vertex v_i to the vertex v_j to obtain the digraph in Fig. 14-47.

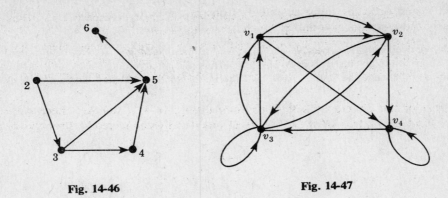

Fig. 14-46   Fig. 14-47

FINITE STATE MACHINES

14.22 Consider a finite state machine M with input symbols a and b and output symbols x, y, z, and the state diagram given in Fig. 14-48. (a) Find the state table for M. (b) Determine the output if the input is the string of symbols $W = aababaabbab$.

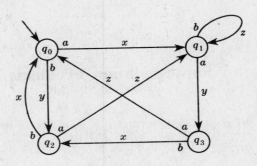

Fig. 14-48

(a)   We label the rows of the table by the four states q_0, q_1, q_2, q_3 and the columns by the input symbols a, b. Using the state diagram, Fig. 14-48, we find the entries in the table as follows. From state q_0, the arrow labeled a goes to state q_1 and is labeled with the output symbol x. Hence q_1, x is put in the table in the position corresponding to the row q_0 and column a. The other entries in the state table below are obtained similarly.

	a	b
q_0	q_1, x	q_2, y
q_1	q_3, y	q_1, z
q_2	q_1, z	q_0, x
q_3	q_0, z	q_2, x

(b)   Beginning at q_0, the initial state, we go from state to state by the arrows which are labeled respectively by the given input symbols:

$$q_0 \xrightarrow{a} q_1 \xrightarrow{a} q_3 \xrightarrow{b} q_2 \xrightarrow{a} q_1 \xrightarrow{b} q_1 \xrightarrow{a} q_3 \xrightarrow{a} q_0 \xrightarrow{b} q_2 \xrightarrow{b} q_0 \xrightarrow{a} q_1 \xrightarrow{b} q_1$$

The output symbols on the above arrows are respectively $xyxzzyzyxxz$.

14.23 Let a and b be the input symbols. Construct a finite automaton M which will accept precisely those strings in a and b which have an even number of a's.

We need only two states, q_0 and q_1. We assume that M is in state q_0 or q_1 according as the number of a's up to the given step is even or odd. (Thus q_0 is an accepting state, but q_1 is nonaccepting.) Then only a will change the state. Also, q_0 is the initial state. The state diagram of M is Fig. 14-49.

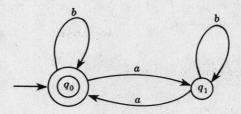

Fig. 14-49

Supplementary Problems

GRAPHS, CONNECTIVITY

14.24 Find the diameter of the graph G with vertices u, v, w, x, y and edges

    (a) $\{u, v\}, \{u, x\}, \{v, w\}, \{v, x\}, \{v, y\}, \{x, y\}$ (b) $\{u, v\}, \{v, w\}, \{w, x\}, \{w, y\}, \{x, y\}$

14.25 Consider the graph in Fig. 14-50. Find: (a) all paths from the vertex A to the vertex H, (b) the diameter of the graph, and (c) the degree of each vertex. (d) Which vertices, if any, are cut points? (e) An edge e in a connected graph G is called a *bridge* if $G - e$, the subgraph obtained from G by deleting the edge e, is disconnected. Which edges, if any, are bridges?

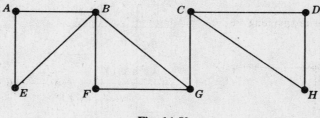

Fig. 14-50

14.26 Show that an edge e is a bridge of a connected graph G if and only if e is contained in no cycle of G. (*Hint*: Use Problem 14.16(a).)

14.27 Consider the multigraphs in Fig. 14-51. Which of them are (a) connected, (b) loop-free, (c) graphs?

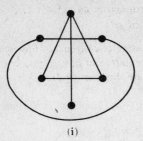

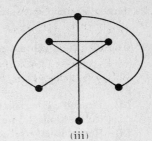

(i)   (ii)   (iii)

Fig. 14-51

14.28 Find all connected graphs with four vertices.

14.29 Draw two 3-regular graphs with eight vertices.

14.30 Determine the diameter of any complete bipartite graph.

14.31 Prove that a graph is bipartite if and only if each of its cycles is of even length.

14.32 Show that any tree is a bipartite graph. (*Hint*: Use Problem 14.31.)

14.33 Consider the following two operations on a graph G: (1) Delete an edge. (2) Delete a vertex and all edges incident on that vertex. Show that every subgraph of a finite graph G can be obtained by a sequence consisting of these two operations.

14.34 Prove that any graph G can be partitioned into maximal disjoint connected subgraphs (i.e. its connected components) by choosing the appropriate equivalence relation on the vertices of G.

TREE GRAPHS

14.35 Figure 14-11(*b*) shows eight of the trees with exactly seven vertices. There are two others; find them.

14.36 Prove that a finite tree (with at least one edge) has at least two vertices of degree 1. (*Hint*: Consider a longest path in the tree.)

14.37 Find the number of spanning trees for the graph of Fig. 14-52.

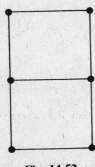

Fig. 14-52

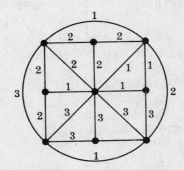

Fig. 14-53

14.38   Find a minimal spanning tree for the labeled graph of Fig. 14-53.

14.39   Consider the algebraic expression

$$\frac{(3x - 5z)^4}{a(2b + c^2)}$$

   (a)   Draw the corresponding ordered rooted tree, using an arrow (↑) for exponentiation, an asterisk (*)
          for multiplication and a slash (/) for division.
   (b)   Rewrite the expression in (i) prefix Polish notation, (ii) postfix Polish notation.
   (c)   Find the scope of each multiplication operation.

14.40   Suppose the payroll record of an employee is organized as follows: 00 Employee, 01 Number, 01 Name,
          02 Last, 02 First, 02 MI, 01 Hours, 02 Regular, 02 Overtime, 01 Rate.   (a) Draw the corresponding tree
          diagram.   (b) Which are the group items?   (c) Which are the elementary items?

DIRECTED GRAPHS

14.41   Consider the directed graph D of Fig. 14-54.

   (a)   Find the indegree and outdegree of each vertex.
   (b)   Find the number of paths from v_1 to v_4.
   (c)   Are there any sources or sinks?
   (d)   Find the matrix M of D.
   (e)   Find the number of walks of length 3 or less from v_1 to v_3.
   (f)   Is D unilaterally connected?   strongly connected?

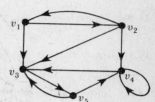

Fig. 14-54

14.42   Let D be the digraph with vertices v_1, v_2, v_3, v_4 corresponding to the matrix

$$M = \begin{pmatrix} 0 & 1 & 1 & 1 \\ 0 & 1 & 1 & 0 \\ 1 & 1 & 0 & 2 \\ 1 & 0 & 0 & 0 \end{pmatrix}$$

   (a) Draw the diagram of D.   (b) Find the number of walks of length 3 from v_1 to (i) v_1, (ii) v_2, (iii) v_3, (iv) v_4.
   (c) Is D unilaterally connected?   strongly connected?

14.43   Let R be the relation on $V = \{2, 3, 4, 9, 15\}$ defined by "x is less than and relatively prime to y"
          (compare Problem 14.20).   (a) Draw the diagram of the digraph of R.   (b) Is R weakly connected?
          unilaterally connected?   strongly connected?

14.44   A digraph D is *complete* if for each pair of distinct vertices v_i and v_j either $\langle v_i, v_j \rangle$ is an arc or $\langle v_j, v_i \rangle$ is an
          arc.   Show that a finite, complete digraph D has a path which includes all vertices.   (This obviously
          holds for nondirected, complete graphs.)   Thus D is unilaterally connected.

14.45   Consider the labeled directed graph in Fig. 14-55.   (a) How many (directed) paths are there from vertex
          s to vertex t?   (b) Find the minimum path from s to t.

FINITE STATE MACHINES

14.46   Consider the finite state machine M with input symbols a, b, c and output symbols x, y, z, and the state
          diagram shown in Fig. 14-56.   (a) Construct the state table of M.   (b) Find the output if the input is the
          string $W = caabbaccab$.

14.47   Let M be the finite state machine with the state table Fig. 14-57.   (a) Draw the state diagram of M
          given that q_0 is the initial state.   (b) Find the output if the input is the string $W = aabbabbbaab$.

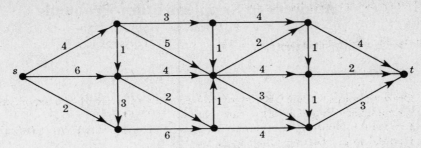

Fig. 14-55

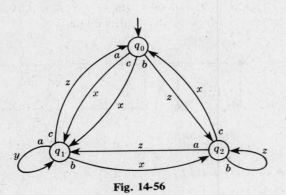

Fig. 14-56

	a	b
q_0	q_1, x	q_2, y
q_1	q_3, y	q_1, z
q_2	q_1, z	q_0, x
q_3	q_0, z	q_2, x

Fig. 14-57

14.48 For each of the machines in Fig. 14-58, with input symbols a and b and output symbols x, y, z, find the output if the input is the string $W = abaabbabbaabaa$.

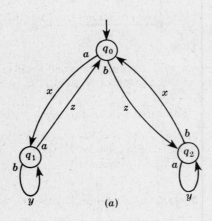

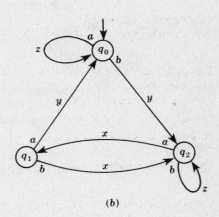

(a)   (b)

Fig. 14-58

14.49 Construct a finite automaton M with input symbols a and b which will accept only those strings in a and b such that the number of b's is divisible by 3.   (*Hint*:   Three states are required.)

14.50 Construct a finite automaton M with input symbols a and b which will accept only those strings in a and b such that *aabb* appears as a substring.   (For example, *ba*(*aabb*)*ba* and *bab*(*aabb*)*a* will be accepted, but *babbaa* and *aababaa* will not be accepted.)

Answers to Supplementary Problems

14.24   (*a*) diam (*G*) = 2, (*b*) diam (*G*) = 3

14.25   (*a*)   There are eight paths:

   (*A*, *B*, *G*, *C*, *H*)   (*A*, *B*, *F*, *G*, *C*, *H*)   (*A*, *B*, *G*, *C*, *D*, *H*)   (*A*, *B*, *F*, *G*, *C*, *D*, *H*)
   (*A*, *E*, *B*, *G*, *C*, *H*)   (*A*, *E*, *B*, *F*, *G*, *C*, *H*)   (*A*, *E*, *B*, *G*, *C*, *D*, *H*)   (*A*, *E*, *B*, *F*, *G*, *C*, *D*, *H*)

   (*b*)   4   (*d*)   *B*, *C*, *G*

   (*c*)   deg (*B*) = 4, deg (*C*) = deg (*G*) = 3; others have degree 2.   (*e*)   {*C*, *G*}

14.27   (*a*) (iii),   (*b*) (i) and (iii),   (*c*) (iii)

14.28   See Fig. 14-59.

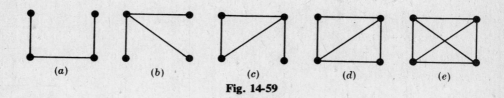

Fig. 14-59

14.29   The two 3-regular graphs shown in Fig. 14-60 are distinct graphs since (*b*) has a 5-cycle but (*a*) does not.

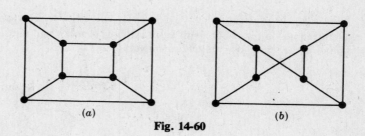

Fig. 14-60

14.30   diam ($K_{1,1}$) = 1; all others have diameter 2.

14.34   Let *u* ~ *v* if *u* = *v* or if there is a path from *u* to *v*.   Show that ~ is an equivalence relation.

14.35   See Fig. 14-61.

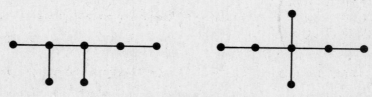

Fig. 14-61

14.37   fifteen

14.38   weight 12

14.39 (*a*) See Fig. 14-62.

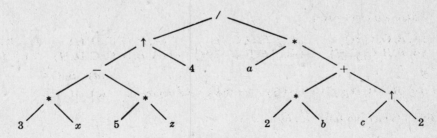

Fig. 14-62

(*b*) (i) $/ \uparrow - * 3\, x * 5\, z\, 4 * a + * 2\, b \uparrow c\, 2$

    (ii) $3\, x * 5\, z * - 4 \uparrow a\, 2\, b * c\, 2 \uparrow + * /$

(*c*) $3x,\, 5z,\, a(2b + c^2),\, 2b$

14.40 (*a*)   See Fig. 14-63.

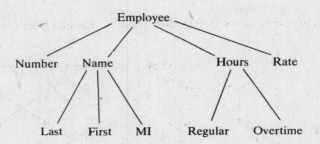

Fig. 14-63

(*b*)   Name, Hours

(*c*)   Number, Last, First, MI, Regular, Overtime, Rate

14.41 (*a*)   Indegrees: 1, 1, 4, 3, 1.   Outdegrees: 2, 3, 1, 2, 2.

(*b*)   Three: (v_1, v_2, v_4), (v_1, v_3, v_5, v_4), $(v_1, v_2, v_3, v_5, v_4)$.   (*c*) No.

(*d*) $M = \begin{pmatrix} 0 & 1 & 1 & 0 & 0 \\ 1 & 0 & 1 & 1 & 0 \\ 0 & 0 & 0 & 0 & 1 \\ 0 & 0 & 1 & 1 & 0 \\ 0 & 0 & 1 & 1 & 0 \end{pmatrix}$.   (*e*) 5.   (*f*) Unilateral, but not strong.

14.42 (*a*) See Fig. 14-64.   (*b*) (i) 3, (ii) 5, (iii) 4, (iv) 4.   (*c*) Unilateral and strong.

14.43 (*a*) See Fig. 14-65.   (*b*) Only weakly connected.

14.44   *Hint*: Suppose $(v_1, v_2, \ldots, v_m)$ is a longest path in D and does not include the vertex u. Then $\langle u, v \rangle$ and $\langle v_m, u \rangle$ are not arcs since, if they were, the path could be extended. Hence $\langle v_1, u \rangle$ and $\langle u, v_m \rangle$ are arcs. Let k be the smallest integer such that $\langle v_k, u \rangle$ and $\langle u, v_{k+1} \rangle$ are arcs. Then $(v_1, \ldots u, v_{k+1}, \ldots, v_m)$ is a longer path.

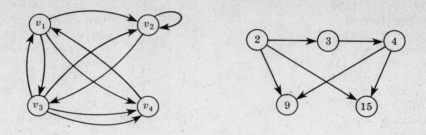

Fig. 14-64   Fig. 14-65

14.45 (a) 62, (b)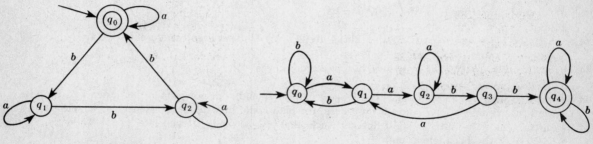

14.46 (a)

	a	b	c
q_0	q_1, x	q_2, z	q_1, x
q_1	q_1, y	q_2, x	q_0, z
q_2	q_1, z	q_2, z	q_0, x

(b) *xyyxzzzxyx*

14.47 See Fig. 14-66. (b) *yyzyzxzxyz*

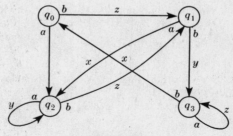

Fig. 14-66

14.48 (a) *xyzxyyzzyyxxz*, (b) *zyxyyzxxzxyyxy*

14.49 See Fig. 14-67.

14.50 See Fig. 14-68.

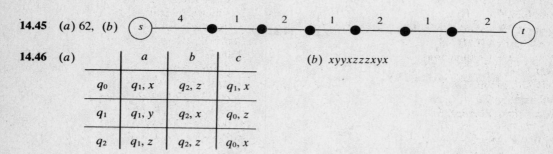

Fig. 14-67   Fig. 14-68

Index

Catalog

If you are interested in a list of SCHAUM'S
OUTLINE SERIES send your name
and address, requesting your free catalog, to:

SCHAUM'S OUTLINE SERIES, Dept. C
McGRAW-HILL BOOK COMPANY
1221 Avenue of Americas
New York, N.Y. 10020

SCHAUM'S OUTLINE SERIES

Each outline includes basic theory, definitions and hundreds of
carefully solved problems and supplementary problems with answers.

ACCOUNTING, BUSINESS & ECONOMICS

Accounting I, 2nd Ed.
Accounting II, 2nd Ed.
Advanced Accounting
Advertising
Bookkeeping & Accounting
Introduction to Business
Introduction to Business
   Organization & Management
Business Statistics
College Business Law
Cost Accounting I
Cost Accounting II
Development Economics
Financial Accounting
Intermediate Accounting I
International Economics
Macroeconomic Theory
Marketing
Mathematics for Economists
Mathematics of Finance
Microeconomic Theory, 2nd Ed.
Personal Finance & Consumer Economics
Principles of Economics
Quantitative Methods in Management
Statistics and Econometrics
Tax Accounting

BIOLOGY

Genetics, 2nd Ed.

CHEMISTRY

College Chemistry, 6th Ed.
Organic Chemistry
Physical Chemistry

COMPUTERS

Boolean Algebra
Computer Science
Computers and Programming
Data Processing
Digital Principles
Discrete Mathematics
Essential Computer Mathematics
Microprocessor Fundamentals
Programming with Basic, 2nd Ed.
Programming with Fortran

EDUCATION, PSYCHOLOGY & SOCIOLOGY

Child Psychology
Introduction to Psychology
Psychology of Learning
Introduction to Sociology
Test Items in Education

ELECTRONICS & ELECTRICAL ENGINEERING

Basic Circuit Analysis
Basic Mathematics for
   Electricity and Electronics
Electric Circuits
Electric Machines and
   Electromechanics
Electromagnetics
Electronic Circuits
Electronic Communication
Electronics Technology
Feedback and Control Systems
Laplace Transforms
Transmission Lines

ENGINEERING

Acoustics
Advanced Structural Analysis
Basic Equations of Engineering
Continuum Mechanics

Descriptive Geometry
Dynamic Structural Analysis
Introduction to Engineering Calculations
Engineering Mechanics, 3rd Ed.
Fluid Dynamics
Fluid Mechanics & Hydraulics
Heat Transfer
Lagrangian Dynamics
Machine Design
Mechanical Vibrations
Operations Research
Reinforced Concrete Design
Space Structural Analysis
State Space & Linear Systems
Static and Strength of Materials
Strength of Materials, 2nd Ed.
Structural Analysis
Theoretical Mechanics
Thermodynamics

ENGLISH

Engilsh Grammar
Punctuation, Capitalization, & Spelling

FOREIGN LANGUAGES

French Grammar, 2nd Ed.
German Grammar, 2nd Ed.
Italian Grammar
Spanish Grammar, 2nd Ed.

MATHEMATICS & STATISTICS

Advanced Calculus
Advanced Mathematics
Analytic Geometry
Basic Mathematics
Calculus, 2nd Ed.
College Algebra
Complex Variables
Differential Equations
Differential Geometry
Elementary Algebra
Review of Elementary Mathematics
   (including Arithmetic)
Finite Differences & Difference Equations
Finite Mathematics
First Year College Mathematics
Fourier Analysis
General Topology
Group Theory
Linear Algebra
Mathematical Handbook
Matrices
Modern Algebra
Modern Elementary Algebra
Modern Introductory Differential Equations
Numerical Analysis
Plane Geometry
Probability
Probability & Statistics
Projective Geometry
Real Variables
Set Theory & Related Topics
Statistics
Technical Mathematics
Trigonometry
Vector Analysis

PHYSICS & PHYSICAL SCIENCE

Applied Physics
College Physics, 7th Ed.
Earth Sciences
Modern Physics
Optics
Physical Science
Physics for Engineering
   and Science

ISBN 0-07-037990-4